THE INTERNATIONAL LAW OF DISASTER RELIEF

Disasters can strike often and with unexpected fury, resulting in devastating consequences for local populations that are insufficiently prepared and largely dependent on foreign aid in the wake of such catastrophes. International law can play a significant role in the recovery after inevitable natural disasters; however, without clear legal frameworks, aid may be stopped, delayed, or even hijacked – placing the intended, suffering recipients in critical condition.

This edited volume brings together experts, emerging scholars, and practitioners in the field of international disaster law from North America, Japan, New Zealand, and Australia to analyze the evolution of international disaster law as a field that encompasses new ideas about human rights, sovereignty, and technology. Chapters focus on specific natural disasters like Hurricane Katrina, Cyclone Nargis, and Typhoon Haiyan in addition to volcanic and earthquake activity, wildfires, and desertification. The book begins a dialogue on the profound implications of the evolution of international law as a tool for disaster response.

David D. Caron is the dean of the Dickson Poon School of Law at King's College London and the C. William Maxeiner Emeritus Distinguished Professor of International Law at the University of California, Berkeley. He currently serves as a member of the World Economic Forum's Global Agenda Council on the Rule of Law, the U.S. Department of State Advisory Committee on Public International Law, and the board of editors of the *American Journal of International Law*.

Michael J. Kelly is associate dean and a professor of law in the School of Law at Creighton University. He is president of the U.S. national chapter of L'Association Internationale du Droit Pénal, co-chair of the American Bar Association's Task Force on Internet Governance, and a member of the President's Advisory Committee on Global Engagement for the American Association of Law Schools. He is the author or coauthor of four books and more than thirty articles and book chapters and continues to serve as a contributing editor to the online legal newspaper *JURIST*.

Anastasia Telesetsky is an associate professor of law in the College of Law at the University of Idaho, where she teaches public international law and international environmental law for the natural resources and environmental law program. She has expertise in climate change index insurance and food security and development. Her scholarship has been published in numerous American, European, and Asian law reviews. She is a member of the international environmental law section of the American Society for International Law.

The International Law of Disaster Relief

Edited by

DAVID D. CARON

Dean and Professor of Law
Dickson Poon School of Law, King's College London

MICHAEL J. KELLY

Associate Dean and Professor of Law
Creighton University School of Law

ANASTASIA TELESETSKY

Associate Professor of Law
University of Idaho College of Law

CAMBRIDGE
UNIVERSITY PRESS

CAMBRIDGE
UNIVERSITY PRESS

32 Avenue of the Americas, New York, NY 10013-2473, USA

Cambridge University Press is part of the University of Cambridge.

It furthers the University's mission by disseminating knowledge in the pursuit of education, learning, and research at the highest international levels of excellence.

www.cambridge.org
Information on this title: www.cambridge.org/9781107061316

First published 2014

Printed in the United States of America

A catalog record for the publication is available from the British Library.

Library of Congress Cataloging in Publication Data
The international law of disaster relief / edited by David D. Caron, Michael J. Kelly, Anastasia Telesetky.
pages cm
ISBN 978-1-107-06131-6 (Hardback)
1. Disaster relief–Law and legislation. 2. Emergency management–Law and legislation.
I. Caron, David D., editor of compilation. II. Kelly, Michael J., 1968– editor of compilation.
III. Telesetky, Anastasia, editor of compilation.
K1980.I587 2014
344.05′348–dc23 2014003455

ISBN 978-1-107-06131-6 Hardback

David D. Caron – For those who need not have suffered

Michael J. Kelly – For my son, Kruz Michael, who should grow up in a safer world

Anastasia Telesetsky – For my husband Matt and for each and every person whose quiet lives have been and will be instantaneously transformed in the face of a disaster

Contents

Editors and Contributors

Editors

David D. Caron is dean of the Dickson Poon School of Law at King's College London. Dean Caron was previously the C. William Maxeiner Distinguished Professor of Law at the University of California, Berkeley. He is currently a member of the World Economic Forum's Global Agenda Council on the Rule of Law, a member of the Executive Council of the American Bar Association Section on International Law, a member of the U.S. Department of State Advisory Committee on Public International Law, co-director of the Law of the Sea Institute, co-director of Berkeley's Miller Institute on Global Challenges and the Law, and a member of the board of editors of the *American Journal of International Law*. Dean Caron was president of the American Society of International Law from 2011 to 2012.

Michael J. Kelly is a professor of law and associate dean for Faculty Research and International Programs at Creighton University School of Law. Professor Kelly is president of the U.S. national chapter of L'Association International du Droit Pénal, a Paris-based society of international criminal law scholars, judges, and attorneys founded in 1924 that enjoys consultative status with the United Nations. He also serves as co-chair of the American Bar Association's Task Force on Internet Governance, a contributing editor of the legal news site *Jurist*, and a member of the president's Advisory Committee on Global Engagement for the American Association of Law Schools.

Anastasia Telesetsky is an associate professor of law at the University of Idaho College of Law. Professor Telesetsky was previously a private attorney, notably representing Ethiopia before the Ethiopia-Eritrea Claims Commission at the Permanent Court of Arbitration. She was also a Bosch Fellow in Germany where she worked for the German Foreign Ministry promoting international food security and assisted in drafting guidelines on implementation for the "Right to Food." As a

Fulbright Fellow and a Berkeley Human Rights Center Fellow, Professor Telesetsky collaborated with communities in the Philippines and Papua New Guinea on developing culturally appropriate legal solutions to environmental protection problems. Her interest in international disaster law is based on her publications proposing the use of environmental insurance as an adaptation mechanism for climate change. She is a member of the American Society of International Law and the IUCN World Commission for Environmental Law.

Contributors

Kirsten Nakjavani Bookmiller is a professor of government and political affairs at Millersville University in Pennsylvania, where she also directs the Center for Disaster Research and Education's Global Partnerships Initiative, a transnational network of academic institutions and international organizations dedicated to advancing best practices in the disaster management field. Her teaching areas include the international humanitarian system, public international law, global governance, and human rights.

Claire Clement is an international law adviser at the British Red Cross Society, where her duties include providing assistance and advice on matters of international humanitarian law to a wide range of stakeholders both within and outside the United Kingdom. She previously worked in the International Humanitarian Law Program at the Australian Red Cross, where she was responsible for training the Australian armed forces, government departments, and nongovernment organizations in international humanitarian law. She holds a master's degree in International Peace and Security from King's College London.

Dug Cubie is a lecturer in the faculty of law at University College Cork, Ireland. He has more than ten years professional experience in refugee protection and humanitarian assistance, including working for the United Nations High Commissioner for Refugees in the United Kingdom, Belgium, Nepal, and the Republic of Congo; the Irish Refugee Council; the International Organization for Migration; the Irish Red Cross in Ireland; and the Immigrant Services Society of British Columbia in Vancouver. He holds an LLM from the University of Cambridge and a PhD from University College Cork, Ireland.

Daniel A. Farber is the Sho Sato Professor of Law at the law school at the University of California, Berkeley. He is also the co-director of the Center for Law, Energy, and the Environment. Professor Farber serves on the editorial board of Foundation Press and is a member of the American Academy of Arts and Sciences and a Life Member of the American Law Institute. He has authored and coauthored eighteen law books including *Disaster Law and Policy* and *Disaster and the Law: Katrina and Beyond.*

Paul Govind is a lecturer at Macquarie University School of Law's Centre for International and Environmental Law and editor of the *Macquarie Journal of International and Comparative Environmental Law*. He holds an LLM from the University of Sydney. His research areas include international and domestic climate change law with a focus on adaptation and issues of vulnerability, socio-economic development, and climate justice.

Catherine Gribbin is the policy advisor on humanitarian issues and emblem use at the Canadian Red Cross. She received her LLM from Dalhousie University and is a member of the Nova Scotia Barristers' Society.

Brian R. Israel is an attorney-adviser in the Office of the Legal Adviser for the Bureau of Oceans and International Environmental and Scientific Affairs, U.S. Department of State, where he is currently responsible for legal issues relating to the Arctic, outer space, international science and technology cooperation, Antarctica, and the extended continental shelf. In this capacity he serves as the U.S. representative to the Legal Subcommittee of the United Nations Committee on the Peaceful Uses of Outer Space. He also teaches international law as an adjunct professor at George Mason University School of Law and is a graduate of the University of California, Berkeley, School of Law, where he was editor-in-chief of the *Berkeley Journal of International Law*.

Akiko Ito is the chief of the secretariat, Convention on the Rights of Persons with Disabilities within the United Nations. She is also a departmental focal point for women for the Department of Economic and Social Affairs of the United Nations. She has lectured and published extensively on issues concerning disabilities and human rights with a focus on the rights of minorities and other disadvantaged groups. Prior to her current post, she worked in the Legal Affairs Section of the United Nations Drug Control Programme in Vienna, Austria. She holds an MA in International Relations from the University of Chicago and an LLM from Boalt Hall School of Law, University of California, Berkeley.

Ilario Maiolo is acting director of the Advocacy Policy and International Humanitarian Law Unit of the Canadian Red Cross. He is responsible for promoting international humanitarian law for the society and has worked with a number of key stakeholders including government departments, the armed forces, and academia. He was a consultant for the government of Canada for the Department of Foreign Affairs and International Trade. He holds a master's degree from the University of Geneva in International Humanitarian Law.

Kentaro Nishimoto is an associate professor of international law at the Graduate School of Law, Tohoku University. He is also a part-time lecturer at the Graduate School of Public Policy at the University of Tokyo. In addition, he is a member of the University of Tokyo Ocean Alliance, a trans-disciplinary network of ocean research and education at the University of Tokyo. He received his PhD in law from the Graduate School of Law and Politics at the University of Tokyo.

Ibironke T. Odumosu-Ayanu is an assistant professor at the University of Saskatchewan. She formerly served as a consultant for the United Nations University (UNU) on a United Nations Convention on Trade and Development/UNU project on the rule of law and good business practices in zones of conflict. She is a barrister and solicitor of the Supreme Court of Nigeria and holds a PhD in law from the University of British Columbia.

Catherine Shanahan Renshaw is an associate of the Sydney Centre for International Law at the University of Sydney where she researches, writes, and teaches about human rights and democracy in Southeast Asia. She serves on the editorial board of the *Journal of South Asian Studies* and between 2008 and 2010 was director of an Australian Research Council project on national human rights institutions in the Asia Pacific region, based at the Australian Human Rights Centre at the University of New South Wales. She is admitted to practice as a lawyer in the Supreme Court of New South Wales and the High Court of Australia.

Imogen Saunders is a lecturer at Australian National University College of Law where she teaches courses including international trade law. She previously practiced as a barrister and solicitor in Western Australia. She received her PhD from Australian National University.

Katie Sykes is an assistant professor at the Thompson Rivers University in British Columbia, Canada. She is a graduate of the LLM programs at Dalhousie University and Harvard University. Her research focuses on international law and food, in particular the tensions between different international norms in the areas of trade, environment, human rights, and animal welfare. She was formerly an associate in the New York office of Cleary Gottlieb Steen & Hamilton LLP, a leading international law firm.

Yukari Takamura is a professor in the Graduate School of Environmental Studies at Nagoya University. She has a master's degree from Hitotsubashi University in Japan. Her research focuses on international law, jurisprudence, environmental impact assessment, and environmental policy. She is an active member of the Japanese Society for International Law and the Japanese Association of World Law. She is the director of the Japanese Association for Environmental Law and Policy.

Amelia Telec is a senior legal officer in the Office of International Law in the Australian Attorney-General's Department. In that role, she has advised the Australian government on a range of international law issues, including the law on the use of force, international humanitarian law, human rights law, and refugee law. She completed her LLM in international law at the Australian National University.

Emika Tokunaga is a PhD researcher in the School of International Public Policy at Osaka University in Japan. She was previously a visiting researcher at the Asiatic Research Institute at Korea University and the Modern East Asia Research Center of the Institute for Area Studies at Leiden University in the Netherlands.

Toshiya Ueki has been a professor of international law at the Tohoku University School of Law in Sendai, Japan, since 1999. Since 2006, he has been the executive vice-president at Tohoku University. He is a graduate of the University of Tokyo Faculty of Law and has been actively involved as a member of the International Law Association's Committee on Accountability of International Organizations and the International Law Association's Study Group on Responsibility of International Organizations.

Foreword

Fumio Kishida
Minister for Foreign Affairs of Japan

History is replete with disasters of all kinds. Natural disasters, in different forms, have affected countries on all continents. Many countries along the Pacific Rim, including Japan, have observed frequent and sometimes violent natural disasters such as floods, volcanic activities, earthquakes, and tsunamis. I find it particularly meaningful that the works on the international law of disaster relief by scholars from the "Four Societies of International Law" – the American Society of International Law, Australian and New Zealand Society of International Law, Canadian Council on International Law, and the Japanese Society of International Law – have been compiled in a publication to lead the international debate in this field.

Responding to a catastrophic disaster requires an international effort, and some form of coordination is necessary. It is not hard to imagine how this gives rise to a host of legal issues, and they need to be addressed effectively in order to minimize human and economic losses. The many aspects of international legal issues dealt with in this volume are all practically important and require urgent attention from both academics and practitioners.

Japan has relied on what nature can offer it to sustain its survival and prosper. Our people have valued living in harmony with nature, and such notions are perhaps manifested in many aspects of traditional Japanese culture. Nonetheless, nature is not always benevolent; our people have been tested through the ages in the face of natural disasters – typhoons, volcanic eruptions, earthquakes, and tsunamis.

The Great East Japan Earthquake of March 11, 2011, spawned a series of massive tsunamis that struck many places along the eastern coast of northern Japan, claiming more than 15,000 lives. Three years after the disaster, reconstruction activities have progressed amid a number of remaining challenges in recovering the affected areas. While the entire nation is continuing its utmost efforts, we still remember how we were truly heartened by the expression of solidarity and support from all corners of the world in the face of this disaster.

We were reminded of the bond of peoples in tackling difficulties, and of our role to respond in kind for the world.

The issues discussed in this book concern the foundation of such international cooperation, and I sincerely hope that it will provide a basis for further enhancing the effectiveness of international efforts to disaster relief. The value of this publication is felt close to my heart particularly with our recent experience of the Great East Japan Earthquake, and, I believe, to many readers around the world who have been affected by a natural disaster in one way or another. I look forward to seeing further collaboration among the legal scholars and experts from the Four Societies and beyond.

FIGURE 1 **"The Great Wave off Kanagawa."** Color print from woodblock, Edo Period (printed between 1826–1833), No. 21 from the series *Thirty-Six Views of Mount Fuji* by Katsushika Hokusai.

Preface

David D. Caron

The international law societies of Australia and New Zealand, Canada, Japan, and the United States (the "Four Societies"), seeking to promote an international network of young scholars, together undertook shortly after the new millennium to hold international conferences biannually at sites within each of the societies. These conferences have exceeded the expectations of all concerned. They have promoted excellence, created the personal ties that enrich and inform great scholarship, and opened numerous opportunities for future collaboration.

This volume flows from the Fourth International Four Societies Conference, and thus marks the completion of the first cycle of the Four Societies Project. The first conference was held in Wellington, New Zealand, in 2006; the second in Edmonton, Canada, in 2008; and the third in Awaji-shima, Japan, in 2010. The fourth and final conference of the cycle was hosted by the American Society of International Law (ASIL) at the University of California, Berkeley, in 2012.

ASIL proposed the topic of disasters as the focus for the fourth biannual conference on March 16, 2011, in the wake of the terrible tsunami that struck the Japanese islands. Then serving as president of the ASIL, I wrote to Professor Masaharu Yanagihara, president of the Japanese Society of International Law and professor of international law at Kyushu University:

> Amidst the continuing tragedy that began with last Friday's earthquake and tsunami, I write to express the heartfelt concern of the American Society of International Law for the people of Japan and for its colleagues in the Japanese Society of International Law. A catastrophe, like lightning, illuminates the world unexpectedly. And in the moment of clear vision that accompanies a catastrophe, we all feel deeply the fragility we share.

With this sense of solidarity in mind, we proceeded to propose that the focus of the 2012 Four Societies meeting be "Disasters and International Law," a proposal to which all of the societies quickly agreed.

Few aspects of the future are certain. Yet, it is certain that catastrophes, attended by widespread suffering, are a part of our collective future. No one will be surprised to wake tomorrow to learn of an earthquake, an accident at a nuclear power plant, or the desperate plight of persons fleeing chaos. Data indicate that a significant natural disaster occurs on average once a week. Every three weeks, there is a disaster that exceeds the response capacities of the country most affected. Damages inflicted by disasters kill one million people each decade and leave millions more homeless. Economic damages from natural disasters have tripled in the last thirty years. All of these statistics tend to increase with a growing population and it is argued by many that climate change will increase the intensity of some types of disasters.

Curiously, however, the international legal order addressing this certain future is, in comparison to other far less certain areas of international affairs, both relatively undeveloped in practice and unexamined in the academic legal literature. Efforts from the bottom up by groups of states develop haphazardly. Driven today by one disaster, tomorrow by another, the ad hoc incoherence of legal and institutional response mirrors the randomness of the catastrophes humanity encounters. The resulting fragmented ad hoc array of responses leaves many holes in the collective effort and often leads assistance amidst an emergency to be as likely a matter of luck as of planning. Simultaneously, the top-down efforts by international organizations and their member states to comprehensively confront this challenge easily spin into ever-broader discussions, sweeping in so many initiatives and efforts that the sheer weight of the agenda seems to impede progress. Similarly, the scholarly attention to catastrophes, with notable exceptions, is diffuse.

With this situation in mind, the Four Societies invited proposals addressing this multifaceted subject. This impressive volume represents the collected resulting papers. They form a balanced study illuminating and progressively advancing the global collective response to this pressing issue. I am delighted to observe that the Four Societies have undertaken a second cycle of international conferences, and it is heartwarming to see young scholars in the first cycle often taking the lead on the next. Scholarship usually is the work of a single person who puts pen to paper. But what that person writes, how that person's writing was made possible, and, indeed, who that person is depends very much on the communities that have nurtured and supported him or her. The Four Societies have done a great service and this insightful volume is proof of that.

FIGURE 2 **Editors and Contributors to the International Law of Disaster Relief.** Photographed at the 4th Biannual Four Societies Symposium at Berkeley Law, University of California, Berkeley (September 2012), courtesy of James Block.

Acknowledgments

This volume would not have been possible without the support of the Four Societies of International Law. Additionally, the travel and research support of the home universities, governments, and organizations of the participants was critical to the success of this important endeavor. Special mention must be made of the Millar Institute for Global Challenges and Law at Berkeley Law at the University of California, Berkeley, and its administrator Karen Chin, for hosting the symposium conference that culminated in this volume; the American Society of International Law (ASIL) and its executive director Elizabeth Anderson for ASIL's cosponsorship and logistical assistance; Creighton University School of Law for the support that funded valuable editorial and research assistance from Rachel Breger and Frank Ginsbach; and the University of Idaho College of Law for the clerical support of Ridley Williams.

As with any edited volume, the success of this endeavor rests chiefly with the level of attention and timeliness of the contributors. This project benefited immensely from a dedicated group of contributors who responded quickly and accurately to editorial comment and direction. The editors are grateful to the contributors for their hard work in producing this book, to the many individuals who are not named authors but who made these scholarly contributions possible by supporting the authors, and to the editorial staff of Cambridge University Press for publishing it with a view toward impacting the global discussion on framing the international law of disaster relief.

Introduction

Michael J. Kelly

The rise of humanity is inextricably linked with the ability to harness nature – whether by agriculture, hunting, exploration, settlement, or civilization. But the natural world is not so easily yoked. Nature is powerful. Despite all the efforts of humanity, the Earth remains a dynamic planet. As such, dynamic events like earthquakes, tsunamis, storms, hurricanes, wildfires, and drought are visited upon us time and again. Indeed, catastrophe is a part of human history, filled with surprise episodes that become historical endnotes such as the Pompeii eruptions and the 1755 tidal wave that hit Lisbon.

As unsettled as we are by catastrophes, humans understandably attempt to respond in an effort to reduce human suffering and environmental loss. Legal frameworks help coordinate such responses. Increasingly, collaborative efforts are undertaken by state and non-state actors to address pre- and post-catastrophe conditions. Some of these have led to the creation or revision of national law; others have resulted in sui generis international frameworks. Both have profound implications for the evolution of international law as a tool for disaster response.

Given the scientific and increasingly dire forecast for new and challenging catastrophes stemming from climate change, there is a growing need for cooperative international response to disaster relief. Some responses may come in the form of a yet to be fully articulated international disaster law. The international law of disaster relief is not well developed. It exists in unsystematic state practice, partial codification, and best practices guidance documents. Significant gaps remain.

Consequently, there is an urgent need to substantiate and solidify this law. Human action, incorrectly taken, or inaction can easily exacerbate already challenging circumstances in the wake of disasters. Tons of foreign relief aid can sit moldering on docks as government agencies wrangle over visa and import issues while people starve. Critical medicines can be intercepted by warring rebel groups if intergovernmental and non-governmental groups are not protected in their disaster relief efforts. Millions of dollars in construction equipment to build up levees, drain flooded areas, or rescue people from collapsed buildings can be irrevocably damaged if something as simple as local fuel additives are not compatible with engines.

But human activity can also contribute to the occurrence of disasters as well. As an Organization of American States 1991 Report on natural hazards and disasters noted:

> Although humans can do little or nothing to change the incidence or intensity of most natural phenomena, they have an important role to play in ensuring that natural events are not converted into disasters by their own actions. It is important to understand that human intervention can increase the frequency and severity of natural hazards. For example, when the toe of a landslide is removed to make room for a settlement, the earth can move again and bury the settlement.
>
> Human intervention may also cause natural hazards where none existed before. Volcanoes erupt periodically, but it is not until the rich soils formed on their eject are occupied by farms and human settlements that they are considered hazardous. Finally, human intervention reduces the mitigating effect of natural ecosystems. Destruction of coral reefs, which removes the shore's first line of defense against ocean currents and storm surges, is a clear example of an intervention that diminishes the ability of an ecosystem to protect itself. An extreme case of destructive human intervention into an ecosystem is desertification, which, by its very definition, is a human-induced "natural" hazard.

Understanding this is the key to developing effective vulnerability reduction measures: If human activities can cause or aggravate the destructive effects of natural phenomena, they can also eliminate or reduce them.[1]

Of course, the world did not fully appreciate in 1991 the negative impacts of human activity on climate change. Now, more than twenty years later, science has proven a direct correlation between industrial pollution and global warming and discovered another direct correlation between global warming and more intense and more frequent weather-related natural disasters.[2] By definition, the wealthiest nations produce the most gases contributing to climate change.[3] These states also have the largest capacity to respond to natural disasters in the developing world. As the chief contributor to worsening weather, the industrialized world has a clear

1 Department of Regional Development and Environment Executive Secretariat for Economic and Social Affairs Organization of American States (OAS), "Natural Hazard Management in Integrated Regional Development Planning," (1991): see chapter 1, accessible at: www.oas.org/dsd/publications/unit/oea66e/begin.htm#Contents.

2 Stephanie Paige Ogburn, "Climate Change Exacerbates Some Extreme Weather: Global Warming Played a Role in Half of 2012's Litany of Extreme Weather Events, from Heat Waves to Storm Surges," *Scientific American* (September 6, 2013).

3 Intergovernmental Panel on Climate Change (IPCC), Working Group I Contribution to the IPCC Fifth Assessment Report, *Climate Change 2013: The Physical Science Basis* (Summary for Policymakers), September 27, 2013, available at: http://news.bbc.co.uk/2/shared/bsp/hi/pdfs/27_09_13_ipccsummary.pdf.

moral duty to not only provide disaster relief, but also to take the lead politically in crafting a workable legal framework for the delivery of that aid.

Law seeks to anticipate issues and resolve them before such damage can happen by providing essential guideposts to institutional and human behavior. However, in the law of international disaster relief, sensitive issues of sovereignty and local cooperation are perhaps more prominent than in other fields of law and may stymie successful codification efforts. The sensitivity becomes particularly acute if local states suffering the effects of catastrophes somehow contribute to further damage. Yet the urgency of legal development in this area remains.

As with any inchoate area of law, efforts to determine the most effective and efficient legal regimes are essential. This volume comprises that effort for this subject. As a new field of study for a very old problem, there are as yet few scholarly contributions. The American Society of International Law (ASIL) on behalf of the Four Societies of International Law (ASIL, the Japanese Society of International Law, the Australian and New Zealand Society of International Law, and the Canadian Council of International Law) convened a symposium in September 2012 at the University of California, Berkeley on the international law of disasters. Scholars with both academic backgrounds and international organization experience shared innovative papers on the law of international disaster prevention and disaster response.

The chapters in this book cover a span of significant topics for delineating vital new subfields of international disaster law, including disaster relief as a general international legal principle, international funding for disaster prevention, the changing relationship between sovereignty and disaster relief, the relationship between humanitarian law and disaster law, the need for harmonization of laws for international search and rescue teams, and the growing role of law in facilitating geographic information system (GIS) technology in responding to international disasters.

Notably, Professor Daniel A. Farber from the University of California, Berkeley, one of the early scholars to approach this field, provides a commentary essay on the evolution of disaster law and Ms. Akiko Ito from the UN Secretariat on the Convention on Disability provides a commentary essay on the relationship between disaster and the Convention, focusing on how disasters specially affect the disabled. These essays demonstrate that the content of the international law of disaster relief matters not just for the relations between states but also for community risk assessment and already vulnerable individuals.

It is the intent of the editors and contributors that this work become not only an essential component of the global discussion on disaster relief occurring in capitals around the world, but also a standard reference for agencies providing disaster relief as well as a thought-provoking inquiry that will spur further academic research to drive the formation, integration, and adoption of progressive laws designed for the betterment of mankind.

PART I

The Legal Theory of International Disaster Relief

FIGURE 3 **The Eruption of Mount Vesuvius in 79 A.D.** Painting by Pierre-Jacques Volaire (1777), courtesy of the North Carolina Museum of Art.

1

International Law and the Disaster Cycle

Daniel A. Farber

This chapter introduces the "disaster cycle" as a framework for organizing national and international emergency responses. Each stage in the cycle of disaster – mitigation, emergency response, insurance/liability compensation, rebuilding – is part of society's risk management portfolio that will include both pre-disaster preparation and post-disaster reconstruction. A holistic review of disaster response as reflected in the disaster cycle is essential if states are to plan effectively for future disasters.

THE NEED FOR DISASTER LAW[1]

This volume reflects an increasing recognition of the role that the legal system plays in disaster prevention, response, and recovery.[2] Attention to these issues expanded after two dramatic events: Hurricane Katrina in 2005 and the Indian Ocean tsunami of 2004.

Attention to disaster issues has been renewed by the Japanese tsunami and Fukushima nuclear accident. On March 11, 2011, in an event that is now known in Japan as 3/11, a 9.0 magnitude earthquake struck off the east coast of Japan, about a hundred miles east-northeast of Fukushima and two hundred miles northeast of Tokyo.[3] The earthquake also triggered a large tsunami that overwhelmed seawalls

[1] This article was originally given as the Keynote Address at the conference on "Disasters and International Law" at Berkeley Law School, September 27, 2012.

[2] These issues are the subject of Daniel A. Farber, Jim Chen, Robert R.M. Verchick, and Lisa Grow Sun, *Disaster Law and Policy* 2nd ed. (New York: Aspen Publishers, 2010). For a perspective on disaster law in a major developing country, see Michael Faure and Andri Wibisana, eds., *Regulating Disasters, Climate Change and Environmental Harm: Lessons from the Indonesian Experience* (Northampton: Edward Elgar Publishing, 2013).

[3] USGS: Earthquake Hazards Program, Magnitude 9.0 – Near the East Coast of Honshu, Japan, http://earthquake.usgs.gov/earthquakes/eqinthenews/2011/usc0001xgp/

and contributed to massive destruction.[4] As a direct result of the earthquake and tsunami, more than 15,000 people were killed and 340,000 displaced.[5] About forty-six minutes after the quake, the first waves of an enormous tsunami reached the Fukushima Daiichi power station. The immediate result was to shut down emergency generators, resulting in loss of coolant and eventually in loss of containment and radioactive releases.

All three of these events revealed a disturbing lack of preparation to handle disasters. Building on earlier efforts, such events have catalyzed a growing community of scholars around the world focusing on disaster law. The importance of this field is obvious. Considering just a single country, the United States, disaster costs have averaged $12 billion per year over the past half-century.[6] From 1980 to 2004, the United States experienced sixty-two weather events causing a billion dollars or more in damages.[7] Averages are misleading, however: Hurricane Katrina alone caused about $100 billion in direct damage.[8] Losses are highly skewed, with the top 20 percent accounting for 80 percent of the damages.[9] These mega-disasters pose unique challenges to legal systems, not just in the United States but around the world. This emerging legal field seeks to inform and improve disaster-related decision making, as evidenced by a spate of recent books[10] and a rapidly expanding number of law review articles.[11]

When national governments lack the ability to respond to large- scale disasters, the international system as a whole is challenged to respond. Such responses are an expression of human solidarity. They may also be based on a recognition of the international impacts that large-scale disasters may have, due to their economic repercussions, effects on migration, and impacts on political stability.

4 "Seawalls Offered Little Protection Against Tsunami's Crushing Waves," *New York Times*, March 13, 2011, www.nytimes.com/2011/03/14/world/asia/14seawalls.html?pagewanted=all.

5 Katherine Harmon, "Japan's post-Fukushima earthquake health woes go beyond radiation effects," *Scientific American*, March 7, 2012, www.nature.com/news/japan-s-post-fukushima-earthquake-health-woes-go-beyond-radiation-effects-1.10179.

6 Carolyn Kousky, "Informing Climate Adaptation: A Review of the Economic Costs of Natural Disasters, Their Determinants, and Risk Reduction Options" (Resources for the Future Discussion Paper No. 12-28), Washington, D.C., 2012. Kousky discusses the data and methodological issues involved in such estimates in detail. Ibid., 11–13.

7 James Miskel, *Disasters Response and Homeland Security: What Works, What Doesn't* (Palo Alto: Stanford University Press, 2008), 25.

8 Ibid., 99.

9 Kousky, "Informing Climate Adaptation" 16.

10 Farber et al., *Disaster Law and Policy*; John Nolon and Daniel Rodriguez, *Losing Ground: A Nation on Edge* (Environmental Law Institute, 2007); Robert R. M. Verchick, *Facing Catastrophe: Environmental Action for a Post-Katrina World* (Cambridge: Harvard University Press, 2010); Nan D. Hunter, *The Law of Emergencies: Public Health and Disaster Management* (Oxford: Butterworth-Heinemann, 2009).

11 We can get some sense of the expansion from a Westlaw search for "flood insurance," "levees," "oil spill," "forest fire," and "natural disaster." For 2000–2005, the search produced 23 documents; for 2007–2012, the search produced 131 documents (search of JLR database on August 1, 2012). A search for "Hurricane Katrina" in the same database on August 1, 2012 produced 3,997 documents, of which 128 had the term in their titles.

Indeed, businesses are increasingly concerned about disruption of global supply chains due to disasters.[12]

Thus, international as well as domestic law becomes relevant to disaster issues. Although some disasters can be handled adequately at the national or even local levels, others exceed local capacities. As we will see, the current legal regime is underdeveloped, characterized by large gaps and a general lack of hard legal standards. But increasing global connectedness, combined with rising levels of global risk due to climate change, is likely to lead to increasingly global structures for disaster management.

THE DISASTER CYCLE AS AN ORGANIZING PRINCIPLE FOR DISASTER LAW

In making sense of the confused body of disaster law and the body of related scholarship, the most useful organizing principle is a simple idea called the disaster cycle, which is described in detail in the section that follows. The disaster cycle delineates a set of strategies for managing disaster risk, including mitigation, emergency response, compensation, and rebuilding, with rebuilding completing the circle by including (or failing to include) mitigation measures:[13]

THE CYCLE OF DISASTER LAW

Using the disaster cycle as a framework has the advantage of putting the emergency response, which usually gets the lion's share of attention, into context. Each stage of the cycle of disaster – mitigation, emergency response, insurance/liability compensation, and rebuilding – is part of society's risk management portfolio. The disaster cycle also highlights the interconnections between risk management strategies. Failures of risk mitigation often contribute to the impact of the later disaster, turning an otherwise manageable event into a more serious risk to life or property, or amplifying what would otherwise be a less significant risk to calamitous proportions. Post-emergency compensation mechanisms and rebuilding are as important as the emergency response in determining the severity of the impact on many victims.

The role of the international community may vary, depending on the situation and the stage of the disaster cycle. There are also likely to be variations depending on the status of the victim country. Developed countries are likely to be less often in need of international assistance than developing ones, but large developing countries may have a better ability to spread risks than small countries. Local capacity

[12] See United Nations Office for Disaster Risk Reduction, *Global Assessment Report on Disaster Risk Reduction 2013: From Shared Risk to Shared Value: the Business Case for Disaster Risk Reduction* (2013) www.preventionweb.net/files/1037_hyogoframeworkforactionenglish.pdf.

[13] Farber et al., *Disaster Law and Policy*, 3.

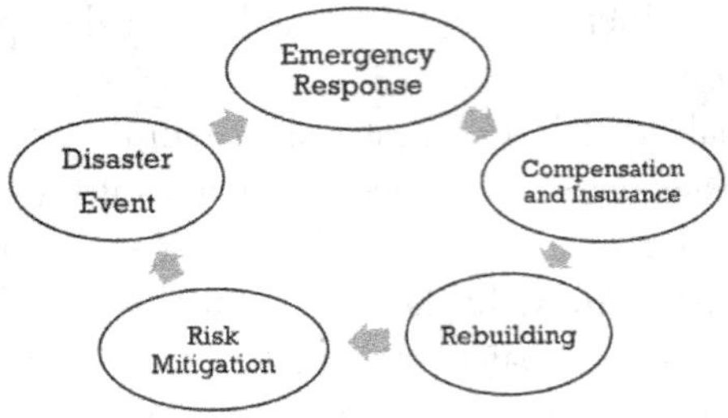

FIGURE 4 **The Cycle of Disaster Law**

building may be as important as direct assistance. The scale of the disaster is also clearly a relevant factor.

The remaining sections of this chapter work through the phases of the disaster cycle, followed by some brief closing thoughts. Institutionally, risk reduction, disaster response, financial compensation, and rebuilding may involve different international institutions and procedures. But none of them can be considered in isolation from the others in mapping disaster management strategies.

MITIGATION OF DISASTER RISKS

If we begin consideration of the disaster cycle in the lower corner of Figure 4, the first issue to consider is pre-disaster risk mitigation. Mitigation efforts attempt to lessen the potential impact or likelihood of disaster events before the fact, and successes and failures at this stage can spell the difference between a routine, manageable disruption and a major catastrophe. Hurricane Katrina would have been a much different story if the levees had not collapsed; an early warning system would have changed the impact of the Southeast Asian tsunami.

The damage wrought by a natural event – extreme weather or earthquakes, for example – is linked with human agency and manipulation of the natural environment, both at the site of the disaster itself and more generally with anthropogenic climate change. It is almost a cliché in the field that there is no such thing as a truly natural disaster. Physical "phenomena are a necessary component of risk, but they are only the starting point in addressing safety concerns" – calculating and planning for disaster risk must account for "acts of nature, … weaknesses of human nature, and … side effects of technology."[14]

Mitigation of disaster risks takes place across a broad range of settings. For instance, the Fukushima disaster has led to arguments to strengthen international regulation of nuclear energy.[15] Similarly, international law may be useful in

[14] Daniel A. Farber et al., "Reinventing Flood Control," *Tulane Law Review* 81 (2006): 1085, 1090.

[15] See Emily Benz, "Lessons from Fukushima: Strengthening the International Regulation of Nuclear Energy," *William & Mary Environmental Law & Policy Review* 37 (2013): 845.

addressing new and potentially high-risk technologies.[16] The growing interest of the international community on risk reduction was evidenced by the UN General Assembly's 1999 adoption of the International Strategy for Disaster Reduction, with the UN establishing an office on disaster risk reduction, the UNISDR, to implement the strategy.[17]

The greatest opportunity for international law to mitigate disaster risks involves climate change. With rare exceptions, recent years rank at the top of the list of the warmest global temperatures,[18] and depending on future emissions and climate sensitivity, the world will ultimately end up 2–7°C warmer than it is today.[19] Temperature change in the Arctic will be about twice as large.[20]

Even warming of 2°C, which may be the best we can hope for, would leave the earth warmer than it has been in millions of years.[21] The World Bank recently examined the more serious outcomes with greater (but not at all implausible) warming:

> The 4°C scenarios are devastating: the inundation of coastal cities; increasing risks for food production potentially leading to higher malnutrition rates; many dry regions becoming dryer, wet regions wetter; unprecedented heat waves in many regions, especially in the tropics; substantially exacerbated water scarcity in many regions; increased frequency of high-intensity tropical cyclones; and irreversible loss of biodiversity, including coral reef systems.
>
> And most importantly, a 4°C world is so different from the current one that it comes with high uncertainty and new risks that threaten our ability to anticipate and plan for future adaptation needs.[22]

The latest report of the Intergovernmental Panel on Climate Change (IPCC) can only heighten concerns about climate impacts.[23] Without mitigation efforts, the IPCC projects that global temperatures could rise 4°C by the end of this century which reaffirms the temperature change discussed in the World Bank's report.[24] Depending on mitigation efforts and the sensitivity of the climate system, sea level rise could range from about a quarter of a meter to a full meter by 2100.[25] In addition, dry regions will likely get drier and wet regions wetter.[26] "Extreme

[16] Grant Wilson, "Minimizing Global Catastrophic Existential Risks From Emerging Technologies Through International Law," *Virginia Environmental Law Journal* 31 (2013): 307 (discussing bioengineering, nanotechnology and artificial intelligence).

[17] See www.unisdr.org/who-we-are/mandate.

[18] David Archer and Stefan Rahmstorf, *The Climate Crisis: An Introductory Guide to Climate Change* (New York: Cambridge University Press, 2010), 43.

[19] Ibid., 129. [20] Ibid., 133. [21] Ibid., 225.

[22] World Bank, *Turn Down the Heat: Why a 4°C Warmer World Must be Avoided* (Washington, D.C.: World Bank, 2012): v. World Bank researchers have begun to consider ways to increase resilience. See Anne T. Kuriakose et al., "Climate-Responsive Social Protection," *Developmental Policy Review* 31 (2013): 31.

[23] Intergovernmental Panel on Climate Change, *Climate Change 2013: The Physical Science Basis – Summary for Policymakers* (2013).

[24] Ibid., Figure SPM7(a). [25] Ibid., Figure SPM.9. [26] Ibid., 17.

precipitation events over most of the mid-latitude land masses and over wet tropical regions will very likely become more intense and more frequent" as well.[27]

In coming years, extreme events such as fires, floods, and heat waves will become more widespread.[28] Adaptation to these impending changes poses serious challenges.[29] "Extreme events such as floods and drought cause extensive damage to many parts of society, and thus a critical issue for adaptation is the degree to which frequency, intensity, and persistence of extreme events change."[30] International law, especially in the form of the United Nations Framework Convention on Climate Change (UNFCCC) negotiation process, is critical to the effort to head off these increased disaster risks.

International law now recognizes a duty to engage in adaptation efforts.[31] According to the Cancun Agreements adopted at the 2010 UNFCCC Conference of Parties, "Enhanced action on adaptation ... should follow a country-driven, gender-sensitive, participatory and fully transparent approach, taking into consideration vulnerable groups, communities and ecosystems..."[Part II, Para.12]. Moreover, the Conference of the Parties "[i]nvites all Parties to enhance action on adaptation ... taking into account their common but differentiated responsibilities and respective capabilities by undertaking [specified actions]." [Part II, Para. 14][32]

In ratifying the UNFCCC, developed countries – including the United States – committed themselves to regulating greenhouse gases.[33] Two provisions of the UNFCCC are particularly relevant. First, in Article 3(3) of the Convention, the parties adopt the principle that they should take "precautionary measures to anticipate, prevent or minimize the causes of climate change and mitigate its adverse effects." Second, under Article 4(1)(b), the parties agree to "formulate,

27 Ibid., 16.

28 Ibid., 174; Heidi Cullen, *The Weather of the Future: Heat Waves, Extreme Storms, and Other Scenes from a Climate-Changed Planet* (New York: Harper Collins, 2010). On the flooding issues, see Howard C. Kunreuther and Erwann O. Michel-Kerjan, *At War with the Weather: Managing Large-Scale Risks in a New Era of Catastrophes* (MIT Press, 2011): 11–12 (impact of climate change on catastrophic weather events).

29 These challenges are discussed in Tim Bonyhady, Andrew Macintosh, and Jan McDonald, *Adaptation to Climate Change: Law and Policy* (Sydney: The Federation Press, 2010); U.S. Government Accountability Office, *Climate Change Adaptation: Strategic Federal Planning Could Help Government Officials Make More Informed Decisions*, www.gao.gov/products/GAO-10-113 (2010).

30 William E. Easterling III, Brian H. Hured, and Joel B. Smith, *Coping with Global Climate Change: The Role of Adaptation in the United States*, www.pewclimate.org/global-warming-in-depth/all_reports/adaptation (2004): 17.

31 Adaptation requires society to manage climate impacts using strategies of resistance, adjustment, and retreat. See Robert R. M. Verchick, "Adapting to Climate Change While Planning for Disasters: Footholds, Rope Lines, and the Iowa Floods," *BYU Law Review* (2011): 2203, 2209.

32 FCCC/CP/2010/7/Add.1

33 See United Nations Framework Convention on Climate Change, May 9, 1992 (New York), FCCC/Informal/84, GE.05-62220 (E) 200705, last modified 2013, http://unfccc.int/essential_background/convention/status_of_ratification/items/2631.php.

implement, publish and regularly update ... measures to mitigate climate change by addressing anthropogenic emissions by sources." In particular, developed countries such as the United States commit themselves to adopting "national policies and take corresponding measures on the mitigation of climate change, by limiting its anthropogenic emissions of greenhouse gases and protecting and enhancing its greenhouse gas sinks and reservoirs."[34] Even without a specific agreement on mitigation measures, the UNFCCC obligates nations to make a good faith effort to address emissions, individually and together.

International law and institutions will play a key role in mitigating the risks of climate change, for the simple reason that this global problem requires a global response. However much bottom-up national and even local actions may contribute to the solution, in the end, international efforts will be required to ensure cooperation. To the extent that climate change is a threat multiplier for disasters, reducing emissions is a risk reducer for natural disasters.

Of course, the role of the international system is not limited to the pre-disaster phase of the emergency cycle. It continues very prominently into the next phase, the emergency response. International participation in that stage is more visible, if only because emergency response efforts attract media attention.

THE EMERGENCY RESPONSE

Disasters are far from being a new issue for the international system. International disaster relief dates back at least to the great earthquake that destroyed Lisbon in 1755.[35] Three years later, the great international law scholar Emer de Vattel recognized a duty to provide humanitarian aid as part of international law.[36]

Thus, at the beginning of the modern era, the need for the international community to respond to major disasters was already recognized. But after this valiant start, progress in creating the legal foundation for such responses has been halting. What has resulted is a patchwork with major gaps.

Efforts to construct international legal regimes on disasters began in the late nineteenth century.[37] The Red Cross called for application of the Geneva Convention's protection of war casualties to victims of natural disasters.[38] By 1921, the Red Cross recommended an international convention on disasters, which led to the creation of an International Relief Union (IRU) – which alas did not survive World

34 The UNFCCC also contemplates international cooperation. Article 4(2)(a) provides that "Parties may implement such policies and measures jointly with other Parties and may assist other Parties in contributing to the achievement of the objective of the Convention."

35 Alejandra de Urioste, "When Will Help Be on the Way? The Status of International Disaster Response Law," *Tulane Journal of International and Comparative Law* 15 (2006): 18, 183.

36 David P. Fidler, "Disaster Relief and Governance After the Indian Ocean Tsunami: What Role for International Law?," *Melbourne Journal of International Law* 16 (2005): 458.

37 Ibid., 462. 38 Ibid.

War II.[39] The IRU also suffered from the absence of a financing mechanism.[40] Thus, what could have provided the skeleton for a multilateral disaster management framework instead atrophied and decayed.

The period after World War II saw the creation of bilateral disaster relief treaties and some regional efforts.[41] These treaties were originally narrow in scope, but became broader over time, covering issues such as responsibility and coordination, access of relief workers and equipment, and costs.[42] The IRU's successor was the United Nations Disaster Relief Coordinator,[43] later subsumed into the Office for the Coordination of Humanitarian Affairs (OCHA).[44] According to its website, OCHA has thirty offices around the world and a staff of 1,900.[45]

There are also some regional conventions and some specialized dealing with specific areas such as telecommunications and aviation after natural disasters.[46] For instance, the Tampere Convention covers the use of telecommunications resources in disaster relief.[47]

The Sphere Project has been one focal point for the effort to address international relief standards.[48] The Sphere Project began in 1997 by the International Federation of Red Cross and Red Crescent Societies (IFRC) and a group of nongovernmental organizations (NGOs).[49] Sphere provides a list of minimum standards, suggested practical actions, and key indicators to determine whether a standard has been met.[50] One of its key concerns is protection of vulnerable groups such as children.[51]

Moving from this NGO-oriented effort to the governmental level, there are some signs that international standards may yet crystallize. In January 2005, government officials met in Kobe, Japan to discuss disaster issues, a meeting shaped by the

39 Fidler, "Disaster Relief and Governance," 463; Urioste, "When Will Help," 184.

40 Fidler, "Disaster Relief and Governance," 464. In addition, the IRU failed to provide guidelines or standards for work in the field. Urioste, "When Will Help," 184.

41 Fidler, "Disaster Relief and Governance," 464; Urioste, "When Will Help," 183.

42 Urioste, "When Will Help," 188.

43 This office was intended as an information clearinghouse rather than an active provider of relief in the field. Urioste, "When Will Help," 185.

44 Urioste, "When Will Help," 186.

45 See www.unocha.org/about-us/who-we-are. (Describing OCHA's missions as mobilizing and coordinating effective and principled humanitarian action, advocating for the rights of people in need, promoting prevention of disaster, and facilitating sustainable solutions.) On OCHA's history, see www.unocha.org/about-us/who-we-are/history.

46 Fidler, *Disaster Relief and Governance*, 464; Urioste, "When Will Help," 188–190.

47 Urioste, "When Will Help," 189–190. A 1965 convention covers ships involved in disaster relief. Ibid.

48 Urioste, "When Will Help," 192.

49 See Sphere Project, *Humanitarian Charter and Minimum Standards in Humanitarian Response* (3d ed. and reprint, 2013): 4.

50 Ibid., 7.

51 Ibid., 31. Urioste speculates that the Sphere guidelines may eventually become translated into customary international law. Urioste, "When Will Help," 192.

occurrence of the Southeast Asian tsunami only days before.[52] One outcome was the Hyogo Framework for Action (HFA), which set priorities for the following ten years.[53] Consideration of the post-2015 period is currently underway.[54]

Yet, despite these areas of progress, the Red Cross still sees a "yawning gap" in international law regarding disaster response.[55] Considering the length of time that disaster relief has been on the international agenda, the amount of progress has not been impressive.

Some of the problems can be seen in the massive international response to the Southeast Asian tsunami, which involved more than 200 NGOs and a dozen governments. In the resulting confusion, communication was poor, efforts were duplicated, and lack of shared expertise led to failures in assistance.[56] For instance, because of competition between groups in the same area, information sometimes was not shared; some groups obtained information from village heads who did not take account of the needs of women; and resources were misallocated, such as the shipment of winter clothing to Sri Lanka.[57]

Some hints on how to proceed might be found in domestic law. We conventionally think of international law as starkly different from domestic law because of the lack of a global sovereign. Yet, in federalist systems, similar issues of coordination between multiple, semi-independent governments may arise. Although the international context for disaster response is undoubtedly different from the domestic setting, the differences may not be as large as they appear – at least, if the United States is any example. National sovereignty is a concern at the international level, as state-government sovereignty is at the local level.

The Stafford Act, the U.S. law governing emergency response, is careful to respect the authority of state governments. Under sections 401 and 501 of the Stafford Act,[58] the president cannot declare an emergency or national disaster without the request of the state governor. Thus, the emphasis is on responding to state requests, not on federal initiative.

The only exception is under section 501(b), which allows the president to provide federal emergency assistance without a state request when the president "determines that an emergency exists for which the primary responsibility for response rests with the United States because the emergency involves a subject area for which, under the Constitution or laws of the United States, the United States exercises preeminent responsibility and authority."[59] Even then, the president is required to consult with the governor of the affected state if practical. This authority was first invoked by President Bill Clinton after the Oklahoma City

52 Farber et al., *Disaster Law and Policy*, 403.

53 Ibid. For the text of the Hyogo Framework, see www.preventionweb.net/files/1037_hyogoframeworkforactionenglish.pdf.

54 See www.preventionweb.net/english/hyogo/post-hfa/.

55 Urioste, "When Will Help,"186.

56 Urioste, "When Will Help," 194–195.

57 Ibid.

58 42 U.S.C. §§ 5170, 5191.

59 42 U.S.C. § 5191(b).

bombing of a federal office building.[60] (Admittedly, the exact scope of this exception has never been clear, but it seems narrow.) Even after an emergency is declared, the federal government provides assistance and coordinates the response, but it cannot directly control local authorities such as police, fire, and medical personnel. As in the international sphere, there is great sensitivity to local prerogatives and severe reluctance to force assistance on unwilling local governments.

Thus, the relationship between states and the federal government is less hierarchical than one might expect. In particular, American states seem nearly as sensitive as national governments about the possibility of unconsented "assistance" being provided in their territories. Not surprisingly, this has also been a source of great controversy at the international level, where sovereignty concerns are even more sensitive.[61] The "responsibility to protect" concept began in the context of mass atrocities, and whether it should extend to natural disasters remains disputed.[62]

Moreover, just as cooperation between nations is important in disasters, so is cooperation between American states. An agreement between states called EMAC, the Emergency Management Assistance Compact,[63] provides for sharing of emergency responders and equipment. All fifty states have become members of EMAC.[64] The EMAC agreement provided valuable interstate assistance during Hurricane Katrina, with almost twenty thousand civilians and nearly fifty thousand National Guard being deployed by other states to Louisiana and Mississippi.[65]

In short, international law and domestic law might have something to teach other about cooperative agreements for disaster relief. This is notable since we often think of the two fields as quite distinct because of the absence of a global sovereign.

COMPENSATION, REBUILDING, AND RESETTLEMENT

After the emergency has passed, insurance, tort law, and government disaster assistance provide ways of spreading and shifting risks. When we cannot (or simply

[60] Farber et al., *Disaster Law and Policy*, 94.

[61] On the dispute under international law, see Jarrod Wong, "Reconstructing the Responsibility to Protect in the Wake of Cyclones and Separatism," *Tulane Law Review* 84 (2009): 219 (arguing that deliberate refusal of needed assistance could under some circumstances constitute a crime against humanity and trigger the "responsibility to protect" as a basis for intervention).

[62] Ibid., 244–246. For an argument in favor of extending the right to protect in the case of Burma, see Alison McCormick, "From Sovereignty to Responsibility: An Emerging International Norm and Its Call to Action in Burma," *Indiana Journal of Global Legal Studies* 19 (2011): 563. Failure to allow humanitarian assistance may be associated in practice with human rights violations or discrimination against particular ethnic groups or regions, blurring the line between disaster issues and human rights violations.

[63] P.L. 104–321, 110 Stat. 3877 (1996).

[64] Farber et al., *Disaster Law and Policy*, 176.

[65] Ibid., 177.

have not) acted to mitigate risks before the event or avoided them through swift response afterwards, these compensation mechanisms help blunt the impact and prepare for reconstruction or rebuilding.

Post-disaster compensation generally takes one of three forms: private insurance, government aid, or the tort system.[66] The first of these methods of compensation is private insurance, but this is often unavailable or fails to provide prompt and efficient compensation. Problems including the frequent unavailability of separate private insurance for catastrophic risks, exclusion of catastrophic risks from insurance coverage, and the difficulty of handling very large numbers of claims create significant hurdles. The second method of compensation, litigation against responsible parties, also has its limitations, and faces barriers even in developed countries like the United States and Japan. Third, is the possibility of obtaining compensation from the governments or international organizations.[67]

The U.S. flood insurance system provides an interesting case study in government-assisted compensation. The federal government provides insurance against flooding, with more than four million flood insurance policies now outstanding.[68] A recurrent problem has been the treatment of pre-existing structures, which have enjoyed subsidized rates.[69] The program includes mapping of the 100-year flood zones, and to limit moral hazard, requires communities to control development in the flood zone in order to participate in the program.[70]

The point of compensation is to give people the wherewithal to begin to rebuild their lives. For wealthier countries, the process is largely internal, apart from the contribution of the international reinsurance market to risk spreading. Poorer countries need more assistance, which often has come in the form of spontaneous responses to obvious humanitarian crises.

International mechanisms are beginning to emerge to a limited extent to assist countries hit by natural disasters. The International Monetary Fund (IMF) can make special funding available to countries if a disaster disrupts their balance of payments.[71] The IMF has "provided resources in cases of earthquakes, drought,

[66] These issues are discussed in detail, in a multinational comparison, by Robert L. Rabin and Suzanne A. Bratis, "United States," in *Financial Compensation for Victims of Catastrophes: A Comparative Approach*, Michael Faure and Ton Hartlief, eds., (New York: Springer, 2006): 303, 356. The remainder of the paragraph in the text presents some of their findings. For a discussion focused on developing countries, see Michael Faure, "Toward Effective Compensation for Victims of Natural Catastrophes in Developing Countries," in Faure and Wibisana, *Regulating Disasters*, 243–76.

[67] Liability and government-supported insurance can also be combined. See W. Kip Viscusi and Richard J. Zeckhauser, "Deterring and Compensating Oil-Spill Catastrophes: The Need for Strict and Two-Tier Liability," *Vanderbilt Law Review* 64 (2011): 1717 (suggesting a combination of private liability and tax-supported compensation for damages that exceed the defendant's financial capacity).

[68] Farber et al., *Disaster Law and Policy*, 312. [69] Ibid. [70] Ibid., 312, 315–8.

[71] Joseph Gold, "Natural Disasters and Other Emergencies Beyond Control: Assistance by the IMF," *International Lawyer* 24 (1990): 621.

hurricanes, floods, and cyclones."[72] The IMF's sister institution, the World Bank, has also been involved in disaster funding. Examples are provided by the creation of "special purpose funds" by the Bank for reconstruction efforts in Indonesia (after the Southeast Asian tsunami) and Haiti.[73]

If disaster risks escalate as expected during this century, we may need to formalize funding for rebuilding rather than relying on these rather ad hoc measures by the existing international financial institutions. Nations may be understandably reluctant to commit funding in advance rather than making individualized decisions about how much to contribute after each disaster. So long as major disasters are relatively few and far between, this case-by-case determination may be seem attractive. Yet, an increased pace of disasters may strain this approach and make a more systematic, institutionalized rebuilding effort seem preferable.

Rebuilding has significance in its own right, but it is also in some sense just the beginning of a new cycle, inasmuch as it incorporates mitigation efforts for the next disaster down the road.[74] The emergency response may also determine whether evacuees turn into permanent migrants, possibly in areas with lower risk levels, or whether they return to impacted areas and (hopefully) improved mitigation measures.

International human rights law may speak to the issue of rebuilding. For instance, there may be a right to housing that is implicated by destruction of residences.[75] Of course, in most cases, governments already have reasons to provide shelter and rebuild communities where they have the capacity to do so. But some groups may be shortchanged. Discrimination against women or ethnic minorities in the rebuilding process is a definite possibility in some countries and triggers other human rights obligations. Yet, after-the-fact complaints about human rights are unlikely to be as effective as before-the-fact capacity-building so that responders understand appropriate ways of responding to the needs of disadvantaged groups.

Rebuilding should be an opportunity to think about mitigation: changes in land use and infrastructure to reduce the expected harm from disasters. So we are now about to begin the cycle again. If we are unsuccessful in addressing some of the root causes, such as climate change and demographic trends, the next cycle may begin with a worsened level of risk, converting the disaster cycle into a kind of downward spiral.

Many victims of disasters may be unable or unwilling to return to their original locales. International law once again becomes relevant to their status. Migration is

[72] Ibid., 633.

[73] Jenny R. Hernandez and Anne D. Johnson, "A Call to Respond: the International Community's Obligation to Mitigate the Impact of Natural Disasters," *Emory International Law Review* 25 (2011): 1087, 1093.

[74] For a discussion of recovery strategies to increase resilience and reduce vulnerability, see Farber et al., *Disaster Law and Policy*, 348–57.

[75] See Charles W. Gould, "The Right to Housing Recovery After Natural Disasters," *Harvard Human Rights Journal* 22 (2009): 169.

a well-established response to disasters and other adverse events[76] that will increase with climate change:

> [M]any people refer to this class of persons as "climate migrants" or "environmentally displaced persons," and seek protection for them through other legal avenues, primarily through domestic law and proposed expansions to the international climate regime. These efforts, however, have proven controversial and largely unsuccessful, leaving these people with minimal legal protections and reliant on inadequate sources of climate change adaptation assistance for relief.[77]

There have been proposals for an international convention to protect these individuals, at least when they cross national borders.[78] One blind spot involves the treatment of internally displaced persons, who do not have clear protection under existing law.[79] Some writers have called for an international convention that covers internal displacement.[80] If climate change results in the increased degree of displacement that some predict, it may not be possible for the international community to avoid coming to grips with this problem.

CONCLUSION

As discussed, the disaster cycle provides a way of organizing the multitude of issues that compose disaster law. The disaster cycle is not merely a convenient way of ordering events. It is also important because no one stage can be considered in isolation. Rather, each stage can only be properly designed in the context of the other stages.

For instance, the appropriate degree of risk mitigation may depend on expectations about emergency response. If planners assume that a disaster will lead to successful evacuation they will downplay potential risks to life in their efforts to mitigate risks. Thus, the existence of a tsunami warning system could lead planners to allow more development near the coast. Of course, that may counteract the beneficial effects of the warning system, at least to an extent, because more people

76 See Leah P. Boustan, Matthew E. Kahn, and Paul W. Rhode, *Moving to Higher Ground: Migration Response to Natural Disasters in the Early Twentieth Century*, 102 Amer. Econ. Rev.: Papers and Proceedings 238 (2012),

77 Cinnamon P. Carlarne, "Risky Business: The Ups and Downs of Mixing Economics, Security and Climate Change," *Melbourne Journal of International Law* 10 (2009): 439, 464.

78 Benoit Mayer, "The International Legal Challenges of Climate-Induced Migration: Proposal for an International Legal Framework," *Colorado Journal of International Environmental Law and Policy* 22 (2011): 357; Bonnie Docherty and Tyler Giannini, "Confronting A Rising Tide: A Proposal for a Convention on Climate Change Refugees," *Harvard Environmental Law Review* 33 (2009): 349.

79 See Cristiano d'Orsi, "Strengths and Weaknesses in the Protection of the Internally Displaced Persons in Sub-Saharan Africa," *Connecticut Journal of International Law* 28 (2013): 73.

80 David Hodgkinson and Lucy Young, *In the Face of Looming Catastrophe': A Convention for Climate Change Displaced Persons*, CCDP Convention (August 2012), www.ccdpconvention.com/documents/Updated%20treaty%20proposal.pdf.

will be exposed to the tsunami risk. Similarly, the availability of insurance may reduce incentives to mitigate risks, requiring careful integration of tools for addressing the risk mitigation and compensation stages of the disaster cycle.

On the other hand, if disaster responders assume that risk mitigation will be successful, they will make less of an investment in planning for an emergency response, and the response will focus on those risks that could not be effectively mitigated in advance. But care must be used in making those determinations, lest excessive optimism about risk mitigation lead to under preparation for emergency response. In the Fukushima context, overconfidence in reactor safety led to a de-emphasis on emergency planning. Overconfidence in sea walls and levees may have contributed to additional flood deaths in Japan and New Orleans.

Moreover, the disaster cycle is illuminating simply because the loop is closed. It is always tempting to downplay the risk of future disasters after a disaster has taken place. Every disaster renews the cycle, calling for a new round of risk mitigation. But each disaster also creates the opportunity for learning about strategies for disaster risk management by individuals, nations, and the international community.

Disasters pose a problem that cannot be ignored and that will only get worse, so long as our populations are growing, so long as so much of the population lives in disaster-prone areas, so long as climate change continues to spawn weather extremes. When they are too big for nation- states to handle on their own, the issues inevitably spill into international law.

If all countries were equal in terms of risk exposure and ability to respond, creating an international agreement would probably be much easier. But countries are unequally situated, with some much more exposed to risk and some much better endowed with resources and response capacity. Thus, as in many international issues, distributional problems are an obstacle to agreement. One might speculate that the increased prominence of emerging economies might help reduce the polarity between developed and developing countries that may hinder efforts at agreement.

We cannot eliminate disasters from our future. As scholars, we cannot even contribute directly to disaster management. But legal scholarship, such as that presented in this volume, may help society become more resilient and its ability to cope with disaster more robust.

2

Natural Disasters and the Theory of International Law

Toshiya Ueki

When we examine the rules of international law applicable to large-scale natural disasters like earthquakes, tsunamis, hurricanes, floods, and volcanic eruptions, the first thing to clarify is to what situations these rules could be applicable – in other words, the *ratione materie* of these rules. Next, we should also consider the *ratione temporis* of these international rules, namely, applicability in the pre-disaster, in-disaster, and post-disaster phases. The issue of *ratione persone* must also be examined, namely, to whom these international rules are addressed: affected states, individual victims, international relief organizations, NGOs, and non-state entities. Finally, the *ratione loci* of these rules must be determined, in other words, the geographical or territorial sphere of applicability. When we consider the whole structure of these international rules applicable to large-scale natural disasters, at least from a theoretical point of view, a "Kelsenian" approach to law might be regarded as an effective analytical method.

INTRODUCTION

Hans Kelsen (1881–1973) was an Austrian legal philosopher who brought a particular order to the world of law during his academic career at the University of Vienna and, upon fleeing Austria by way of Switzerland during the rise of National Socialism, at the University of California, Berkeley. Kelsen developed his "Pure Theory of Law" in the early twentieth century as a method of rationalizing legal legitimacy. In it, he held that:

> Legal norms necessarily come in systems. There are no free-floating legal norms... Furthermore, legal systems are themselves organized in a hierarchical structure, manifesting a great deal of complexity but also a certain systematic unity. We talk about Canadian law, or German law, etc., not only because these are separate countries in which there is law. They are also separate legal

systems, manifesting a certain cohesion and unity. This systematic unity Kelsen meant to capture by the following two postulates:

(1) Every two norms that ultimately derive their validity from one basic norm belong to the same legal system.
(2) All legal norms of a given legal system ultimately derive their validity from one basic norm.[1]

And when one follows the derivative legal authority of law back up the line to a final resting place, like a constitution, there must at that point rest a presupposition of validity.[2] Kelsen's Pure Theory of Law in general, as well as his theory on international law, was a very strong influence from the 1920s to the mid-1930s not only in Europe but also in Japan.[3] Kelsen's legal theory also held a strong influence on some of the leading Japanese international law scholars.[4] However, these theoretical impacts by Kelsen on Japanese scholars were limited to a very small circle of intelligentsia mainly in universities at that time, and had no impact on the actual diplomatic policy adopted by the Imperial Japanese government during the 1930s.

Even after 1945, the Kelsenian view of international law theory survived in the academic circle of international law in Japan to some degree although Kelsen's theoretical impact or influence on international law theory in the United States, Canada, and Oceania seemed rather limited. Even so, in 1971, a special volume of *California Law Review* was published celebrating the ninetieth anniversary of his birth.[5] In the first paragraph of the preface, Albert A. Ehrenzweig wrote: "For half a century, there have hardly been publications in legal philosophy anywhere, books or articles, that have not praised or opposed the Pure Theory of Law. But whether in praise or opposition, legal philosophers all over the world agree with Roscoe Pound who saw in Hans Kelsen one of the world's outstanding men of the law."[6]

1 Andrei Marmor, "The Pure Theory of Law," in *The Stanford Encyclopedia of Philosophy* (Fall 2010 Edition), ed. Edward N. Zalta available at plato.stanford.edu/entries/lawphil-theory/.

2 Ibid.

3 Martti Koskenniemi describes as follows; "Many other German-language international lawyers participated in the Kelsen *Kreis* from 1911 that grew into the *Vianna School* in the 1920s." Martti Koskenniemi, *The Gentle Civilizer of Nations, The Rise and Fall of International Law 1870–1960* (Cambridge University Press, 2002), 246.

4 Hans Kelsen's main academic works published in 1920s and 1930s, during his chair of professorship at the University of Vienna and in Geneva, *Allgemaine Staatslehre* (Berlin: Verlag von Juliua Springer, 1925) and *Reine Rechslehre* (Vienna: Verlag Franz Deuticke Wein, 1934), were quickly translated into Japanese in 1931 by Shiro Kiyomiya and in 1935 by Kisaburo Yokota, respectively, each of whom was the most eminent scholar in constitutional law and international law in Japan at that time. Hans Kelsen, *Keruzen Ippan Kokkagaku*. Translated by Shiro Kiyomiya. (Tokyo: Iwanami Shoten, 1931). Hans Kelsen, *Keruzen Jyunsui Hogaku*. Translated by Kisaburo Yokota. (Tokyo: Iwanami Shoten, 1935).

5 University of California Berkley, *Essays in Honor of Hans Kelsen, Celebrating the 90th Anniversary of His Birth* (Colorado: Fred B. Rothman & Co., 1971)

6 Ibid., 609.

Kelsen's work has been applied in many fields of law and his ideas about the operation of law in society and its efficacy continue to resonate around the world. This chapter considers the application of Kelsen's ideas to the international law of disaster relief. The importance of this inquiry rests in the reality that disaster relief must be effectuated in specific locations controlled by nation-states. Rarely is disaster relief needed in the areas beyond national jurisdiction. Consequently, the law that operates validly (locally or internationally) to provide for such relief must be located, and thereby legitimized, in some manner.

LEGAL FRAMEWORK ON NATURAL DISASTERS — DID THE KELSENIAN LEGAL STRUCTURE SURVIVE?

As for the international law on natural disasters, when the International Law Commission (ILC) of the United Nations General Assembly started its work on the "Protection of Persons in the Event of Disasters," the reports submitted by Special Rapporteur Eduardo Valencia-Ospina were, seemingly, influenced by the Kelsenian structure of law. The preliminary report as well as the second report submitted by Valencia-Ospina clustered the discussion around three axes: *ratione materiae*, *ratione personae*, and *ratione temporis*.[7] In addition to these three aspects, Valencia-Ospina also pointed out that the question of whether the scope of the report should include *ratione loci* was left open.[8] These four aspects of validity spheres of law mirror the structure of Kelsen's legal theory on international law. In his major monograph on international law Kelsen states:

> International law is a normative order, and a normative order is a system of valid norms. Legal norms regulate human behavior, and human behavior takes place in time and space. Consequently, legal norms have relation to time and space. They are valid for a certain time and for a certain space (territory). Hence we speak of a temporal and territorial sphere of validity of legal norms or a legal order.[9]

According to Kelsen, every legal norm or legal order has its own sphere of validity, namely, temporal, territorial, personal, and material.[10] It is from these four aspects, according to his theory, that every legal rule or legal regime shall be regulated.

When one examines the legal norms of international law applicable to the situation of large-scale natural disasters like earthquakes, tsunamis, hurricanes, floods, and volcanic eruptions, the first thing to clarify under Kelsen's framework

[7] Eduardo Valencia-Ospina, *The Preliminary Report on the Protection of Persons in the Event of Disasters*. UN Doc A/CN 4/598, 2008, 15–21. Eduardo Valencia-Ospina, *Second Report on the Protection of Persons in the Even of Disasters*. UN Doc A/CN 4/615, 2009, 6–11.

[8] Valencia-Ospina, *The Preliminary Report*, 15, par. 43.

[9] Hans Kelsen, *Principles of International Law* (Colorado: Rinehart and Company, Inc., 1952), 93. Hans Kelsen, *Principles of International Law*. Edited by Robert W. Tucker. Vol. 2. Holt, Rinehart, and Winston, Inc., 1966, 177.

[10] Ibid. *First Edition*, 93–202.; Ibid. *Revised Edition*, 177–301.

is the *ratione materie* of such rules – in other words, to what situations would these rules apply? Second, one would examine the *ratione temporis* of these international rules – namely, applicability of the rules on pre-disaster, in-disaster and post-disaster phases. Third, the *ratione persone* of these rules must be considered – namely, to whom are these international rules on disasters addressed? For instance, these rules might be, or shall be, addressed to the affected states, individual victims of natural disasters, and/or international organizations, NGOs, and non-state entities. And fourth, what is the *ratione loci* – in other words, what is the geographical or territorial sphere of applicability of such international rules? Does it extend from the territory, territorial waters, and airspace of the affected states to areas beyond national jurisdictions, like high seas, the deep seabed, and outer space?

When one considers the whole structure of the international rules applicable to the situation of large-scale natural disasters, at least from a theoretical point of view, this Kelsenian approach might be regarded as an effective analytical method. Following the Kelsenian method of analysis, the international law of disaster relief may be fitted into the framework in four distinct spheres.

KELSEN'S FOUR ASPECTS OF THE SPHERE OF VALIDITY

Ratione Materiae *of the Rules*

One must first examine the concept or notion of disaster in the context of international law. The term "disaster" has not yet been clearly defined as a strict legal term in international law. For instance, the Tampere Convention of 1998 defined a disaster as "a serious disruption of the functioning society that poses a significant, widespread threat to human life, health, property or the environment, whether caused by accident, nature or human activity, and whether developing suddenly or as the result of complex, long-term processes."[11]

According to the Association of Southeast Asian Nations' (ASEAN) Agreement on Disaster Management and Emergency Response, disaster means a serious disruption of the functioning of a community or a society causing widespread human, material, economic or environmental losses.[12] On the other hand, the Hyogo Declaration and Hyogo Framework of Action (HFA) 2005–2015 adopted

[11] "Tempere Convention on the Prevention of Telecommunication Resources for Disaster Mitigation and Relief Operations." Tempere, June 18, 1998, Article 1, par. 6. This Convention was entered into forces in January 2005 and there are 46 parties in September 2012.

[12] ASEAN, "ASEAN Agreement on Disaster Management and Emergency Response," *ASEAN Documents Series* 2005, 157–165. This Convention was adopted on July 26, 2005, but not yet been entered into forces. According to Article 33 of this Agreement, the ratifications by all ten member states of ASEAN are necessary in order to enter into force in this Convention.

by the World Conference on Disaster Reduction held in Kobe, Japan in 2005 defined a "hazard" as follows:

> A potentially damaging physical event, phenomenon or human activity that may cause the loss of life or injury, property damage, social and economic disruption or environmental degradation. Hazard can include latent conditions that may represent future threats and can have different origins: natural (geological, hydro-meteorological and biological) or induced by human processes (environmental degradation and technological hazards).[13]

These definitions of disaster and hazard in international documents seem too large and all-encompassing from the legal point of view. Thus, Kelsen would deem the applicability factor as very widespread. A new definition, however, has emerged in the ILC's Draft Articles on Protection of Persons in the Event of Disasters, where a disaster is defined as "a calamitous event or series of events resulting in widespread loss of life, great human suffering and distress, or large-scale material or environmental damage, thereby seriously disrupting the functioning of society."[14]

Ratione Temporis *of the Rules*

Second, one must ask what is the *ratione temporis* of the international rules on disasters? As for the legal rules applicable to the situation on large-scale natural disasters, there are three stages: first, at pre-disaster stage, rules on preventions and predictions of disasters should be applied; second, at the time of disasters, rules for the rescue and humanitarian assistance operations should be applicable; third, at post-disaster stage, rules on rehabilitation and reconstruction phases should be applied. It is also very important, in substance, to ensure smooth coordination between international organizations and NGOs on the one hand and domestic agencies on the other at the time of break out of large-scale natural disasters.

According to the commentary of the ILC's Draft Articles on Protection of Persons in the Event of Disasters, the phrase "in the event of" in the title implies that *ratione temporis* of these draft articles is primarily focused on the immediate post-disaster response and recovery phase including the post-disaster reconstruction phase.[15] Nevertheless, it is generally agreed in the ILC that the draft articles should also, where relevant, cover the pre-disaster phase as relating to disaster risk reduction and disaster prevention and mitigation activities.[16]

[13] United Nations. *Report of the World Conference on Disaster Reduction.* U.N. Doc A/CN. 206/ 6. Resolution 2, n. 2.

[14] Article 3 "Definition of disaster" of the Draft Articles on the Protection of Persons in the event of Disasters provisionally adopted by the Commission, *International Law Commission, Report on the work of its Sixty-Second Session (3 May to 4 June and 5 July to 6 August 2010)*, Official Records of the General Assembly, Sixty-fifth Session, Supplement No.10 (UN Doc. A/65/10), 321, par. 320.

[15] Ibid., 322, par. 331. [16] Ibid.

Ratione Personae *of the Rules*

Third, the *ratione personae* of these rules must be used to determine upon whom they operate. On this point, the victims of disasters and the affected states are the two main subjects of such international rules. Consequently, both natural persons and states as corporate persons are considered. Article 1 of the ILC's Draft Articles states: "The present Draft Articles apply to the protection of persons in the event of disasters." It is obvious, from this article, that *ratione personae* of these draft articles is limited, at first stage, to natural persons affected by disasters.

In order to protect these individual victims of natural disasters, affected states shall cooperate with other states, the United Nations, and other competent intergovernmental organizations.[17] The "duties" of the affected states by natural disasters are provided for in Article 9 and Article 10 of the draft articles. Article 9 provides the affected state's duty to ensure the protection of persons and provision of disaster relief and assistance on its territory, and Article 10 provides the affected state's duty to seek assistance from among other states, the United Nations, other competent intergovernmental organizations, and relevant NGOs, as appropriate.[18] In this regard, therefore, the affected states by natural disasters are also within the scope of *ratione personae* of the ILC's Draft Articles.

It should be also noted, however, that international organizations, NGOs, and other non-state entities would, actually, take very important roles at the time of break out of natural disasters. In this context, the applicability of the rules of Human Rights Law in emergency situations of large-scale natural disasters might be examined. The ILC's Draft Articles note as much in Article 8, which provides for "Human Rights."[19] As for the relationship with international humanitarian law, however, Article 4 states: "The present draft articles do not apply to situations to

[17] Article 5 of the ILC's Draft Articles provides "Duty to cooperate" for the states whose territory is suffered or affected by natural disasters. Article 5 provides as follows: "In accordance with the present draft articles, States shall, as appropriate, cooperate among themselves, and with the United Nations and other competent intergovernmental organizations, the International Federation of Red Cross and Red Crescent Societies and the International Committee of the Red Cross, and with relevant non-governmental organizations." International Law Commission, *Report on the work of its Sixty-Second Session (3 May to 4 June and 5 July to 6 August 2010)*, Official Records of the General Assembly, Sixty-fifth Session, Supplement No.10 (UN Doc. A/65/10), 321, par. 330.

[18] Texts and Commentaries of Article 9 and Article 10 of the Draft Articles are in International Law Commission, *Report on the work of its Sixty-Third Session (26 April to 3 June and 4 July to 12 August 2011)*, Official Records of the General Assembly, Sixty-sixth Session, Supplement No.10 (UN Doc. A/66/10), 261–267, par. 289.

[19] Texts and Commentaries of Article 7 ("Human Dignity") and Article 8 ("Human Rights") of the Draft Articles are in International Law Commission, *Report on the work of its Sixty-Third Session (26 April to 3 June and 4 July to 12 August 2011)*, Official Records of the General Assembly, Sixty-sixth Session, Supplement No.10 (UN Doc. A/66/10), 258–261, par. 289.

which the rules of international humanitarian law are applicable."[20] The provision of this article implies that the ILC's Draft Articles on natural disasters could not be applicable, in principle, to the situation of armed conflict as long as the rules of international humanitarian law could be applied.

Ratione Loci *of the Rules*

Fourth, the *ratione loci*, or territorial or geographical sphere of applicability of such international rules, should be applied not only to the territory, territorial waters, and airspace of the affected states by natural disasters but also to some areas beyond national jurisdictions, like high seas and the deep seabed, which are also governed, albeit more loosely, under the spheres of these international rules.

For instance, as the result of the Great East Japan Earthquake on March 11, 2011, huge amounts of wreckage washed away from large coastal areas along the Pacific Ocean in Japan. This wreckage flowed into a very wide range of the Pacific Ocean. Some portions of wreckage flowed into the high seas and arrived on the other side of the Pacific Ocean in the territorial waters and the Exclusive Economic Zones (EEZs) of the United States and Canada. The existing legal rules for dealing with this wreckage, one of the consequences of the natural disaster, are not so clear. Thus, it could be said that there exist some legal vacuums in the law of the sea and international environmental law on this issue. On the other hand, these phenomena imply that *ratione loci* of the international rules on natural disasters could extend not only to the territory of the affected states by disasters but also to the territory or the areas under the jurisdiction of other states, and also to high seas and the deep seabed.

CONCLUSION

The international law on natural disasters is a new area in international law, and the whole framework of this new area of law itself has not yet been made evident. Therefore, especially from the theoretical point of view, it is essential to clarify the spheres of validity for these new rules as they exist and as they develop into new legal regimes with wider and perhaps deeper applicability. This is the reason why the Kelsenian approach of international law, which might be viewed today as too formalistic and "out-of-date," has its own value and is actually a very effective means to tackle many of the issues presented in the international law of disaster relief.

[20] Text and Commentary of Draft Article 4 ("Relationship with international humanitarian law") is in International Law Commission, *Report on the work of its Sixty-Second Session (3 May to 4 June and 5 July to 6 August 2010)*, Official Records of the General Assembly, Sixty-fifth Session, Supplement No.10 (UN Doc. A/65/10), 327, par. 331.

The Finnish diplomat and member of the ILC, Martti Koskenniemi, describes Kelsen as follows:

> Most international lawyers today know of his work and many continue to admire him as legal theorist. The admiration is almost without exception tempered by the rejoinder that, of course, it was just theory and therefore could never be realized in practice. This is a problematic position: that theory is right but nonetheless useless or even dangerous and that moral goodness lies in our carrying out practices on the pragmatic basis betrays an altogether objectionable admiration of the profession at its most bureaucratic, a most unreflective, most self-perpetuating form of elitism.[21]

As for a newly evolving and developing area in international law, such as international rules on natural disasters, however, Kelsen's formalistic approach for law in general and international law in particular seems a very effective means of theoretical analysis.

[21] Koskenniemi, *The Gentle Civilizer of Nations*, 249.

3

International Disaster Relief Law and Article 38(1)(c) of the Statute of the International Court of Justice: The Forgotten Source of International Law

Imogen Saunders

Disaster mitigation in international disaster relief law operates largely in the area of soft law, with various institutions encouraging the adoption of guidelines and best practice standards by national governments. Current international law regulation of disaster risk reduction is scarce – no conventional obligations require states to mitigate disasters within their own territories. Further, the primarily internal nature of such requirements means customary international law, best used to regulate external state behavior, struggles to govern disaster mitigation. However, when examining the existing international law for state disaster mitigation, a third source of law may be most appropriate – the general principles of law recognized by civilized nations. This chapter explains how general principles as a source of law could be used to develop international law with regard to disaster risk reduction, using a case study of wildfire risk management.

INTRODUCTION

While progress has been made to codify conventional international law, no binding treaties on the international law of disaster relief are currently in force. Likewise, customary international law is less than clear. In the absence of conventional or customary law, states look to general principles of law for guidance in their conduct of international legal relations. This chapter proposes not only that states look to general principles of law for disaster relief, specifically on the question of wildfires, but more particularly to ecological practices of indigenous communities within states that have proven effective in combating natural disasters as equally incumbent general principles.

In so doing, the concept of general principles is greatly expanded, offered more weight in the international legal field as a source of international law, and the voice of native peoples is accorded the resonance it deserves. Particularly in the field of wildfire management, the case study offered in this chapter demonstrates the wisdom of practices by indigenous populations within nation-states to manage such disasters when they occur. If such disaster management practices are available to other affected states via general principles, then so much the better.

WILDFIRES AND DISASTERS

National and international disasters caused by wildfire occur throughout the world. In 2009, Australia suffered its worst known wildfire, as the Black Saturday fires tore through Victoria, devastating bush land and killing 173 people.[1] From 1997 to 1999, massive wildfires burned in Mexico, Brazil, Guatemala, Honduras, Nicaragua, Russia, Tanzania, Kenya, Rwanda, Congo, Senegal, Indonesia, Malaysia, and Papua New Guinea.[2] The smoke from these fires choked airports[3] and caused transnational air pollution in neighboring states.[4]

Some predict that as global temperatures increase, wildfires will become both more common, as fire risk seasons lengthen,[5] and more intense.[6] So called "megafires"[7] have the potential to cause transnational harm if they burn across national borders. The question becomes, in the context of international disaster relief law, can, or indeed does, international law contribute to the mitigation of devastating wildfires. This chapter argues that international law through the application of general principles of law may provide an internationally harmonized framework for managing risks associated with domestic wildfires.

States deal with wildfires in a variety of ways. However, modern thinking encompasses a "fire management" system, where states undertake various activities to decrease the risk and devastation of wildfires.[8] As will be discussed, much of state reaction to wildfire risk occurs at the domestic level – often at the urging of international bodies.

1 Liz T. Williams, "The Worst Bushfires in Australia's History," *Australian Geographic*, www.australiangeographic.com.au/topics/features/2011/11/the-worst-bushfires-in-australias-history/

2 Nicholas Robinson, "Forest Fires as a Common International Concern: Precedents for the Progressive Development of International Environmental Law," *Pace Environmental Law Review*, 2001: 460–461.

3 Ibid., 460–462.

4 Simon S C. Tav, "Southeast Asian Fires: The Challenge for International Environmental Law and Sustainable Development," *Georgetown International Environmental Law*, 1999: 241.

5 C. Giannakopolous, "Climatic Changes and Associated Impacts in the Mediterranean Resulting from 2°C Global Warming," *Global and Planetary Change*, 2009: 213.

6 Yongqiang Liu, John Stanturf, and Scott Goodrick, "Trends in Global Wildfire Potential in a Changing Climate," *Forest Ecology and Management*, 2010: 685.

7 Jerry Williams, "Managing Fire-Dependent Ecosystems: We Need a Public Lands Policy Debate," *Fire Management*, 2004: 8–9.

8 Frederico Cheever, "The Phantom Menace and the Real Cause: Lessons from Colorado's Hayman Fire 2002," *Penn State Environmental Law Review*, 2010: 191.

INTERNATIONAL LAW, CUSTOMARY LAW, AND GENERAL PRINCIPLES

It should be noted from the outset that although wildfire mitigation may involve international cooperation, it is primarily the domestic concern of states. Indeed, this is how the United Nations has approached general disaster relief, focusing the Hyogo Framework for Action on states' domestic measures.[9] Given this *internal* quality to an issue that nonetheless has potential transnational (i.e., *external*) effects, how then can international law respond?

Customary international law traditionally regulates the external behavior of states – how states treat each other, their diplomats, and citizens; acts in the international field.[10] When dealing with such external conduct, the dual requirements of state practice and *opinio juris*[11] are relatively easy to identify: state practice is what states do, and *opinio juris* is how states justify such actions. However, with the expansion of international law, custom started regulating more than just external behavior, but also internal behavior. It is evident that there are two inherent problems with this attempt. First are those areas of inconsistency between what states do and what states say the law is. Norms regarding human rights are the most obvious example of this. The conflict between state practice and views expressed by nations through international bodies[12] has resulted in a new conception of custom. *La coutume sauvage*,[13] or "modern custom,"[14] is based primarily on *opinio juris*, allowing for instant custom through the legislative actions of international organizations.[15] This practice of ignoring the disparity between state practice and *opinio juris* in human rights to establish customary norms has been criticized as an unjustified manipulation of customary law.[16]

However, this disparity is indicative of the bigger problem of attempting to fit custom to primarily internal behavior. Because, unlike with external state-state conduct, there is no natural alignment between the acts of states and their *opinio juris* that such acts are obligatory. For human rights this means the existence of

9 United Nations, "Hyogo Framework for Action." www.unisdr.org/we/coordinate/hfa.

10 Hugh Thirlway, *International Customary Law and Codification*. (Leiden: A.W. Siithoff, 1972).

11 *North Sea Continental Shelf (Federal Republic of Germany v. Netherlands)*. ICJ Rep 3 at 44 (1969).

12 Niels Peterson, "Customary Law Without Custom? Rules, Principles and the Role of States Practice in International Norm Creation," *American University International Law Review*, 2008: 279–280.

13 Bruno Simma and Philip Alston. "The Sources of Human Rights Law: Custom, Jus Cogens and General Principles," *Australian Yearbook of International Law*, 1988–1989: 89.

14 Anthea Elizabeth Roberts, "Traditional and Modern Approaches to Customary International Law: A Reconciliation," *American Journal of International Law*, 2001: 758.

15 Bin Cheng, "United Nations Resolutions on Outer Space: 'Instant' International Customary Law?" *Indian Journal of International Law*, 1965: 45; Robert Kolb, "Selected Problems in the Theory of Customary International Law," *Netherlands International Law Review*, 2003: 122.

16 Simma and Alston, "Sources of Human Rights Law," 89.

often conflicting state practice and *opinio juris*. But even for content for which there is no reason for such contradiction, relying on domestic laws for state practice can be problematic. *Opinio juris* can be hard to find – perhaps in fact nonexistent because states do not tend to justify internal laws on an international level.

It should be noted that traditionally, international law did not concern itself with internal behavior. But as globalization increases, formerly internal matters have increasingly transnational effects. Examples are myriad, and at least include domestic laws regulating cybercrime, financial policies that affect the global economy, and policies and laws regarding natural and man-made disasters. For these areas, states enact domestic legislation but often do not undertake corresponding acts of *opinio juris* at the international level. Indeed, progress in these areas is often concentrated on promulgating domestic legislation, under the auspices of an overseeing international body.[17] However, the focus of such a body is often to gain harmonized domestic laws, not to develop a norm of international customary law.

What then can international law say of those areas of law where there are high levels of domestic legislation and regulation, but no accompanying international rhetoric? The answer lies not in twisting conceptions of customary international law to encompass such areas, but rather in recourse to the general principles of law recognized by civilized nations (general principles).[18] The use of general principles as a source of law allows examination of domestic regulation to find norms of requisite horizontal generality, which can be elevated to customary norms of international law. This chapter examines the use of general principles to articulate global norms that can be applied to international disaster relief law and in particular in prevention and risk reduction of wildfires.

In doing so, three main arguments are set out. First, that international disaster relief law, particularly in relation to preventative or risk reduction measures, does not currently exist in binding conventional or customary form. Second, that general principles may be used in a substantive, formal sense, examining both law and policy of governments and law of chthonic legal systems, or "living law."[19] Third, a general principle with the relevant degree of horizontal generality exists and may operate independently, or may exist within a broader customary norm.

[17] See for example the "Project on Cybercrime" carried out by Microsoft, the Council of Europe and Estonia from 2006 to 2009: Council of Europe, Economic Crime Division of the Directorate General of Human Rights and Legal Affairs. "Project on Cybercrime: Final Report" ECD/567. 2009:1; the work of the International Federation of Red Cross and Red Crescent Societies in preparing "The Model Act for the Facilitation and Regulation of International Disaster Assistance." Model Act on IDRL. www.ifrc.org/en/what-we-do/idrl/model-act-on-idrl/.

[18] "Statute of the International Court of Justice," 1945: ATS 1, Article 38(1)(c).

[19] *Gabčíkovo-Nagymaros Project (Hungary v Slovakia)* [1997] ICJ Rep 7 at 109–110 (Separate Opinion of Judge Weeramantry).

CURRENT INTERNATIONAL REGULATION OF DISASTER RISK REDUCTION LAW

Much of the discussion of international disaster relief law focuses on disaster *response* – whether states are under an obligation to offer assistance following a disaster and whether states can offer aid without the consent of the government of the affected state.[20] There is no convention currently governing these areas (although international humanitarian law will apply in wartime disasters),[21] and the outlook for custom is grim:

> [N]otwithstanding an increased international interest in how states respond to natural disasters, there is no obligation upon states to seek international disaster assistance and no practical right or power to provide assistance to a disaster affected community without the consent of the government of the affected state.[22]

Although disaster response has been historically at the forefront of disaster law discussion, policy discussion is shifting toward consideration of disaster reduction as well.[23] However, much of the international activity surrounding this area is focused on practical governance issues – helping governments implement effective disaster relief policies and strategies.[24] The question of whether there is any norm directing states to engage in specific disaster risk reduction strategies has not been developed. While the "responsibility to protect" doctrine is often argued to be applicable, this is usually in the context of the aftermath of a disaster, rather than in the prevention of one.[25] There is an emerging argument in the European Court of Human Rights that a responsibility to mitigate natural disasters is an aspect of the right to life, as guaranteed in various international law instruments.[26] Much of the law relating to the standards of mitigation is limited to the European context.[27] However, if the more generalized duty of disaster risk reduction is read into the right to life, the general principles of law discussed in the sections that follow may be useful in providing a standard against which to assess such measures.

Apart from this rights-based duty, much international effort and attention has been focused on encouraging domestic measures for disaster risk reduction.

[20] Michael Eburn, "International Law and Disaster Response," *Monash University Law Review*, 2010: 162; see also David Fidler, "Disaster Relief and Governance after the Indian Ocean Tsunami: What Role for International Law," *Melbourne Journal of International Law*, 2005: 458.

[21] Eburn, "International Law and Disaster Response," 162–163. [22] Ibid., 164.

[23] International Strategy for Disaster Reduction. *Living with Risk: A Global Review of Disaster Reduction Intiatives*. Vol. 1. New York, Geneva: United Nations, 2004:ix.

[24] Fidler, "Disaster Relief," 470.

[25] Tyra Ruth Sacchao, "Natural Disasters and the Responsibility to Protect: From Chaos to Clarity," *Brooklyn Journal of International Law*, 2006–2007: 666.

[26] See *Budayeva and others v. Russia*. Application nos. 1533/02, 21166/02, 20058/02, 11673/02, and 15343/02 (ECHR, March 20, 2008).

[27] Dinah Shelton, "Tatar C. Roumanie," *American Journal of International Law*, 2010: 250–251.

The United Nations Office for Disaster Risk Reduction (UNISDR) monitors implementation of the Hyogo Framework for Action (HFA) a ten-year plan aimed at improving disaster risk reduction laws and policies in states.[28] The first priority under the HFA is that countries should "ensure that disaster risk reduction is a national and a local priority with a strong institutional basis for implementation."[29] The HFA is essentially a soft law instrument, and does not create any binding customary norms. But the promotion of domestic legislation raises the question of whether general principles can assist in developing the international law of disaster risk reduction. To do so, some consideration of the source itself is necessary.

GENERAL PRINCIPLES OF LAW

General principles of law, as a source of international law, have always attracted contention. Their introduction into the Statute of the Permanent Court of International Justice (PCIJ) aroused heated discussion between the Belgian delegate and President of the Advisory Committee of Jurists, Baron Eduoard Descamps,[30] and the American delegate, Elihu Root.[31] Root eventually won the day, with his amended version of the source drafted in conjunction with the British delegate Lord Walter Phillimore,[32] being inserted into the PCIJ Statute.[33]

Among commentators, divisions exist as to the fundamental importance of the source itself, as well as what general principles are and how they can be found. While Sir Robert Jennings dismissed general principles in 1982, stating: "I have never been convinced of the supposed importance of 'general principles of law' as a

[28] United Nations, "Hyogo Framework for Action 2005–2015: Building the resilience of nations and communities to disasters," *Extract from the final report of the World Conference on Disaster Reduction*. UN DOC A/CONF.206/6.

[29] United Nations, "Hyogo Framework for Action." www.unisdr.org/we/coordinate/hfa.

[30] A Belgian minister of state, former secretary-general of the Interparliamentary Union, former president of the *Institut de Droit International*, and Nobel Peace Prize nominee in 1901, 1902, and 1903. See Arthur Eyffinger, *The 1809 Hague Peace Conference: "The Parliament of Men, the Federation of the World* (London: Kluwer International, 1999).

[31] The former US secretary of state and former US secretary of war, first president of the Carnegie Endowment for International Peace, a member of the Permanent Court of Arbitration, and the winner of the 1912 Nobel Peace Prize: Christopher Rossi, *Equity and International Law: A Legal Realist Approach to Decision Making*, 1993; Margaret White, "Equity – A General Principle of Law Recognised by Civilized Nations?" *Queensland University of Technology Law and Justice*, (2004): 107–108.

[32] An expert in international, ecclesiastical and admiralty law, the former Lord Justice of the Court of Appeal of the Privy Council (UK), and president of the International Law Association in 1905. See John Sankey and Hugh Mooney, "Phillimore, Walter George Frank, first Baron Phillimore (1845–1929)," *Oxford Dictionary of National Biography*. 2008. www.oxforddnb.com/view/article/35511.

[33] Permanent Court of International Justice: Advisory Committee of Jurists. "Procès-verbaux of the Proceedings of the Committee" Annex 1 at 344 (15th Meeting). 1920.

source of general international law..."[34], other commentators laud it. M. Cherif Bassiouni espoused the view in 1990 that "it is quite likely that 'General Principles' will become the most important and influential source of international law in this decade."[35] While Bassiouni's prediction did not come to pass within that timeline, the time for general principles to be realized as an important source of law is ripe.

General principles are uniquely positioned to respond to rapidly developing areas of law with transnational impact in a way that custom cannot. For these areas, conventional law is often slow to develop, and sometimes impeded by domestic politics or a general lack of domestic impetus. Nonetheless, in these areas, domestic regulation is enacted, often in large volumes in many countries. In disaster law, this domestic legislation is encouraged and supported by international organizations such as the United Nations[36] and the International Federation of Red Cross and Red Crescent Societies (IFRC).[37]

But how are general principles defined? As previously discussed, there are many aspects of the source that have been open to debate, and to address them all here is well beyond the scope of this chapter. Drawing on the development of the source throughout the jurisprudence of the PCIJ and the International Court of Justice (ICJ), it is clear that the source operates by drawing content from domestic legal systems.[38] That content must have a high degree of horizontal (but not necessarily vertical) generality, and must be appropriate in the international context.

Furthermore, horizontality is generally viewed across domestic legal systems at the state-level, not below it. And chthonic practices exist well below state-level law.

[34] Sir Robert Jennings, "The Identification of International Law" in *International Law: Teaching and Practice*, ed. Bin Cheng (London: Stevens, 1982), 4.

[35] M. Cherif Bassiouni, "A Functional Approach to 'General Principles of International Law," *Michigan Journal of International Law*, 1990: 769.

[36] "Hyogo Framework for Action." [37] "Model Act on IDRL."

[38] See for example George Finch, *The Sources of Modern International Law* (Washington: Carnegie Endowment for International Peace, 1937); Michael Virally, "The Sources of International Law" in *Manual of Public International Law*, ed. Max Sorenson (Macmillan: 1968), 144; Georg Schwarzenberger and E. D. Brown, *A Manual of International Law* (Milton: Professional Books, 1976); Michael Akehurst, "Equity and General Principles of Law," *International and Comparative Law Quarterly*, 28 (1976): 814; Michael Bogdan, "General Principles of Law and the Problem of Lacunae in the Law of Nations," *Nordisk Tidsskrift for International Ret*, 1977: 42; Johan G. Lammersammers, "General Principles of Law Recognized by Civilized Nations," in *Essays on the Development of the International Legal Order in Memory of Haro F Van Panhuys*, ed. Frits Kalshoven, Pieter Jan Kuyper, and Johan G Lammers (Springer: 1980), 56–7; Linda Maki, "Genral Principles of Human Rights Law Recognized by All Nations: Freedom from Arbitrary Arrest and Detention," *California Western International Law Journal*, 1980: 275; Godefridus J. H. Hoof, *Rethinking the Sources of International Law* (The Netherlands: Kluwer, 1983); Alain Pellet, "The Normative Dilemma: Will and Consent in International Law-making," *Australian Yearbook of International Law*, 1988–1989: 38; Theodor Meron, *Human Rights and Humanitarian Norms as Customary Law* (Oxford University Press, 1989); Jonathon Charney, "Universal International Law," *American Journal of International Law*, 1993: 535; David Bederman, *The Spirit of International Law* (The University of Georgia Press, 2002); James Hathaway, *The Rights of Refugees under International Law* (Cambridge University Press, 2005).

Indeed, within a federal system, this would be like asking a court to pierce the veil of national law and draw general principles from the practices of Arizona, Texas, and California instead. A point should be made about each of these requirements.

1) *Appropriateness*

Perhaps the most famous formulation of this requirement is found in the judgment of Lord Arnold McNair in the *International Status of South West Africa* case.[39] McNair warned against "importing private law institutions 'lock, stock and barrel.'"[40] This general requirement has been interpreted as requiring that the domestic law be appropriate,[41] or transposable,[42] to be a norm of *international*, as well as domestic law.

Not only is this requirement well established, its application to general principles is not doubted. However, it should be noted that as the scope of international law grows, so too will the field of appropriateness.[43] Growing global interdependence drives the expanse of international law, and increasingly new areas require regulation:

> This big change in the content and span of the law, was not the result of doctrinal enlightenment but of actual necessities arising from the juxtaposition of a larger number of States, all of them pushed by modern economic and technological developments into ever-increasing interdependence in ever more matters; matters which therefore imperatively required regulation by international law.[44]

Disaster risk reduction is such an area. Where previously the actions of states in preventing disasters within their own territories may have been viewed as domestic concerns, such a view cannot be supported today. Disasters themselves may impact other states – both in the initial event, and in the aftermath. Further, international law itself now accepts regulation of domestic activities, particularly where such activities affect the human rights of the domestic population.

39 *International Status of South-West Africa (Advisory Opinion)*. (ICJ Rep 128., 1950).

40 Ibid., 148 (Separate Opinion of Judge McNair).

41 Georg Schwarzenbeger, *International Law as Applied by International Courts and Tribunals*. Vol. 1. (London: Stevens, 1957).

42 Allain Pellet, "Article 38," in *The Statute of the International Court of Justice: A Commentary*, ed. Andreas Zimmerman, Christian Tomuschat, and Karin Oellers-Frahm(Oxford University Press, 2006), 772–3.

43 Wolfgang Friedmann, *The Changing Structure of International Law* (New York: Columbia University Press, 1964). See also Wolfgang Friedman, "The Uses of 'General Principles' in the Development of International Law," *The American Journal of International Law*, 1963: 281.

44 Robert Jennings, "Teachings and Teaching in International Law," in *Essays in International Law in Honour of Judge Manfred Lachs*, ed. Jerry Makarczyk (The Hague: Martinus Nijhoff, 1984), 124.

2) *Vertical and Horizontal Generality*

There are two components to this requirement: first, general principles *do* require horizontal generality; and second, they *do not* necessarily require vertical generality. The first requirement is well-established – although the breadth of comparative studies used to support general principles in international jurisprudence may honor it more in breach than reality.[45] However, there is a discussion among commentators whether general principles must be considered in the context of the rule/principle distinction. Rules are generally considered to have a low vertical generality, while principles are more abstract, and general in concept.[46] Such a requirement is often assumed by commentators, who variously describe general principles as foundational,[47] fundamental,[48] constitutive,[49] and value oriented.[50]

However, throughout the development of general principles, neither the Advisory Committee of Jurists, the PCIJ, nor the ICJ ever drew such a distinction.[51] As Maurice Mendelson, formerly the Rapporteur for the International Law Association International Committee on Formation of Customary (General) International Law argued:

> [A]lthough there is quite a debate among legal theorists as to the difference and hierarchical relation between rules and principles, none of this finds any reflection in the utterances of the ICJ, which tends to treat the two terms as synonymous.[52]

[45] See for example *Lighthouse Case (France v. Greece) (Merits)*. (ser A/B) No 62 (PCIJ, 1934). (Separate Opinion of Judge Séfériadès); *International Status of South-West Africa (Advisory Opinion)*. (ICJ Rep 128., 1950). (Separate Opinion of Judge McNair); *Prosecutor v. Erdemovic, Sentencing Judgement*. Case No IT-96-22-T (November 29, 1996).

[46] Joseph Raz, "Legal Principles and the Limits of Law," *Yale Law Journal*, 1972: 838; Robert A. Alexy, A *Theory of Constitutional Rights* trans. Julian Rivers (Oxford Univeristy Press, 2002).

[47] Bin Cheng, *General Principles of Law as Applied by International Courts and Tribunals* (London: Stevens, 1953).

[48] Jan Hendrik and Willem Verzijl, *International Law in Historical Perspective*. Vol. 1 (Kluwer Law International, 1968).

[49] Michael Bogdan, "General Principles of Law and the Problem of Lacunae in the Law of Nations," *Nordisk Tidsskrift for International Ret*, 1977: 42.

[50] Robert Kolb, "Principles as Sources of International Law (with Special Reference to Good Faith)," *Netherlands International Law Review*, 2006: 29; Peterson, "Customary Law," 288.

[51] See for example Permanent Court of International Justice: Advisory Committee of Jurists, "Proces-verbaux of the Proceedings of the Committee." 14th Meeting, 315 per Mr Ricci-Busatti; 15th Meeting, Annex 2 per Mr. Fernandes; *Panevezys-Saldutiskis Railway (Estonia v. Lithuania) (Merits)*. (PCIJ (ser A/B) No 76 at 40, 1935). (Separate Opinion of Judge Van Eysinga); *Certain Phosphate Lands in Naura (Naura v. Australia) (Preliminary Objections)*. (ICJ, 1992). (Separate Opinion of Judge Shuhabuddeen).

[52] Maurice Mendelson, "The International Court of Justice and the Sources of International Law," in *Fifty Years of the International Court of Justice: Essays in Honor of Sir Robert Jennings*, ed. Vaughan Lowe and Malgosia Fitzmaurice (Cambridge University Press, 1996), 63, 80.

Indeed, the practice of the ICJ and the International Criminal Tribunal for the Former Yugoslavia (ICTY) in particular has been to accept specific rules drawn from domestic systems as forming the content of a general principle. Thus the Trial Chamber of ICTY looked to the definition of rape under the auspices of general principles in *Furundžija*,[53] Judge Dillard of the ICJ sought to include doctrines of frustration and impossibility as a general principle in *Legal Consequences for States*,[54] and the Inter-American Court of Human Rights considered rules on succession and heirs as general principles of law in the *Aloeboetoe* case.[55]

This practice would also operate in relation to general principles of international disaster relief law. It should be noted however that as a matter of practicality, in order to find horizontal generality some degree of vertical abstraction may be required to account for differences among different states and legal systems. However, more generally for international disaster relief law, where states have domesticated a legislative tool like the International Federation of Red Cross' Model Act on Disaster Relief certain rules may possess a high degree of horizontal generality *without* a great degree of abstraction. Assuming they fulfill all other requirements, such rules could be properly considered General Principles, irrespective of their specificity.

Legal Systems

While it is well established that general principles are drawn from domestic legal systems, a question remains as to *which* legal systems should be looked to. Historically, much of the consideration of general principles by the PCIJ and ICJ assessed only common law and civil law countries. ICJ Judge Ammoun criticized this approach in his separate opinion in the *North Sea Continental Shelf* case.[56] Ammoun argued that the term "civilized" had led to a "Concert of Europe,"[57] where previous international courts and tribunals had only looked to European systems of law to draw general principles.[58] Ammoun supported a comparative approach to finding General Principles, but felt it should be across "national rules in all latitudes."[59] Ammoun then looked to the following:

> [I]n the law of Western Europe and of Latin America ... in the common law ... in Muslim law ... by the Koran ... in the teaching of the four great jurisconsults of Islam condensed in the Shari'a ... Chinese law ... Soviet law ... Hindu law ... [and] finally the law of the other Asian countries, and of the African countries.[60]

[53] *Prosecutor v. Furundžija, Judgement*. IT-85-17/1-T (December 10, 1998).

[54] *Legal Consequences for States of the Continued Presence of South Africa in Namibia (South West Africa) notwithstanding Security Council Resolution 276 (1970) (Advisory Opinion)*. (ICJ, 1971). (Separate Opinion of Judge Dillard).

[55] *Aloeboetoe et al – Reparation (Art 63(1) American Convention on Human Rights)*. (IACHR, September 10, 1993).

[56] *North Sea Continental Shelf Cases*. [57] Ibid., 135 (Separate Opinion of Judge Ammoun).

[58] Ibid. [59] Ibid. [60] Ibid., 139–44.

Although this call was made in 1969, more than years later the same point needs to be made. Considerations of general principles *must* look to more diverse systems than simply (European) common and civil law systems. The need to focus on diverse legal systems is supported by the recent spread of jurisprudence examining these ideas, and the general notion of globalization of law recognizes this growing movement.[61]

However, for the purposes of this query, one must go beyond the common law, civil law, religious law, and Asian and African legal systems as set out by Judge Ammoun. H. Patrick Glenn identifies another type of legal systems, the "chthonic legal systems." Glenn uses the term chthonic to identify the traditional or customary legal systems used by indigenous societies.[62] A chthonic legal system is characterized by being the law of people "who live ecological lives by being chthonic, which means that they live in or in close harmony with the earth."[63]

In his separate opinion in the *Gabčíkovo-Nagymaros Project* case,[64] ICJ Judge Weeramantry looked to practices within chthonic systems to understand that extent of sustainable development as a global "principle with normative value." His analysis considered environmental practice in ancient civilizations, focusing on evidence of ancient irrigation systems,[65] as well as customary environmental practices in traditional indigenous cultures.[66] In articulating a general principle that all systems of law that stem from past and present civilizations can be the source of legal rights and obligations, Judge Weeramantry observed:

> Living law which is daily observed by members of the community, and compliance with which is so axiomatic that it is taken for granted, is not deprived of the character of law by the extraneous test and standard of reduction to writing... Moreover, when the Statute of the Court described the sources of international law as including the 'general principles of law recognized by civilized nations' it expressly opened a door to the entry of such principles into modern international law.[67]

Thus, Weeramantry looked to chthonic legal systems and historical practice to support a general principle. He argued that the orality of what he termed "living law" does not deprive chthonic systems of the quality of law:

> By virtue of its representation of the main forms of civilization, this Court constitutes a unique forum for the reflection and the revitalization of those global

61 See for example James Heckman, Robert Nelson, and Lee Cabatingan, *Global Perspectives on the Rule of Law* (Milton Park: Routledge-Cavendish, 2010); William Twining, *General Jurisprudence* (Cambridge University Press, 2009); Abdullahi Ahmed An-Na'im, Abd, "Globalization and Jurisprudence: An Islamic Law Perspective," *Emory Law Journal*, 2005: 25; Paul Schiff, *The Globalisation of International Law* (Aldershot: Ashgate Publishing, 2005); Spencer Zifcak, *Globalisation of International Law* (Aldershot: Ashgate Publishing, 2005); C. G. Weeramantry, *Universalising International Law* (The Netherlands: Martinus Nijhoff, 2004); H. Patrick. Glenn, *Legal Traditions of the World* (Oxford University Press, 2004); William Twining, *General Jurisprudence* (Cambridge University Press, 2009).

62 Glenn, *Legal Traditions*, 59. 63 Ibid., 60.

64 *Gabčíkovo-Nagymaros Project (Hungary v Slovakia) (Merits)* [1997] ICJ Rep 7.

65 Ibid., 98–106. (Separate Opinion of Judge Weeramantry).

66 Ibid., 107–8. 67 Ibid., 109–10.

> legal traditions. There were principles ingrained in these civilizations as well as embodied in their *legal systems*, for legal systems include not merely written legal systems but traditional legal systems as well, which modern researchers have shown to be no less legal systems than their written cousins, and in some respects even more sophisticated and finely tuned than the latter.[68]

This approach should be followed. The inclusion of the term "civilized" in general principles unfortunately echoes the often racist and European-superiority aspects of early international law. Westlake, writing at the turn of the twentieth century, argued that international society "is composed of all the States of European blood"[69] and "understood international law as a tool for ensuring the supremacy of the interests of 'peoples of European blood' over those of the inhabitants of the territories they colonized."[70] Such opinions (though not quite as blatant) continued throughout much of the twentieth century.[71] By removing these limiting associations and looking to *all* legal systems, Courts should recognize an obligation to investigate all evidence of legal practice including practices by chthonic societies.

Such a realization is important when considering disaster mitigation in the context of wildfires. While there is current contemporary practice of risk reduction through various strategies, including prescribed burns,[72] there also exists significant chthonic practice in the same area.[73] By applying an approach such as Judge Weeramantry's that involves examining all practice including chthonic practice, it can be argued that a "principle of normative value" relating to wildfire mitigation may exist.

It is particularly appropriate to consider disaster risk reduction in this way. The UNISDR explicitly links disaster risk reduction with sustainable development initiatives. The *Living with Risk* report states:

> Disaster reduction has emerged as an essential prerequisite for sustainable development. The UN General Assembly includes disaster reduction in its treatment of the sustainable development items in its annual deliberations.[74]

If disaster reduction is viewed as inherently part of sustainable development, resorting to a General Principle of disaster reduction in the context of the

68 Ibid., 109.

69 J. Westlake, *International Law*. Vol. 1 (Cambridge University Press, 1904). Alexander Orakhelashvili, "The Idea of European International Law," *European Journal of International Law*, 2006.

70 Orakhelashvili, "European International Law," 320.

71 See Charles Fenwick, *International Law* 4th ed. (New York: Appleton-Century-Crofts, 1965); Schwarzenberger and Brown, *A Manual of International Law*, 13.

72 Stephen Pyne, *Tending Fire* (Washington: Island Press, 2004).

73 Robin Wall Kimmerer and Frank Kanawha Lake, "The Role of Indigenous Burning in Land Management," *Journal of Forestry*, 2001: 37.

74 Reduction, International Strategy for Disaster. *Living with Risk: A Global Review of Disaster Reduction Intiatives*. New York; Geneva: United Nations, 2004.

established General Principle of sustainable development seems fitting. Indeed, the more general chthonic practices of sustainable development include chthonic practices of fire prevention by burning and other methods.

WILDFIRE RISK REDUCTION

Just as Judge Weeramantry found an independent norm of international law regarding sustainable diversity, it may be that there is a similar norm pertaining to wildfire risk reduction. However, the usefulness of a general principle is not limited to this function. General principles can also be used to find aspects of norms or rules otherwise established – as for example, in the *Furundžija* case, where general principles were used within the existing statutory elements of the crime.[75] Accordingly, although the existence of a responsibility to protect is debated, such a responsibility if established would encompass a responsibility to *prevent*.[76] If such a norm is established, a general principle requiring the use of chthonic sources could inform the content of that responsibility.

CHTHONIC PRACTICES

Although colonial views of indigenous people living without impact on the land persisted for some time,[77] the concept of indigenous land management has been increasingly accepted over the last thirty years.[78] In particular the role of indigenous societies in actively managing wildfires by burning bush to reduce fuel load has been recognized.[79] Such anthropogenic fires are widespread in chthonic systems. Australia is the most fire-prone continent on Earth,[80] and indigenous burning practices of Australian Aborigines are well documented.[81] In traditional Aboriginal culture, people have a spiritual link to the land, which entails custodial responsibility for the land.[82] These custodial responsibilities include deliberate burning of bush, in deliberate applications so as to target dry grass (with a high fire danger).[83]

The beneficial result of such practices becomes apparent when compared with unmanaged land in the central and west Arnhem Land Region where fires "may burn unchecked for months, over tens of thousands of square kilometres."[84]

75 *Prosecutor v Furundžija*.

76 International Commission of Intervention and State Sovereignty. *The Responsibility to Protect* (Ottawa: International Development Research Centre, 2001) 19.

77 Ramona Butz, "Traditional Fire Management: Historical Fire Regimes and Land Use Change in Pastoral East Africa," *International Journal of Wildland Fire*, 2009: 442.

78 Kimmerer and Lake, "Role of Indigenous Burning," 37.

79 D. Yibarbuk et al., "Fire Ecology and Aboriginal Land Management in Central Arnhem Land, Northern Australia: a Tradition of Ecosystem Management," *Journal of Biogeography*, 2004: 208.

80 D. M. J. S. Bowman, Angie Walsh, and L. D. Prior, "Landscape Analysis of Aboriginal Fire Management in Central Arnhem Land, North Australia," *Journal of Biogeography*, 2004: 208.

81 Ibid. 82 Yibarbuk et al., "Fire Ecology," 328. 83 Ibid. 84 Ibid., 329.

Although much of the research in this area focuses on Australia, in part because of amounts of available data,[85] the practice was common throughout other chthonic societies.

Indigenous fire management in North America has shaped landscapes – from the prairie grasses of the Midwest to the Appalachian forests.[86] Traditional practices are now not only understood to have helped prevent catastrophic wildfires,[87] but there are calls for traditional knowledge of indigenous Americans to be integrated into modern wildfire management.[88] Anthropogenic fire use by indigenous groups is also evident in South America: Brazilian indigenous tribes used fire to control the *cerrado* (savannahs).[89] There is evidence of prescribed burns being used in contemporary times by indigenous groups in Bolivia.[90]

There is also evidence of indigenous fire management in Africa. Grassfires often occur in the African savanna where they burn more frequently and extensively than any other region on earth.[91] The Masaai tribe has lived in this area since the mid nineteenth century.[92] Evidence shows that the Masaai used fire to burn savannah grass in small patches early in the dry season to prevent catastrophic fires later in the season.[93] Although evidence is scarce at this stage, there is some support for suggestions of indigenous fire use and management in Cambodia.[94]

CONTEMPORARY PRACTICES

Prescribed burning is a common contemporary practice to prevent, or reduce the intensity of, wildfires. It occurs throughout the United States.[95] Although fire suppression as a policy gained popularity during the nineteenth century, modern policies aim to "restore fire as an essential ecosystem process."[96] Fire suppression

85 Bowman, "Landscape Analysis," 208.

86 Kimmerer and Lake, "Role of Indigenous Burning," 37.

87 Ernesto Alverado et al., "Integration and Application of Traditional Ecological Knowledge and Modern Science for Contemporary Wildland Fire Management in Tribal Lands of North America," in *Food and Agriculture Organisation of the United Nations Fire Management Working Paper*, 2011: 99.

88 Ibid., 99–100.

89 Jayalaxshmi Mistry et al. "Indigenous Fire Management in the Cerrado of Brazil: The Case of Krabo of Tocantins," *Human Ecology*, 2005: 366.

90 Carlos Pintos and Ernesto Alverado, "Lessons Learned on Fire Management in Indigenous Communities of Bolivia," in *Food and Agriculture Organisation of the United Nations Fire Management Working Paper*, 2011: 113.

91 Butz, "Traditional Fire Management," 442.

92 Ibid.

93 Ibid., 446.

94 Andrew Maxwell, "Fire Regimes in North-eastern Cambodian Monsoonal Forests, with a 9300-year Sediment Charcoal Record," *Journal of Biogeography*, 2004: 236.

95 John Loomis et al. "Prescribed Fire and Public Support: Knowledge Gained, Attitudes Changed in Flordia," *Journal of Forestry*, 2001: 18–19.

96 David Ostergren, "Comparing Wilderness Restoration and Fire Policy in Three Federal Agencies: Variations on a Theme in Northern Arizona," *Journal of Land, Resources and Environmental Law*, 2005–2006: 267.

policies led to a fire "crisis," with a buildup of fuel load resulting in "economically and ecologically disastrous wildfires."[97] Prescribed burning is now embraced as a legitimate and necessary policy to reduce fuel load buildup.[98]

Prescribed burning also occurs throughout Australia,[99] and is credited with lower incidences of wildfires in states with a higher prevalence of burning.[100] Prescribed burning also occurs in South Africa, and it is suggested it should be used on gorse shrub land in Spain.[101]

CONCLUSION

Although a more thorough analysis of domestic laws and policies regarding fire prevention strategies needs to be undertaken, the theoretical ground is now set to see a common practice of fire prevention, shared by those who are responsible for the land. For chthonic societies, such responsibility rests with the tribes and people themselves, and there is evidence that controlled fire has been consistently used to prevent and mitigate wildfires. For contemporary, nonchthonic legal systems, governments engage in fire prevention strategies. For those countries often affected by wildfires, prescribed burning is common and gaining traction.

How then does this operate in the context of international law? First, as acceptance of indigenous anthropogenic fire practices becomes more widespread, combined with increasing uses of prescribed burning in current times, indigenous fire practices may reach a level similar to the practices that informed Judge Weeramantry's understanding of a "normative principle of sustainable development." An emerging norm could require that governments be actively involved in disaster risk reduction, and specifically engage in prescribed burning to prevent and reduce the risk of wildfires. In viewing the actions of governments as the potential basis for a General Principle rather than a norm of custom, problems with a lack of *opinio juris* are overcome. Indeed, like much disaster risk reduction, fire prevention strategies are often not carried out on a federal level, but are left to state or even local authorities.

This is not an area where a traditional conception of customary law – custom as regulating state to state behavior – fits easily. It is, however, a situation where a

97 Sara Elizabeth Jensen, "Policy for Wildland Fire Management: Principles, Incentives, and Conflicts," *Natural Resources Journal*, 2006: 959–960.

98 Ibid., 968.

99 R. A. Bradstock et al., "Bushfire Risk at the Urban Interface Estimated from Historical Weather Records: Consequences for the use of Prescribed Fire in the Sydney Region of South-eastern Australia," *Journal of Environmental Management*, 1998: 260.

100 Rick Sneeuwiagt, "Prescribed Burning: How Effective is it in the Control of Large Forest Fires," *Victoria Royal Commission*. www.royalcommission.vic.gov.au/getdoc/e9e5a53e-5406-4f94-977f-7ed0ffaf46/WIT.135.001.0099.

101 M. J. Baeza et al., "Factors Influencing Fire Behavior in Shrublands of Different Stand Ages and the Implications for Using Prescribed Burning to Reduce Wildfire Risk," *Journal of Environmental Management* 65 (2002): 200.

common practice has occurred over tens of thousands of years, in countries, and different parts of the world, and such a practice should be considered a general principle. As localized practice rather than state-level practice, however, the notion of "general principle" would need to be tweaked to include consideration of practices below state-level. Even so, more evidence of contemporary domestic laws and policies, however, would have to be considered to be able to state such a conclusion with certainty.

The second, more conservative, view is to consider a general principle in this context as informing the content of a broader, more general customary norm. If a general duty to prevent "harm" can be established, then a specific duty to take actions to prevent and mitigate wildfires can be shown by reference to evidence collected under a general principle that requires states to consider chthonic sources.

Finally, for those fires that cause transboundary harm, such as air pollution, a general principle might assist in determining state responsibility. It is well established that transboundary environmental harm is governed by international law.[102] However, this is traditionally expressed as a negative duty, encompassing what states must not do: "[N]o State has the right to use or permit the use of its territory in such a manner to cause injury by fumes in or to the territory of another or the properties or persons therein...[103]"

In the context of the *Trial Smelter Case*, the pollution in question was clearly caused by the operation of the smelter by a Canadian company.[104] There, the pollution migrated across the border and caused great harm inside the United States. But what of an event that a government has no control over? Principle 21 of the *Stockholm Declaration* states:

> States have, in accordance with the Charter of the United Nations and the principles of international law ... the responsibility to ensure that activities within their jurisdiction or control do not cause damage to the environment of other states or of areas beyond the limits of national jurisdiction.[105]

Would wildfires, especially if lit by lightning, be an activity within a state's jurisdiction or control? Even if it is, such strict liability for transboundary harm is not universally accepted. Negligence-type analysis of transboundary harm cases in international law often require that the act causing the harm must be attributable to the state.[106] Could a wildfire (particularly one that is accidental – or even one ignited by lightning) ever be attributable to a state? This question has been specifically addressed in relation to the Southeast Asian fires of 1997 by

[102] *Trail Smelter Case (United States of America v. Canada)*. (1949). 3 RIAA 1905.

[103] Ibid., 1965. [104] Ibid., 1917.

[105] United Nations Environment Programme. *Declaration of the United Nations Conference on the Human Environment*. Stockholm: United Nations, 1972.

[106] Devereaux McClatchey, "Chernobyl and Sandoz one Decade Later: The Evolution of State Responsibility for International Disasters 1986–1996," *Georgia Journal of International and Comparative Law*, 1995–1996: 677.

Alan Tan.[107] Tan applied the Articles on State Responsibility to the fires and concluded that the acts of private Indonesian companies engaging in "slash and burn" practices could not be directly attributed to Indonesia.[108]

The usefulness of a general principle becomes apparent here. If a state *fails* to comply with an established general principle regarding fire prevention (and perhaps in particular prescribed burning), then this could be an omission in international law. Questions of attribution are thus made simple – instead of trying to trace back the actions of private actors to the state, noncompliance with the general principle is clearly an action of the state itself.

General principles, like conventional and customary law, are currently state-centered. Consequently, the view advocated here – to consider practices of indigenous populations as potential contributors to international law – requires a theoretical leap. But it is a leap well worth taking. Subunits within nation states sometimes have much to offer in areas of international law that bear directly on matters of urgent import, like containing wildfires.

107 Alan Khee-Jin Tan, "Forest Fires of Indonesia: State Responsibility and International Liability," *International and Comparative Law Quarterly*, 1999: 826.

108 Ibid., 831–832.

4

Evolution of International Disaster Response Law: Toward Codification and Progressive Development of the Law

Emika Tokunaga

International Disaster Response Law (IDRL) is one of the emerging fields of international law. While there is no comprehensive multilateral legal framework on international disaster response, there are two recent major international developments that focus on the legal aspects of disaster response: First, the Protection of Persons in the Event of Disasters project undertaken by the United Nation's International Law Commission (ILC) in 2007; and second, the Disaster Law Programme initiated by the International Federation of Red Cross and Red Crescent Societies (IFRC) in 2001 as the world's largest humanitarian organization. This chapter focuses on the central concepts of IDRL. Key questions raised here include: What is the definition of disaster? What is the responsibility of affected states? And, what is the content of the consent requirement that an affected state must undertake to receive external assistance?

INTRODUCTION

Focusing on the ILC project for the Protection of Persons in the Event of Disasters and the Disaster Law Programme initiated by the IFRC, this chapter considers the evolution of IDRL in the context of international human rights and humanitarian law. The chapter begins its exploration in 2007, when the ILC included the topic of Protection of Persons in the Event of Disasters in its program of work and appointed Eduardo Valencia-Ospina as Special Rapporteur.[1] From 2008 until 2012, Valencia-Ospina submitted five reports to the ILC, which were considered at their annual sessions. The Sixth Committee, which deals with international legal matters, subsequently considered the reports and the debate thereon held in the ILC at each session since 2008.

[1] The topic of the protection of the persons in the events of disasters was proposed at the fifty-eighth session in 2006 by the Working Group on the Long term Programme of Work of the Commission.

Then, the analysis moves to consider the latest developments in the IFRC's project. The IFRC began its activities on IDRL in 2001, focusing on technical assistance, capacity building, advocacy, dissemination, and research. There are two important points of progress: First, Guidelines for the Domestic Facilitation and Regulation of International Disaster Relief and Initial Recovery Assistance (IDRL Guidelines), which were adopted in 2007; and second, the Model Act for the Facilitation and Regulation of International Disaster Relief and Initial Recovery Assistance (Model Act) adopted in 2011. In 2012, the IFRC changed its project name from the International Disaster Response Laws, Rules and Principles (IDRL) Programme to the Disaster Law Programme in response to the expressed interests of national societies as well as its evaluation of humanitarian needs.

These twin disaster response projects yield important preliminary answers to vital questions in the new field of IDRL. For example, what is the scope and definition of disaster? What is the responsibility of the affected state? What is the requirement of the consent of the affected state to external assistance? What is the duty to cooperate in the context of disaster response? The following sections track the progressive development of IDRL through the drafting phase toward final adoption.

THE ILC: PROTECTION OF PERSONS IN THE EVENT OF DISASTERS

Considerations in 2008

Valencia-Ospina prepared the preliminary report[2] on the topic of the Protection of Persons in the Event of Disasters and submitted it to the sixtieth session of the ILC in 2008. The report outlined the scope of the topic as well as legal issues concerned. These included the following:

(a) the possibility of the adoption of the rights-based approach to the project;
(b) the appropriateness of the three axes of the topic, including *ratione materiae*, *ratione personae*, and *ratione temporis*;
(c) the consideration of all types of disasters, whether natural or man-made, and of the issues revolving around the various phases of a disaster, namely the pre-, in-, and post-disaster phases, which corresponded to concepts of prevention, response, and rehabilitation;
(d) the exclusion of the situation applied to international humanitarian law from the topic, for instance armed conflict, because of the existence of other well-established standards of the conduct;

[2] A/CN.4/598.

(e) whether the concept of the protection of persons was the comprehensive one that encompassed response, relief, and assistance for the victims affected by the disaster;
(f) whether to include both issues of the right to humanitarian assistance and the responsibility to protect; and
(g) the decision as to the final form of the work at the early stage, without prejudice to the outcome of further debates, among others.

At the sixtieth session in 2008, the ILC considered the preliminary report.[3] At the same time, the ILC had before it the memorandum[4] prepared by the secretariat of the ILC upon the request of the Commission at the previous session in 2007.[5] It focused on natural disaster and provided the existing legal instruments and texts applicable to the topic.

These suggestions by Valencia-Ospina were welcomed as a whole in the discussion, especially the adoption of the rights-based approach to the project and the inapplicability of the project to existing international humanitarian law. On the other hand, the ILC gave general preference to response and assistance in the aftermath of a disaster over prevention during the pre-disaster phase at least at the early stage of the consideration of this topic.[6]

While states agreed to permit the Special Rapporteur to investigate the Responsibility to Protect (R2P) as part of the ILC project,[7] there remained some negative opinions regarding R2P considerations. For example, some argued that it was doubtful whether the largely political and moral concept existed as a legal concept since the legal parameters had yet to be developed.[8]

Regarding the final form of the work on the topic, the ILC could not reach a conclusion. Some members contended that it was premature to take a decision on the final form.[9] Other members noted that the ILC's work would largely be in the area of progressive development rather than codification and therefore the goal of the project would be to lay down a framework of legal rules, guidelines, or mechanisms that would facilitate practical international cooperation in disaster response, such as a framework convention.[10]

At the sixty-third session in 2008, the Sixth Committee considered the preliminary report of the Special Rapporteur and the debate within the ILC.[11] The

[3] Official Records of the General Assembly, Sixty-third Session, Supplement No. 10 (A/63/10), paras. 214–264. See also Arnold, Pronto,. "Consideration of the Protection of Persons in the Event of Disasters by the International Law Commission." *ILSA Journal of International and Comparative Law*, 2009: 449–457.
[4] A/CN.4/590 and Add.1 to 3.
[5] Official Records of the General Assembly, Sixty-second Session, Supplement No. 10 (A/62/10), par. 386.
[6] Official Records of the General Assembly, Sixty-third Session, Supplement No. 10 (A/63/10), par. 239.
[7] Ibid., pars. 247 and 249. [8] Ibid., pars. 248–249. [9] Ibid., par. 257. [10] Ibid.
[11] A/CN.4/606, pars. 78–88.

Committee expressed general support for the view of the Commission in the adoption of the rights-based approach to the topic, as well as the exclusion of the situation applied to international humanitarian law from the subject matter involved. However, it was proposed that there were a variety of other ideas to examine concerning the legal issues around the scope of the topic, since the work of the topic was at an early stage.

With respect to the right to humanitarian assistance and the responsibility to protect, there were both affirmative and negative reactions as to whether these rights and responsibilities should be included in the work of the project. In the view of the Sixth Committee, it seemed difficult to reach a concurrence of opinion. With respect to the final form of the work, many states welcomed the Commission's decision to present draft articles for further consideration without prejudice to the final form of its work.[12] Instead of creating a legally binding instrument, some states noted that it would be of more practical value to develop non-binding guidelines, a set of principles, and practical tools for states and others engaged in disaster relief.[13]

Considerations in 2009

In the second report, Valencia-Ospina[14] identified the following four points to be focused on by the ILC in the earliest phases of discussion:

(1) the definition of disaster;
(2) the appropriate implication of the protection of persons in the event of disasters;
(3) the extent of the scope between the affected and nonaffected States as well as among those States with relevant actors for disaster relief and assistance;
(4) which disaster phase was to be considered for the topic.

He elaborated on the three draft articles on the scope of the ILC project (Article 1), the definition of disaster (Article 2), and the duty to cooperate (Article 3) within the context of the three axes of *ratione materiae*, *ratione personae*, and *ratione temporis*. At the sixty-first session of the Commission in 2009,[15] Valencia-Ospina

[12] India (A/C.6/63/SR.23, par. 21), Republic of Korea (A/C.6/63/SR.23, par. 24). Spain (A/C.6/63/SR.23, par. 37), Japan (A/C.6/63/SR.23, par. 42), El Salvador (A/C.6/63/SR.23, par. 55), Malaysia (A/C.6/63/SR.23, par. 77), Poland (A/C.6/63/SR.24, par. 56), and Argentina (A/C.6/63/SR.24, par. 65).

[13] Hungary (A/C.6/63/SR.20, par. 31), Germany (A/C.6/63/SR.22, par. 60), India (A/C.6/63/SR.23, par. 21), United Kingdom (A/C.6/63/SR.23, par. 64), and United States of America (A/C.6/63/SR.23, par. 86).

[14] A/CN.4/615 and A/CN.4/615/Corr.1.

[15] Official Records of the General Assembly, Sixty-Fourth Session, Supplement No. 10 (A/64/10), pars. 147–183.

also pointed out that it was not necessary to include R2P in further ILC discussions, concurring with the view of the UN Secretary-General[16] that R2P did not apply to natural disasters.[17]

Several members expressed support for the second report. The positive opinions for the second report included: the adoption of the rights-based and need-based approach;[18] the non-applicability of the responsibility to protect to this project;[19] the preference for discussion of the rights and obligations of states vis-à-vis each other rather than the rights and obligations of states vis-à-vis persons in need of protection;[20] and the consideration of the duty to cooperate in the context of the protection of persons in the event of disasters.

Valencia-Ospina's suggestions in the second report, which received support from some states, also raised a number of concerns from states. With regard to the adoption of a rights-based approach, some members noted that "rights" were considered legal concepts, but "needs" referred to a particular factual situation, not a legal notion.[21] Moreover, a reference to an approach based on "needs" rather than "rights" could end up with the derogation of human rights in emergency situations.[22]

Regarding the proposed definition of disaster, while concurring with the existence of actual loss in line with the 1998 Tampere Convention, some states proposed that the occurrence of certain types of events should qualify as a disaster under the draft articles. On the duty to cooperate contained in Article 3, some states suggested that international assistance be considered complementary to and not a substitute for the primary obligation of affected states in a disaster situation.[23] However, some members remained concerned that the scope of the duty to cooperate was not sufficiently clarified.[24]

In response to the reactions of the various states, the Commission referred the proposed three articles to the drafting committee who made some revisions and sent the five draft articles on scope (Article 1), purpose (Article 2), definition of disaster (Article 3), relationship with international humanitarian law (Article 4), and duty to cooperate (Article 5), back to the plenary of the ILC.

At the sixty-fourth session of the Sixth Committee, there was some support for adoption of the rights and needs-based approaches, and the exclusion of the responsibility to protect.[25] The state parties also welcomed the draft articles adopted by the ILC.

However, while there was support for the definition of disaster as suggested by the Valencia-Ospina and the ILC, some delegations sought to include the situation of complex emergencies, both natural disaster and armed conflict, occurring

[16] A/63/677, par. 10 (b). [17] A/64/10, par. 156. [18] Ibid., pars. 159–160.
[19] Ibid., par. 164. [20] Ibid., par. 155. [21] Ibid., par. 161. [22] Ibid. [23] Ibid., par. 171.
[24] Ibid., par. 177. [25] A/CN.4/620, pars. 40–59.

simultaneously. Finally, on the protection of persons, there was support for establishing the principles in relation to disaster relief, including humanity, neutrality, impartiality, sovereignty, and non-intervention, among others.

Considerations in 2010

Valencia-Ospina, in the third report,[26] focused on the following two points:

(1) the principles that inspire the protection of persons in the event of disaster, in its aspect related to persons in need of protection,"[27] namely humanity, neutrality, impartiality, and human dignity; and
(2) the primary responsibility of the affected State, due consideration to sovereignty and non-intervention.

In his view, the state's role in disaster response included two aspects. One is the internal aspect; a state must provide humanitarian assistance and direct, control, coordinate, and supervise relief operations under its territorial jurisdiction during all disaster phases including initiation and termination of disaster relief and assistance. The other is the external aspect, which holds that humanitarian aid should be provided only with the consent of the affected state. Valencia-Ospina therefore proposed three further draft articles on humanitarian principles in disaster response (Article 6), human dignity (Article 7), and primary responsibility of the affected state (Article 8).

At the sixty-second session, while the Commission adopted the five draft articles as provisionally adopted in the drafting committee, it considered the third report of the Special Rapporteur.[28] Although general support was expressed for the proposed humanitarian principle in disaster response, concern was noted for each principle proposed. Many members were also concerned about assigning primary responsibility for disaster response to the affected state.

Regarding neutrality, there were a variety of opinions. Some members pointed out that it was not appropriate to refer to the principle, since it relates traditionally to armed conflict.[29] Other members suggested that the principle of neutrality should be replaced with that of non-discrimination.[30] With regard to the impartiality, several members doubted that the requirement of the proportionality applied, which is one of the elements of impartiality explained by Valencia-Ospina.[31] They also noted in this regard that the linkage of the principle involved the needs of affected persons as well as economic considerations relating to a state's capability of providing assistance.[32]

[26] A/CN.4/629. [27] A/CN.4/615, par. 71.
[28] Official Records of the General Assembly, Sixty-Fifth Session, Supplement No. 10 (A/65/10), pars. 290–329.
[29] Ibid., par. 311. [30] Ibid. [31] Ibid., par. 312. [32] Ibid.

Concerning the primary responsibility of the affected state, there was support for the requirement of consent by the affected state established in the proposed draft Article 8, paragraph 2.[33] However, there were divergent concerns as to its formulation.[34] For example, some suggested that the draft article should be restated for sovereignty and non-intervention, so as to reflect the right of affected states to decide whether to require or refuse international assistance.[35] It was also noted that balance among sovereignty, human rights, and the needs of persons to be protected must be achieved since the topic focus was on the protection of persons in the event of disasters.[36] Other members stressed that primary responsibility did not imply exclusive responsibility.[37]

Following the discussion, the Commission referred the three articles proposed by the Special Rapporteur in his third report to the drafting committee. The committee provisionally adopted the four draft articles: humanitarian principles in disaster response (Article 6); human dignity (Article 7); human rights (Article 8); and role of the affected state (Article 9). Paragraph 2 of draft Article 8 was also referred to the committee, but not elaborated on during the meeting because of lack of time. Meanwhile, the Commission adopted draft Articles 1 through 5 together with the accompanying commentary in the plenary.

At the sixty-fifth session of the Sixth Session in 2010, support was expressed for draft Articles 1 to 5 together with commentaries, as well as draft Articles 6 to 9 as provisionally adopted in the drafting committee.[38] In particular, support was expressed for inclusion of humanitarian principles in the draft articles as provisionally adopted by the drafting committee, including humanity, neutrality, impartiality, and human dignity, among others. The principle of non-discrimination was also endorsed. Nevertheless, concern was expressed that the principle of neutrality could cause unnecessary confusion and complications, since it is closely related to armed conflict and presupposed the existence of two opposite parties.

As for affected states, many states concurred with draft Article 9 adopted by the drafting committee based on paragraph 1 of draft Article 8, as proposed by the Valencia-Ospina in his third report. It was noted that the version adopted by the drafting committee properly asserted in the text that the duty of affected states was to ensure the protection of persons and the provision of disaster relief and assistance on their territory, and that the primary role of such states included the direction, control, coordination, and supervision of such relief and assistance. Support was also expressed for the reference to an affected state's "duty" rather than "responsibility." However, there was some concern that it was not clear what the content of such a duty for affected states amounted to in legal terms and what it would entail in practice.

[33] Ibid., par. 322. [34] Ibid., pars. 316–324 [35] Ibid., par.316. [36] Ibid.
[37] Ibid., par. 318. [38] A/CN.4/638, pars. 75–95.

Considerations in 2011

In his fourth report, Valencia-Ospina[39] focused on the following three issues: (1) the responsibility of the affected state to seek assistance where its national response capacity is exceeded; (2) the duty of the affected state not to arbitrarily withhold its consent to external assistance; and (3) the right to offer assistance in the international community.

While stressing the core principles of state sovereignty and non-intervention, Valencia-Ospina proposed that in determining the appropriate response when a disaster overwhelmed the national capacity of an affected state, the state must take all appropriate and necessary measures to ensure the protection of persons under its jurisdiction, including the acceptance of international assistance, on the basis of the duty to cooperate.[40] He also suggested that there arose the duty to seek international assistance in cases where national incapability was demonstrated, on the one hand, and that it was appropriate to frame the duty to "seek" rather than to "request" assistance in this context, on the other hand.[41]

With regard to the duty of the affected state not to arbitrarily withhold its consent to external assistance, he noted that consent is the expression of the willingness of the sovereign of affected states to allow the activities of non-affected states or other relevant actors on its territory, which include the right of affected states to refuse such disaster relief and assistance.[42] However, in his view, such consent was not unlimited when the authorities concerned were unable or unwilling to provide humanitarian assistance to meet the essential needs of the persons concerned, with full respect for their rights.[43] Therefore, he suggested that affected states should not be able to arbitrarily withhold their consent to external assistance.[44]

Concerning the right to offer assistance in the international community, Valencia-Ospina emphasized that the protection of persons in the event of disasters was also in the interest of non-affected states.[45] While confirming the primary role of the affected state, he also noted that the offer of humanitarian assistance by the international community in the event of disasters was the practical manifestation of solidarity, which was inherent in any global matter while confirming the primary role of the affected state.[46]

Consequently, he suggested three further draft articles in his report: duty of the affected state to seek assistance (Article 10); duty of the affected state not to arbitrarily withhold its consent (Article 11); and right to offer assistance (Article 12).

At the sixty-third session of the ILC, the Commission considered the fourth report and the proposed articles.[47] While support was expressed for these draft

39 A/CN.4/643 and A/CN.4/643/Corr.1. 40 A/CN.4/643, pars. 28–50.
41 Ibid., pars. 43–44. 42 Ibid., pars. 51–52. 43 Ibid., pars. 52–77. 44 Ibid., par. 76.
45 Ibid., par. 81. 46 Ibid., pars. 78–109.
47 Official Records of the General Assembly, Sixty-sixth Session, Supplement No. 10 (A/66/10), pars.264–287.

articles, there was concern in particular with the term "arbitrarily" in draft Article 11 and the ensuing formulation in Article 12. Regarding Article 11, it was pointed out that it was not appropriate to use "arbitrarily" in the context of offer of assistance in the international community and that the implication was not clear and therefore should be replaced with "unreasonably."

As for Article 12, some members suggested that the provision should more clearly elucidate the circumstances under which an affected state could reject offers of assistance.[48] Some also noted that offers of assistance could be different within the international community, especially offers from non-affected states and other relevant humanitarian actors; therefore Article 12 should be clarified.[49] It was also stressed that in interpreting Article 12, nothing should imply permission to interfere in the internal affairs of the affected state.[50] Moreover, the reference to "rights" should be clarified, since offering assistance offered by the international community was one of the parts of international coöperation, not an assertion of rights.[51]

Finally, the Commission referred Articles 10 to 12 to the drafting committee, requesting elaboration on the duty of the affected state to seek assistance (Article 10); and consent of the affected state to external assistance (Article 11). These were clearly the two most politically charged provisions. The Commission thereafter adopted Articles 6 to 9, as provisionally adopted in the drafting committee at the last session, together with commentaries. Meanwhile, the drafting committee failed to adopt draft Article 12 at this session, owing to the lack of time.

At the sixty-sixth session, the Sixth Committee considered the fourth report of the Special Rapporteur and the text of draft articles and their commentaries adopted by the Commission.[52] In addition, most states also focused on the question of whether the duty to cooperate included a duty for states to provide assistance when requested by the affected state. The Commission asked governments in the plenary session of the committee to give them some guidance in light of the Commission's view that states had a duty to cooperate with the affected state in disaster relief matters.[53] While general support was expressed for the adopted draft articles, there were divergent opinions on each draft article.

On draft Article 5, some expressed the need to formulate a completely new one.[54] Concerning the ILC's question as to whether the duty to cooperate included a duty on states to provide assistance when requested by the affected state, most doubted there was such a duty.[55]

48 Ibid., par. 279. 49 Ibid., par. 280. 50 Ibid., par. 281. 51 Ibid.

52 A/CN.4/650, pars.28–39.

53 Chairman of the International Law Commission (A/C.6/66/SR.21, par. 44).

54 Cuba (A/C.6/66/SR.24, par. 26) and Malaysia (A/C.6/66/SR.24, par. 120).

55 Mexico (A/C.6/66/SR.18, par. 55 and A/C.6/66/SR.22, par. 21), Slovenia (A/C.6/66/SR.20, par. 12), Singapore (A/C.6/66/SR.21, par. 76), Italy (A/C.6/66/SR.21, par. 91), Colombia (A/C.6/66/SR.22, par. 28), Germany (A/C.6/66/SR.23, par. 28), United Kingdom (A/C.6/66/SR.23, par. 45), Netherlands (A/C.6/66/SR.23, par. 48) (supporting the Special Rapporteur's earlier

Draft Article 9 on role of the affected state received general support.[56] It was also suggested that it would be appropriate to consider the addition of a third paragraph on the affected state's duty toward the international community as a whole, since inaction could have terrible consequences not only within a state's own territory, but within that of its neighbors.[57]

Concerning draft Article 10 on the duty of the affected state to seek assistance, many delegations welcomed establishing the legal obligation of the affected state to seek assistance, not merely a moral or political duty.[58] However, others emphasized that imposition of such a duty constituted infringement of the sovereignty of states as well as of international cooperation and solidarity, and that there was no basis in customary international law and state practice.[59]

With respect to draft Article 11, most states concurred with the codification of the fundamental principle that the affected state must give its consent to external assistance in accordance with state sovereignty.[60] However, some expressed concern for imposing such a legal obligation, which could undermine the current practice of international cooperation and solidarity.[61]

understanding of the duty to cooperate), Spain (A/C.6/66/SR.23, par. 50), Malaysia (A/C.6/66/SR.24, par. 120), and Ireland (A/C.6/66/SR.25, par. 21). See also Singapore (A/C.6/66/SR.21, par. 76).

[56] European Union (A/C.6/66/SR.21, par. 55), United States of America (A/C.6/66/SR.21, par. 69), Colombia (A/C.6/66/SR.22, par. 27), France (A/C.6/66/SR.23, par. 38), China (A/C.6/66/SR.23, par. 42), Argentina (A/C.6/66/SR.25, par. 10), India (A/C.6/66/SR.25, par. 13), Romania (A/C.6/66/SR.25, par. 17), Ireland (A/C.6/66/SR.25, par. 21), and Algeria (A/C.6/66/SR.25, par. 31).

[57] Romania (A/C.6/66/SR.25, par. 17).

[58] Slovenia (A/C.6/66/SR.20, par. 11), European Union (A/C.6/66/SR.21, par. 56), Finland (on behalf of the Nordic States) (A/C.6/66/SR.21, par. 60), El Salvador (A/C.6/66/SR.22, par. 12), Colombia (A/C.6/66/SR.22, par. 27), Czech Republic (A/C.6/66/SR.23, par. 19), Chile (A/C.6/66/SR.24, par. 8), India (A/C.6/66/SR.25, par. 13), Romania (A/C.6/66/SR.25, par. 18), Ireland (A/C.6/66/SR.25, par. 21), Egypt (A/C.6/66/SR.25, par. 36), and International Federation of Red Cross and Red Crescent Societies (A/C.6/66/SR.25, par. 41).

[59] Austria (A/C.6/66/SR.23, par. 23), Israel (A/C.6/66/SR.23, par. 33), France (A/C.6/66/SR.23, par. 38), China (A/C.6/66/SR.23, par. 42), Netherlands (A/C.6/66/SR.23, par. 48), Greece (A/C.6/66/SR.24, par. 25), Cuba (A/C.6/66/SR.24, par. 26), Russian Federation (A/C.6/66/SR.24, par. 37), Islamic Republic of Iran (A/C.6/66/SR.24, par. 50), Portugal (A/C.6/66/SR.24, par. 66), Indonesia (A/C.6/66/SR.24, par. 70), Republic of Korea (A/C.6/66/SR.24, par. 82), Thailand (A/C.6/66/SR.24, par. 90), Malaysia (A/C.6/66/SR.24, par. 114), Pakistan (A/C.6/66/SR.25, par. 7), Argentina (A/C.6/66/SR.25, par. 10), Algeria (A/C.6/66/SR.25, par. 33), and Sri Lanka (A/C.6/66/SR.27, par. 19).

[60] European Union (A/C.6/66/SR.21, par. 56), Finland (on behalf of the Nordic States) (A/C.6/66/SR.21, par. 60), El Salvador (A/C.6/66/SR.22, par. 13), Colombia (A/C.6/66/SR.22, par. 27), Czech Republic (A/C.6/66/SR.23, par. 19), Austria (A/C.6/66/SR.23, par. 24), Israel (A/C.6/66/SR.23, par. 33), France (A/C.6/66/SR.23, par. 39), Niger (A/C.6/66/SR.23, par. 54) (but noting that it could cause delay in cases where rapid reaction was needed), Chile (A/C.6/66/SR.24, par. 9), India (A/C.6/66/SR.25, par. 13), Romania (A/C.6/66/SR.25, par. 19), and Egypt (A/C.6/66/SR.25, par. 36).

[61] China (A/C.6/66/SR.23, par. 42), Russian Federation (A/C.6/66/SR.24, par. 37), Portugal (A/C.6/66/SR.24, par. 66), and Pakistan (A/C.6/66/SR.25, par. 7).

With respect to draft Article 12 proposed by the Special Rapporteur, some agreed only with the general premise articulated in the draft article and urged a limitation on its applicable scope and conditions, without undermining the principle of non-interference in the internal affairs of the affected state.[62]

The effort at the UN is continuing and, while progress has been made, significant agreement has yet to be achieved across the spectrum of related interests. Nevertheless, one can identify the broad outlines of emerging international disaster relief law from the UN process.

IFRC: DISASTER LAW PROGRAMME

The second effort to codify the international law of disaster relief has been undertaken by the IFRC. With long experience in the codification of international humanitarian law, the IFRC is perhaps better equipped than the UN to undertake this task from a legal and process standpoint. However, the IFRC is an NGO and, as such, does not enjoy the same level of global representation as the UN.

Aim of the IFRC Programme

The IFRC is the largest humanitarian organization in the world, with a focus on humanitarian relief and assistance in four core areas: promoting humanitarian values, disaster response, disaster preparedness, and health and community care. It initiated the project of International Disaster Response Laws, Rules and Principles (IDRL) Programme in 2001.[63] The twin goals of the program were to explore the role of law in the response to disasters and to strengthen legal preparedness of governments for disasters. The IFRC has been addressing this project through its activities, including technical assistance, capacity building, advocacy, dissemination, and research, among others, by collaborating with the National Red Cross and Red Crescent Societies in their individual capacities.

There have been two significant developments over this ten-year project: promulgation of the Guidelines for the Domestic Facilitation and Regulation of International Disaster Relief and Initial Recovery Assistance (IDRL Guildelines) adopted in 2007,[64] and promulgation of the Model Act for the Facilitation and Regulation of International Disaster Relief and Initial Recovery Assistance (Model Act) in 2011.[65] The IDRL Guidelines aim to support governments and strengthen

[62] Poland (A/C.6/66/SR.21, par. 86), Mexico (A/C.6/66/SR.22, par. 20). (noting that the exercise of such a right was subject to two constraints), Austria (A/C.6/66/SR.23, par. 25), Chile (A/C.6/66/SR.24, par. 10), and Romania (A/C.6/66/SR.25, par. 19).

[63] See "International Federation of Red Cross and Red Crescent Societies (IFRC)." www.ifrc.org/en/what-we-do/idrl..

[64] See IFRC. "IDRL Guidelines." http://www.ifrc.org/en/what-we-do/idrl/idrl-guidelines.

[65] See IFRC. "Model Act for the Facilitation and Regulation of International Disaster Relief and Initial Recovery Assistance." www.ifrc.org/en/what-we-do/idrl/model-act-on-idrl.

their legal preparedness systems on international disaster response operation. The Model Act has been developed as a tool to assist states by incorporating the IDRL Guidelines into national legal systems. A final version of the Model Act was due at the end of 2012.

In early 2012, the IFRC renamed its project the Disaster Law Programme so as to cover legal issues on disaster risk reduction and recovery as well.[66] The current name addresses the reduction of human vulnerability at all disaster phases, so that the IFRC could reflect its evolving focus in response to the expressed interests of national societies as well as its evaluation of humanitarian needs.

Development of the IFRC Programme

In 2007, on the basis of the findings from the project and a series of high-level regional forums held in 2006 and 2007, the 30th International Conference of the Red Cross and Red Crescent Societies unanimously adopted a set of operational guidelines for the facilitation and regulation of international disaster relief, called IDRL Guidelines.[67]

These guidelines are a set of recommendations to governments, not a treaty.[68] They focus on national legal preparedness by providing guidance to states in the area of disaster laws and plans, so that they can improve their regulatory frameworks in international disaster relief operations and initial recovery assistance. These guidelines refer to a number of existing international instruments, both binding and nonbinding, including UN General Assembly resolutions 46/182 of 1991 and 57/150 of 2002, the draft Convention on Expediting the Delivery of Emergency Assistance of 1977, and the HFA of 2005, among others.

The guidelines include sections on scope, purpose, and definitions, as well as the following five parts: core responsibilities (part I); early warning and preparedness (part II); initiation and termination of international disaster relief and initial recovery assistance (part III); eligibility for legal facilities (part IV); and legal facilities for entry and operations (part V).

As for the scope and definition, the IDRL Guidelines define disaster as follows: it "means a serious disruption of the functioning of society, which poses a significant, widespread threat to human life, health, property or the environment, whether arising from accident, nature or human activity, whether developing suddenly or as the result of long-term processes, but excluding armed conflict." This definition

[66] See IFRC. "About the Disaster Law Programme." www.ifrc.org/en/what-we-do/idrl/about-idrl.

[67] *Introduction to the Guidelines for the Domestic Facilitation and Regulation of International Disaster Relief and Initial Recovery Assistance*, Geneva: The International Federation of Red Cross and Red Crescent Societies, 2011.

[68] *Annotations to the Draft Guidelines for the Domestic Facilitation and Regulation of International Disaster Relief and Intitial Recovery Assistance*, Geneva: The International Federation of Red Cross and Red Crescent Societies, 2007.

is drawn from Article 1(6) of Tampere Convention. The guidelines do not intend to apply to situations of armed conflict, a disaster that occurs during armed conflict, and complex emergencies. In this regard, the IFRC insists that these situations should be covered by settled and widely-adopted international humanitarian law and that there is a practical difference between disaster and armed conflict in the regulation of external relief, noting that the "[c]onflict settings are inherently more politically sensitive than pure disaster situations."[69]

The first part on core responsibilities involves responsibilities of affected states and assisting actors, additional responsibilities of all states, and responsibilities concerning diversion and the intended use of resources.[70] Regarding the responsibility of the affected state, the IDRL Guidelines indicate that affected states have the primary responsibility to ensure disaster risk reduction as well as disaster relief and recovery assistance in their territory, and that they should seek international and/or regional assistance to address the needs of affected persons where a disaster situation exceeds national coping capacities. They also acknowledge that affected states are in the best position to coordinate, regulate, and monitor disaster relief and recovery assistance in their jurisdiction on the basis of sovereignty.

The second part of the IDRL Guidelines on early warning and preparedness comprise three elements: early warning; legal, policy, and institutional frameworks; and regional and international support for domestic capacity.[71] On early warning, the guidelines point out that all states should have procedures in place to facilitate the expeditious sharing of information about disasters. This includes identifying emerging hazards that are likely to cause disasters so as to minimize transboundary impacts and maximize the effectiveness of any international assistance. The guidelines stress the significance of legal preparedness for disaster risk reduction at all disaster phases, and therefore suggest that states adopt comprehensive legal policies and institutional frameworks on disaster prevention, mitigation, and preparedness as well as disaster relief and recovery.

The third part of the guidelines aim at initiation and termination of disaster relief as well as the specific initiation of military relief. While military organizations are often the most able to quickly respond to catastrophe, the guidelines emphasize that such responses should be initiated or terminated only with the consent of the affected state in principle.[72] The IDRL Guidelines also urge the affected state to decide whether or not to request and consent to disaster relief or initial recovery quickly and then notify its decision to the relevant assisting states and/or actors promptly.

The fourth part, concerning eligibility for legal facilities, focuses on preparedness of legal frameworks of affected states and transit states for assisting actors, whether states or humanitarian organizations.[73] Transit and affected states are encouraged to grant assisting states the legal facilities described in part five as to their disaster relief or initial recovery assistance.

[69] Ibid., 6–7. [70] Ibid., 13–21. [71] Ibid., 21–24. [72] Ibid., 24–26. [73] Ibid., 26–29.

Part five contains the following nine issues in the area of legal facilities for entry and operations: (1) personnel, (2) goods and equipment, (3) special goods and equipment, (4) transport, (5) temporary domestic legal status, (6) taxation, (7) security, (8) extended hours, and (9) costs. These sections seek to reduce legal and administrative barriers to the exportation, transit, importation, and re-exportation of goods and equipment, as well as establish expedited procedures for temporary recognition of professional qualifications and for granting visa and work permits for disaster relief and initial recovery personnel working in assisting states and humanitarian organizations.[74] For example, as regards special goods, the IDRL Guidelines provide for entry of medical equipment, telecommunications and information technology equipment, and vehicles imported by assisting states during operation of disaster relief and assistance. The guidelines encourage affected states to grant temporary recognition to foreign registration in disasters.

With respect to the Model Act, the IFRC has enhanced the pilot version that was intended to assist lawmakers with incorporating the recommendations of the IDRL Guidelines into national legislation as a reference tool for the facilitation and regulation of international disaster relief and initial recovery assistance.[75] This intensive work was welcomed by the 31st International Conference of the Red Cross and Red Crescent in 2011. To develop the Model Act, the IFRC has been collaborating with the UN Office for the Coordination of Humanitarian Affairs (OCHA) and the Inter-Parliamentary Union (IPU) with support from World Customs Organization and other relevant organizations and agencies, so as to produce a final version.

EVOLUTION OF INTERNATIONAL DISASTER RESPONSE LAW: TOWARD CODIFICATION AND PROGRESSIVE DEVELOPMENT

While there have been many recent developments contributing to IDRL, the field of IDRL also reflects codification of existing law. Although there is not a comprehensive multilateral legal framework focused on international disaster response, there are several relevant regulations dispersed in existing international, regional, and bilateral frameworks.[76]

International disaster relief law as a distinct field of law has experienced the same successes and failures toward codification and progressive development as many other fields of international law. Nevertheless, both the ILC and the IFRC have remained involved in the codification of IDRL over a long period. Indeed,

[74] Ibid., 29–43.

[75] IFRC, "Model Act for the Facilitation and Regulation of International Disaster Relief and Initial Recovery Assistance (with commentary),": 7. www.ifrc.org/PageFiles/125656/MODEL%20ACT%20INGLES.pdf.

[76] David Fisher, *Law and Legal Issues in International Disaster Response: A Desk Study* (Geneva: International Federation of Red Cross and Red Crescent Societies, 2007): 25.

the International Committee of the Red Cross (ICRC) was founded in the mid-1800s with part of its mission to protect persons in the event of natural disasters. This mission continued through the formation of the International Red Cross and Red Crescent Movement at the end of 1900s. At the intergovernmental level, the Convention and Statute Establishing an International Relief Union (IRU Convention) were adopted by the League of Nations, as a the first attempt to establish an intergovernmental organization on disaster relief, but the IRU Convention did not work effectively. The next attempt to move the IRU forward resulted in the failure of the draft Convention on Expediting the Delivery of Emergency Assistance of 1977, as prepared by the Office of the UN Disaster Relief Coordinator in the UN Economic and Social Council. There was, however, an increase in the number of sector and regional treaties, as well as "soft law" instruments on international disaster response during the 1980s.[77]

It was in this context of limited success and repeat failure that the projects of the ILC and IFRC were undertaken to codify IDRL. The ILC started to consider their topic on Protection of Persons in the Event of Disasters in 2007. On the other hand, the IFRC initiated their work on the Disaster Law Programme in 2001. While both seek similar certainties with respect to international regulation, it is appropriate to stress the difference between these two ongoing projects. The ILC aims to formulate the first international legal document dealing with general aspects and scopes in the disaster relief and assistance with full respect for human rights and needs of persons concerned. The IFRC, on the other hand, explores ways to enhance new non-binding guidelines and models which strengthen domestic legal preparedness for disasters, making use of the analysis of existing international and national norms as well as instruments relating to disaster relief and assistance.

While these two projects encompass great efforts at codification, they are occurring with very little input from the legal academy. Even though IDRL encompasses the breadth of public international law, legal scholars have been slow to engage in developing this field. To date, the codification efforts tend to have been undertaken by the legal practice community and tend to reflect an intent only to accomplish what is possible politically in the context of disaster response under international law.[78] The key questions that remain include: What is the definition of disaster? What is the responsibility of the affected state? Is there a consent requirement for the receipt of external assistance? What is the content of the duty to cooperate?

77 Ibid., 25–8.

78 David Caron, "Addressing Catastophes: Conflicting Images of Solidarity and Separateness" in *Les Aspects Internationaux des Catastrophes*, ed. David Caron and Charles Leben (The Hague: The Hague Academy of International Law, 2001), 3–29 at 29.

ASSESSMENT OF IDRL COMPONENTS

Definition of Disaster

With regard to the definition of disaster, the ILC relies on definitions offered by the Tampere Convention and the IDRL Guidelines. The elements for the legal definition of disaster suggested by the ILC are the following four points: (1) the existence of an event, not mere occurrence of the one; (2) two causation elements, including the calamitous event that results in widespread loss of life, great human suffering and distress or large-scale material or environmental damage, and the nature of the event that result in seriously disrupting the functioning of society; (3) the possibility of great human suffering and distress; and (4) large-scale material or environmental damage.[79]

The IFRC and ILC take somewhat different approaches to applying disaster response principles to situations arising under international humanitarian law. The IFRC excludes from their project armed conflict as well as disaster that occurs during armed conflict, namely complex emergencies. On the other hand, the ILC leaves space to deal with the situation of complex emergencies in draft Article 4 on the relationship with international humanitarian law, which provides that "the present draft articles do not apply to situations to which the rules of international humanitarian law are applicable." According to the drafting committee, if the draft articles exclude the applicability to complex emergencies because of the coexistence of armed conflict, it would undermine the principle of protection of persons in the event of disaster.[80]

Responsibility of the Affected State

Concerning the responsibility of the affected state, the ILC and the IFRC share the following three legal elements, as premised on the principles of sovereignty and non-intervention: (1) the duty to ensure the protection of persons on its territory; (2) the duty to ensure the provision of disaster relief and assistance on its territory; and (3) the primary role in the direction, control, coordination, and supervision of such relief and assistance.[81]

In this regard, while the Special Rapporteur suggested using the term "responsibility," which seemed to be derived from the first part of the IDRL Guidelines on responsibilities of affected states and assisting actors, the ILC decided to use the

79 See Official Records of the General Assembly, Sixty-Fifth Session, Supplement No. 10 (A/65/10) pars. 325–326.

80 Marcelo Vázquez-Bermúdez, "Statement of the Chairman of the Drafting Committee on Protection of Persons in the Event of Disasters." July 31, 2009, 14.

81 See Official Records of the General Assembly, Sixty-sixth Session, Supplement No. 10 (A/66/10), pars. 259–261.

terms "role" and "duty" instead. The Commission indicated in its commentary that the term "responsibility" could "give arise to confusion given its use as a term elsewhere in the Commission's work."[82] They also noted that "role" was used in UN General Assembly resolution 46/182 of 1991, which affirmed *inter alia* that an affected state had "the primary role in the initiation, organization, coordination, and implementation of humanitarian assistance within its territory."[83] However, this resolution uses the term as soft law, though it is one of the significant documents on disaster response. Even if the ILC explored the legally binding aspects of these terms in their work, it still would be insufficient to explain the usefulness of the term as a legal term. It would also be necessary to clarify the differences among each concept of "responsibility," "role," and "duty," so as to identify each meaning as well as any activating threshold that triggered a duty to act.

The consent of affected states to external assistance is a corollary of the state sovereignty. UN General Assembly resolution 46/182 of 1991 notes that "humanitarian assistance should be provided with the consent of the affected country and in principle on the basis of an appeal by the affected country."[84] Other instruments reaffirm the principle of consent by affected states, including the Tampere Convention, the ASEAN Agreement on Disaster Management and Emergency Response of 1995, and IDRL Guidelines, among others.

However, the consent of affected states to external assistance is not exclusive. It should be recalled that the main duty of affected states is to ensure the protection of persons and provision of disaster relief and assistance to the victims under their jurisdiction and in accordance with the principle of non-discrimination. If a disaster exceeds the capacity of an affected state to cope with the adverse effects, the state has the duty to seek external assistance as a complement to their capacities. Affected states also should not unreasonably refuse offers from non-affected states on disaster response. As the Commission indicated in its commentary, these ideas derive from the obligations of international human rights law[85] as well as the duty of cooperation[86] and solidarity.

This point gives rise to the question of whether the duty to cooperate includes a duty on states to provide assistance when requested by the affected state, as implied by the Commission in the plenary of the Sixth Committee.[87] In this regard the Commission's view is that states have a duty to cooperate with the affected state

[82] Ibid., (2) of Commentary of Draft Article 9.
[83] Ibid., (3) of Commentary of Draft Article 9.
[84] U.N.General Assembly (A/RES/46/182) Strengthening of the Coordination of Humanitarian Emergency Assistance of the United Nations
[85] Official Records of the General Assembly, Sixty-sixth Session, Supplement No. 10 (A/66/10), (1), (5) and (6) of Commentary of Draft Article 10, and (4) and (5) of Commentary of Draft Article 11.
[86] Ibid., (1), (9) and (11) of Commentary of Draft Article 10.
[87] See Chairman of the International Law Commission (A/C.6/66/SR.21, par. 44).

in disaster relief. Although most delegations did not concur with the idea, they only voiced their stance and did not explain specifically.[88] This question will most certainly be raised again in the plenary of the ILC as well as the Sixth Committee.

It is also important to consider the requirements associated with refusing external humanitarian assistance. According to the Valencia-Ospina, the requirements include two elements: the lack of a clear need for assistance; and the offer for assistance does not comport with humanitarian principles, including humanity, neutrality, and impartiality. The ILC decided to use the term "arbitrariness" as one of the requirements. They note in their commentary that this term "directs attention to the basis of an affected State's decision to withhold consent" and "the determination of whether the withholding of consent is arbitrary must be determined on a case-by-case basis."[89] However, the term "arbitrary" remains ambiguous and thereby could leave considerable discretion to the affected states.

Finally, in addition to those points, it is must be stressed that adverse effects from disaster, man-made or otherwise, tend to disproportionately impact the most vulnerable within a population. Women, children, people with disabilities, older persons, and minority groups have consistently fared the worst as disasters occur.[90] In many cases vulnerable people already suffer from discrimination pre-disaster because of the existence of traditional customs and typically negative notions associated with them. International law requires generally that all components of a population be treated equally and that discrimination not occur, even in the event of a disaster. All appropriate means, including legislative, administrative, and judicial measures, must be taken without discrimination of any kind, in accordance of international law.

CONCLUSION

There is no comprehensive international legal framework in the area of international disaster response. In many states there is no system in place to facilitate the urgent assessment and facilitation for disaster relief and assistance. This lack of law adversely affects persons impacted by disaster. For example, the human rights situation of victims in Japan's Triple Disaster was very severe. Many people are still suffering from the negative impact caused by the slow time of recovery. One of the

88 See Mexico (A/C.6/66/SR.18, par. 55 and A/C.6/66/SR.22, par. 21), Slovenia (A/C.6/66/SR.20, par. 12), Singapore (A/C.6/66/SR.21, par. 76), Italy (A/C.6/66/SR.21, par. 91), Colombia (A/C.6/66/SR.22, par. 28), Germany (A/C.6/66/SR.23, par. 28), United Kingdom (A/C.6/66/SR.23, par. 45), Netherlands (A/C.6/66/SR.23, par. 48) (supporting the Special Rapporteur's earlier understanding of the duty to cooperate), Spain (A/C.6/66/SR.23, par. 50), Malaysia (A/C.6/66/SR.24, par. 120), and Ireland (A/C.6/66/SR.25, par. 21). See also Singapore (A/C.6/66/SR.21, par. 76).

89 See A/66/10, (7) and (8) of Commentary of Draft Article 11, pars. 264-287.

90 See Erica Harper, *International Law and Standards Applicable in Natural Disaster Situations* (International Development Law Organization, 2009).

reasons for this slow recovery time was that Japan has not prepared the law on emergency disaster response and accepted external assistance adequately and effectively.

Disasters stemming from the Indian Ocean tsunami, Cyclone Nargis in Myanmar, and the Haiti earthquake are also still fresh in our minds. Affected states and other assisting actors struggle to save affected people in relief operations, while planning for and responding to the risks and consequences of future disasters and minimizing the damage of current disasters. Because various legal issues remain unresolved and there is no sign of cessation of disasters, it is has become more urgent that we construct a comprehensive international legal framework in this area.

PART II

The Law of International Disaster Relief: From Local to Global

FIGURE 5 **San Francisco burned to the ground in the wake of the 1906 earthquake.** Source: Panoramic photographs, Library of Congress, Reproduction number LC-DIG-ppmsca-07823 DLC.

5

International Disaster Response Laws, Rules, and Principles: A Pragmatic Approach to Strengthening International Disaster Response Mechanisms

Claire Clement

In recent decades, there has been a growing recognition of the role of domestic and international law in ensuring efficiency and effectiveness in international disaster response. When disaster strikes, an absence of appropriate domestic legislation, policies, and guidance can lead to unnecessary delays and otherwise hinder the delivery and coordination of incoming relief from abroad. Mindful of these difficulties, the 30th International Conference of the Red Cross and Red Crescent in 2007 adopted Guidelines for the Domestic Facilitation and Regulation of International Disaster Relief and Initial Recovery Assistance (IDRL Guidelines). This chapter explains how, despite their nonbinding nature, the IDRL Guidelines, and considerable efforts undertaken to encourage their use, have made important contributions to the development of best practices in the regulation of international disaster relief. Various projects undertaken by the National Red Cross and Red Crescent Societies have benefited disaster response processes and mechanisms at the national and regional levels from a "bottom-up" approach. Given the difficulties in encouraging the development of specialized international law in the area of disaster response, and understanding that many of the major legal obstacles to the provision of effective disaster relief could be addressed by facilitating changes at the domestic level, such a bottom-up approach is both a pragmatic solution and well suited to these issues.

INTRODUCTION

In recent decades, there has been a growing recognition of the role of laws, both international and national, in ensuring efficiency and effectiveness in international responses to situations of disaster.[1] When such a disaster strikes, an absence of

[1] For the purposes of this paper, the term "disaster" is defined as in the Guidelines for the Domestic Facilitation and Regulation of International Disaster Relief and Initial Recovery

appropriate domestic legislation, policies, and guidance can lead to unnecessary delays and otherwise hinder the delivery and coordination of incoming relief from abroad. While many emergencies will not require or receive cross-border assistance, the serious impact caused by major disasters that affect more than one state, or otherwise overwhelm domestic capacities, serves to remind the world of the need to guard against complacency.[2] The need for effective systems of international support in such circumstances can be expected to continue to grow.

Unlike international humanitarian law in times of armed conflict, there is no universal, specialized body of international law that applies to situations of natural or technological disaster. Rather, international disaster response is primarily governed by domestic legislation, policies, and plans, as well as a number of bilateral and even fewer multilateral agreements. Consequently, a variety of actors, both state and non-state, have over time turned their attentions to exploring suitable means of addressing the perceived gaps in this field.

Some such efforts have attempted to identify and address the common legal and regulatory barriers that may serve to hinder efforts to deliver disaster relief,[3] while others have paid increasing attention to the identification of rights and obligations in such circumstances (and some have endeavored to combine both approaches). One initiative, focused primarily on the technical aspects of international disaster relief, was instigated more than a decade ago by the International Federation of Red Cross and Red Crescent Societies (IFRC), which established an area of research commonly referred to as the Disaster Law Programme.

Mindful of the difficulties in encouraging the development of specialized international law in the area of disaster response, and believing that many of the major legal obstacles to the provision of effective disaster relief could be addressed by

Assistance (paragraph 2.1) (IDRL guidelines): "a serious disruption of the functioning of society, which poses a significant, widespread threat to human life, health, property or the environment, whether arising from accident, nature or human activity, whether developed suddenly or as the result of long-term processes, but excluding armed conflict." Annotated IDRL Guidelines available at http://www.ifrc.org/PageFiles/41203/annotations.pdf.

2 The 2011 earthquake and tsunami affecting Japan, major flooding in Pakistan in 2010, the protracted food crisis in the Sahel, and the 2010 earthquake in Haiti all serve as sobering reminders.

3 For practical examples of how gaps in states' domestic legal frameworks have caused challenges in the provision of incoming disaster assistance, see Victoria Bannon and David Fisher, "Legal Lessons in Disaster Relief from the Tsunami, the Pakistan Earthquake, and Hurricane Katrina," *ASIL Insights* (American Society of International Law) 10, no. 6 (March 2006). For example, a year after the tsunami in Indonesia, hundreds of containers of tsunami relief aid were reportedly still with customs authorities outside Jakarta. Similarly, in Sri Lanka many food items perished before they could be distributed and other items, such as tents and body bags, were no longer needed, due to delays in official inspections. Such difficulties may not only be caused by gaps in the domestic frameworks in affected countries, but may equally be generated by assisting actors providing unsuitable or inappropriate assistance. For example, the United Kingdom flew nearly 400,000 "ready meals" to the United States in the wake of Hurricane Katrina, which were subsequently quarantined in accordance with U.S. health regulations.

facilitating changes at the domestic level, the IFRC began to explore the possibility of developing a set of practical guidelines primarily for use by states, but also of relevance to aid organizations and other actors. Following years of research and consultation, the IDRL Guidelines were eventually adopted in 2007 by the 30th International Conference of the Red Cross and Red Crescent.

This chapter argues that, in spite of their nonbinding nature, the IDRL Guidelines have made a significant contribution to the development of best practice in the regulation of international disaster relief by facilitating tangible though limited improvements to a number of domestic and regional response mechanisms. Projects undertaken in a number of countries, where National Red Cross or Red Crescent Societies have worked with their respective governments to examine domestic disaster response frameworks in the light of the IDRL Guidelines, attest to the many benefits of this so-called bottom-up approach.

The development of international law is a slow and often fractious process. To date, efforts to promote the need for universally applicable, legally binding standards in situations of disaster have been largely unsuccessful. This chapter takes a pragmatic approach to the problem of strengthening the quality and efficiency of international disaster response by focusing on the adoption of a bottom-up approach. Taking into account states' primary responsibility in responding to disasters on their territory and the largely domestic nature of most disaster response activities, a bottom-up approach provides states with a measure of flexibility that allows for appropriate solutions tailored to specific national circumstances.

The main goal of the IDRL Guidelines is to strengthen domestic laws, policies, and institutions so that states are prepared both to expedite and coordinate international disaster assistance when it is needed. In the longer term, and as the increasing utilization of the IDRL Guidelines at a regional level perhaps indicates, it is possible that the strengthening of domestic frameworks may influence positively future efforts to regulate international disaster assistance at the global level.

THE DEVELOPMENT OF INTERNATIONAL DISASTER LAWS: A TROUBLED PATH

The field of international disaster assistance is characterized by a lack of international law; in particular, binding treaty law. Consequently, despite an increasing tendency by some academics and practitioners to refer to a body of "international disaster response law" or even "international disaster law," it is not possible to assert that, at this juncture, any such coherent body of law exists.[4]

The processes and mechanisms by which disaster relief is delivered and distributed across national borders are generally determined by states' domestic laws and

[4] The British Red Cross, for example, has continually maintained its preference for use of the term "disaster laws" rather than "disaster law," for reasons of accuracy and clarity.

regulations, as well as by a number of bilateral and regional agreements, although many gaps in such national and regional frameworks remain. Overall, in spite of these gaps, the sovereign responsibility of a state to provide assistance to those affected by disasters on its territory, and the corresponding duty of other states not to intervene in such circumstances without the consent of the affected state, are well recognized.[5]

Unlike the field of international humanitarian law, where the universally recognized 1949 Geneva Conventions form a backbone of specialized rules aimed at regulating behavior in situations of armed conflict, no similar comprehensive and binding treaty framework applies to situations of natural disaster. While several attempts have been undertaken to develop such an instrument, these efforts have remained largely unsuccessful. It is not the purpose of this chapter to set out a comprehensive list of all such initiatives, in particular as these have been documented extensively in other literature.[6] However, the most notable efforts are set out in the sections that follow.

THE INTERNATIONAL RELIEF UNION AND BEYOND

One of the first attempts to develop a specific framework for international disaster response was undertaken in the 1920s under the auspices of the League of Nations, leading to the establishment of the short-lived International Relief Union.[7] Following this failure,[8] further serious efforts to establish an international legal framework for disaster relief were not undertaken until the 1970s, when the UN Disaster Relief Organization (UNDRO), the predecessor organization to the UN Office for the Coordination of Humanitarian Affairs (OCHA), proposed a draft Convention on Expediting the Delivery of Emergency Assistance. The draft convention sought to solve a number of logistical issues commonly faced in the delivery and distribution of international assistance, as well as to tackle core duties, such as the responsibilities of the state receiving such assistance. Despite referral to the Economic and Social Council in 1984, the draft convention was never taken up by the UN General Assembly.[9]

[5] International Law Commission. "Third Report on the Protection of Persons in the Event of Disasters." 2010, 23.

[6] For example, see: David Fisher, *Law and Legal Issues in Disaster Response: A Desk Study*. International Federation of Red Cross and Red Crescent Socities, 2007.

[7] The Convention and Statutes Establishing an International Relief Union (IRU) was adopted in 1927, following an original proposal by the Red Cross. The Convention entered into force in 1932. The IRU was designed to be a centralized operational agency, channelling international funds and support in disaster settings, among other tasks. It did not have notable success in regard to relief actions during its brief existence, and ultimately failed due to a lack of funding. See Peter Macalister-Smith, "The International Relief Union of 1932," *Disasters* (2007), 147.

[8] The International Relief Union did not have notable success and very quickly failed due to a lack of funding. Ibid.

[9] Fisher, *Law and Legal Issues*, 4.

In 2007, a comprehensive desk study of laws pertaining to international disaster relief was completed by the IFRC. Noting the absence of any centralized framework, it concluded that: "International law on disaster relief has thus developed in a fragmented manner, often through the inclusion of disaster related provisions in treaties of a more general character or through declarations or resolutions in various forums."[10] In addition, it was recognized that a growing number of instruments and initiatives (both of a binding and nonbinding nature) seek to address such difficulties at the regional level.[11]

A range of theories has been offered to explain the failure of the international community to agree upon a comprehensive international regime for international disaster response activities to date. One argument is that, due to the political process by which international law is developed, such rules may simply be fundamentally ill-suited to this context. In general, international law is typically developed where states' interests converge on the need to regulate sovereignty, a classic example being states' relative willingness to create binding international rules to regulate their behavior in armed conflict. In contrast, in the context of international disaster response, the interests of both the affected state and assisting states favor the maintenance and strengthening of states' sovereign role, and underscore the need to preserve the principle of nonintervention.[12] This reality, it is contended, "does not stimulate the robust development of international law."[13]

[10] Ibid., 6. One notable example (in a relatively sparse field) of an international treaty specifically related to international disaster response is the 1998 Tampere Convention on the Provision of Telecommunication Resources for Disaster Mitigation and Relief Operations. The Tampere Convention governs the provision and availability of communications equipment during disaster relief operations. Other examples related to customs law include the 1999 Kyoto Convention on the Simplification and Harmonization of Customs Procedures and the 1990 Istanbul Convention on Temporary Admission, both of which include annexes addressing the need to expedite customs clearance of relief consignments.

[11] Some such regional instruments take the form of binding treaties (e.g., the Inter-American Convention to facilitate Disaster Assistance, the Association of Southeast Asian Nations (ASEAN) Agreement on Disaster Management and Emergency Response) and some entail the establishment of practical coordination mechanisms (e.g., the EU Civil Protection Mechanism). None of these instruments has the universal scope envisaged by the Draft Convention on Expediting the Delivery of Emergency Assistance.

[12] Regarding states' interests in international disaster response, the debates in the context of the ILC's work on the Protection of Persons in the Event of Disasters are instructive. For example, the ILC's query to UN member states in 2011 as to whether international law imposes a general obligation on states to offer disaster assistance when requested elicited generally negative responses. The question as to whether a state is under a duty to seek assistance when a disaster situation exceeds its own capacities received contradictory (both negative and positive) responses.

[13] In particular, disasters have historically been regarded by states as episodic events that do not systematically affect their interactions with other states (unlike warfare or trade), or necessarily form a broader part of their foreign policy. David Fidler, "Disaster Relief and Governance After the Indian Ocean Tsunami: What Role for International Law?" *Melbourne Journal of International Law* 6 (2005): 458.

This inherent collision of interests between states receiving aid, states needing aid, and individuals is readily apparent in the most recent attempt to develop binding rules on international disaster assistance at the international level. In 2007 the International Law Commission (ILC) began to develop draft articles on the Protection of Persons in the Event of Disasters.[14] Although the final form and purpose of the draft articles is still unclear because only fifteen draft articles have so far been provisionally adopted by the drafting committee,[15] the ILC continues to grapple with the challenge of balancing the need to address the common technical and regulatory barriers to international relief with its more rights-based aspects.[16] In relation to the latter, the question of whether and to what extent persons are entitled to international relief in the context of natural disasters and the potential obligations of states in such circumstances has proved particularly vexing in the context of the ILC's work.

RIGHTS-BASED APPROACHES VERSUS TECHNICAL APPROACHES

Although increasingly of interest to scholars and practitioners, efforts to examine cross-border disaster relief through the lens of human rights have gained little traction at the international level from states even though a number of aid organizations have begun to utilize rights-based approaches in their humanitarian programming. In particular, a general right to receive humanitarian assistance, purportedly established by combining various accepted norms of international human rights law,[17] has not received widespread support to date.[18] States appear reluctant to

[14] The decision to include this topic in the ILC's long-term program of work was taken under General Assembly Resolution 62/66 of December 6, 2007. For a full summary and progress report if the ILC's work on this topic, see International Law Commission, *Protection of Persons in the Event of Disasters*. http://untreaty.un.org/ilc/guide/6_3.htm.

[15] The most recent draft articles (5 bis, 12, 13, 14 and 15) were provisionally adopted in July 2012. See the Statement of the Chairman of the Drafting Committee at untreaty.un.org/ilc/sessions/64/ProtectionOfPersonsDCstatement%282012%29.pdf

[16] For further discussion, see Dug Cubie, "An Enchanted Tool? Humanitarian Assistance and the ILC Draft Articles on the Protection of Persons in the Event of Disasters," *Irish Yearbook of International Law*. 2009–2010. 121. Draft Articles 12 and 13 (provisionally adopted by the drafting committee in July 2012) elaborate a right of states, the United Nations, and other inter-governmental organizations to offer assistance to the affected state. In turn, the affected state may place conditions on the provision of external assistance in accordance with its national laws.

[17] It is argued, for example, that the combined effect of several rights enshrined in the 1966 International Covenant on Economic, Social and Cultural Rights, including, among others, the right to an adequate standard of living, the right to be free from hunger, and the right to health, serve to engender a related yet distinct right to humanitarian assistance.

[18] For example, Yoram Dinstein notes that "there is no clear-cut right under existing international law to humanitarian assistance in peacetime, not even when natural disasters strike." Yoram Dinstein, "The Right to Humanitarian Assistance," Working Paper, *Naval War College Review (Autumn 2000)* 77–92. On the other hand, Fisher argues that "existing human rights instruments imply a right to assistance in situations of crisis and a certain obligation on states

create specific customary norms in the area of humanitarian assistance, aside from some limited provisions of specific application to situations of armed conflict.[19]

A second human rights-based approach to disaster assistance endeavors to make use of the relatively recent doctrine of the Responsibility to Protect (R2P), a concept that was endorsed by the UN General Assembly in 2005 but continues to elicit varying levels of acceptance within the international community.[20] In the event of serious violations of human rights or international humanitarian law occurring within a state's territory, the R2P doctrine creates a valid basis for various forms of collective intervention (including, as a last resort, military intervention) in a state's internal affairs.[21] It is unlikely that R2P, in its current limited form[22] can reasonably be reformulated to extend to situations where the provision of international disaster relief is perceived as being intentionally and unreasonably barred by a state.[23] At least one relatively recent example demonstrates the reticence of the international community to consider implementation of the doctrine of R2P in such circumstances.[24]

to seek international support if their own means are insufficient to address humanitarian needs." See David Fisher, "Domestic regulation of international humanitarian relief in disasters and armed conflict: a comparative analysis," *International Review of the Red Cross*, (June 2007) 89:2, 345–372 at 349.

19 Even where they exist in international humanitarian law, such provisions are limited in scope. For a full analysis of such provisions, see Yoram Dinstein, *The Right to Humanitarian Assistance*, Working Paper, *Naval War College Review* (Autumn 2000), 77–92.

20 Carlo Focarelli, "The Responsibility to Protect Doctrine and Humanitarian Intervention: Too Many Ambiguities for a Working Doctrine," *Journal of Conflict and Security Law* 2 (2008), 191–213

21 The UN General Assembly's 2005 World Summit Outcome Document included the following statement on R2P ((Resolution 60/1, 2005) par. 139):
"...we are prepared to take collective action, in a timely and decisive manner, through the Security Council, in accordance with the Charter, including Chapter VII, on a case-by-case basis and in cooperation with relevant regional organizations as appropriate, should peaceful means be inadequate and national authorities manifestly fail to protect their populations from genocide, war crimes, ethnic cleansing and crimes against humanity."

22 The 2005 narrow formulation of R2P accepted by the UN General Assembly has four triggers: genocide, war crimes, crimes against humanity, and ethnic cleansing. Harrington argues that, not only does the attempted invocation of R2P in situations of natural disaster potentially undermine the various international cooperative efforts to address challenges in international disaster response, but its expansion into such areas also risks damaging the doctrine's own utility and strength. See Joanna Harrington, "R2P and Natural Disasters," in *The Routledge Handbook of the Responsibility to Protect*, ed. W.A. Knight and F. Egerton. (New York: Routledge, 2012), 141–51.

23 For further discussion of the various challenges regarding the applicability of R2P to situations of natural disaster, see Milena Costas Trascasas, "Access to the Territory of a Disaster-Affected State," in *International Disaster Response Law*, ed. Marco Gestri, Gabriella Venturini, and Andrea de Guttry (The Hague: T.M.C Asser Press, 2012), 221–50.

24 During the aftermath of Cyclone Nargis, which devastated Myanmar in May 2008, affecting 2.4 million people and causing the deaths of approximately 30,000, the possibility of invoking the R2P doctrine in relation to the Myanmar government placed severe restrictions on incoming international assistance. This perceived blocking of international disaster relief was raised in some quarters and led to considerable debate as to whether the R2P doctrine

An international or "top-down" approach that focuses solely on technical matters relating to international disaster relief and avoids the more contentious matters of rights and obligations can also face major challenges in gaining vital support from states. In part, this is due to the fact that addressing technical obstacles to disaster relief delivery may require the enactment or amendment of domestic regulations and procedures to allow for cross-border delivery in areas such as immigration, customs, transport, and telecommunications. These jurisdictional areas can be of particular sensitivity for many national authorities who are reluctant to yield powers to actors from other states.

Further, the somewhat negative practical experience of a number of states in receiving international disaster assistance can undermine efforts to seek support for the development of binding international rules of a technical nature. For example, a number of challenges in institutional cooperation and coordination between international responders and the domestic authorities of an affected state emerged following the introduction of the UN cluster system.[25] A recent evaluation of cluster coordination found evidence of a continuing failure of international responders to engage with national authorities sufficiently:

> In their current implementation, clusters largely exclude national and local actors and often fail to link with, build on, or support existing coordination and response mechanisms... As a result, the introduction of clusters has in several cases weakened national and local ownership and capacities.[26]

On the basis of these and similar findings, a 2011 report concluded that, in practice, "the relationship between government coordination systems and those set up by international agencies is often tense, and coordination problems are common."[27]

could be invoked in this circumstance, to facilitate military intervention for the purpose of delivering humanitarian aid. This proposal was problematic on a number of levels. It was not clear that the suffering of the population in Myanmar in the wake of Cyclone Nargis, although severe, constituted one of the serious crimes for which the doctrine of R2P may be invoked. There were also practical difficulties in potentially delivering aid by force, threats to humanitarian principles, and a lack of overall consensus on and support for R2P. Ultimately, a diplomatic solution was sought, utilizing regional partners including ASEAN. See, for example, Joanna Harrington, "R2P and Natural Disasters" and also Doctors Without Borders, *Myanmar – Responsibility to Protect?* www.doctorswithoutborders.org/publications/article.cfm?id=2740&cat=ideas-opinions.

[25] Introduced in 2005, the cluster approach seeks to make humanitarian assistance more effective by introducing a system of sectoral coordination with designated lead organizations. Julia Steets et al., "Cluster Approach Evaluation II: Synthesis Report," *IASC Cluster Approach Evaluation 2nd Phase* (Groupe URD and the Global Public Policy Institute, 2010), 8.

[26] Ibid., 9.

[27] Paul Harvey and Adele Harmer, "Building Trust: Challenges for National Authorities and International Aid Agencies in Working Together in Times of Disasters," Background Paper for the International Dialogue on Strengthening Paternships in Disaster Response: Bridging National and International Support, *Humanitarian Outcomes*, (September 2011), 32.

Efforts to develop binding international norms in the area of international disaster assistance may also be undermined by fundamental perceptions of bias in the nature of such assistance. In particular, there is a prevailing notion of international disaster relief as a largely "Western" undertaking. In spite of emerging trends pointing to a shifting balance in the international humanitarian system, for example seen through the increasing provision of humanitarian assistance by "nontraditional" donors including China, Russia, Saudi Arabia, and Brazil.[28] Nonwestern state actors are unlikely to voice their support for the creation of new international laws in disaster relief even of a purely technical nature, where these are perceived as being incapable of representing their particular needs and interests.

SOFT LAW APPROACHES AND DEVELOPMENT OF THE IDRL GUIDELINES

In the absence of formally binding international rules, there exists a wide range of guidelines, operational guidance, resolutions, recommendations, and best practice, all aimed at addressing the various and multiple challenges of international disaster response. Given their nonbinding nature, there are obvious limitations to the utility and acceptance of such "soft" instruments. However, they may usefully fill gaps in existing laws on international disaster response, where such assistance is sought. For example, UN member states have agreed to a range of soft measures in order to strengthen their ability to respond to natural disasters, as well as to reduce their effects.[29]

Soft instruments are able to address sensitive or controversial issues that may face more resistance in the process of development of international laws. One pertinent example is the extensive "Operational Guidelines on the Protection of Persons in Situations of Natural Disasters," issued by the Inter-Agency Standing Committee (IASC)[30] in January 2011.[31] These guidelines seek in particular to address issues of

[28] For a comprehensive analysis of such trends see Andrea Binder, Claudia Meier, and Julia Steets, "Humanitarian Assistance: Truly Universal? A Mapping Study of Non-Western Donors," *Global Public Policy Institute*, 12 (August 2010).

[29] One such well-known example relevant to the field of disaster risk reduction is the Hyogo Framework for Action, adopted in 2005 by the World Conference in Disaster Reduction. The Hyogo Framework for Action emphasizes the sovereign role of states in developing disaster risk reduction strategies for their own populations, while stressing the need for cooperation at the international level. See Conference Proceedings. *World Conference on Disaster Reduction*. Kobe, Hyogo, Japan, 2005. See www.unisdr.org/2005/wcdr.

[30] The Inter-Agency Standing Committee is a forum of key humanitarian agencies (both UN and non-UN) established in 1992 to promote better inter-agency coordination in disaster assistance.

[31] The Operational Guidelines state that, while they are primarily aimed at non-state actors, they may also "be useful for those governmental actors, in particular disaster management institutions, who are primarily responsible for providing protection and humanitarian assistance to affected persons. Furthermore, they may also inform national laws and policies." The Brookings-Bern Project on International Displacement, "IASC Operational Guidelines on

human rights in natural disasters. However, they are primarily intended to assist international and nongovernmental humanitarian agencies, rather than affected states themselves.[32] Further, similar principles to those set out in the IASC Operational Guidelines may be found in guidance on themes that cut across, or otherwise bear some relevance to, situations of disaster. For example, the 1998 Guiding Principles on Internal Displacement, which are intended to be used by both state and nonstate actors, emphasize "the responsibility of sovereign governments and the complementary role of international assistance."[33]

An alternative approach to addressing the common problems faced in international disaster relief was launched in 2001, when the Council of Delegates[34] of the International Red Cross and Red Crescent Movement launched a body of work which eventually became known as the Disaster Law Programme. .[35] Led by the IFRC, high-level representatives of governments, National Red Cross and Red Crescent Societies, and other humanitarian actors held a series of formal regional fora from 2006 to 2007. The outcomes of these fora, in combination with a body of extensive research on the subject, influenced the development of the IDRL Guidelines.

the Projection of Persons in Situations of Natural Disasters," *The Brookings-Bern Project on International Displacement* (January 2011), 8.

32 Other examples of guidance primarily aimed at humanitarian agencies in disaster response, but which may have wider relevance for government and other actors include the Code of Conduct for the International Red Cross and Red Crescent Movement and NGOs in Disaster Relief, and the Sphere Project's Humanitarian Charter and Minimum Standards in Humanitarian Response.

33 Elizabeth Ferris and Kei Hakata, "Beyond Disaster: A Call for Japanese Leadership in International Disaster Response Law," Brookings Institute, (2011) www.brookings.edu/research/opinions/2011/03/22-japan-leadership-ferris. The UN Human Rights Commission took note of the Guiding Principles on Internal Displacement (Resolution 1998/50, 1998) and has since continued to encourage their use by states and other relevant actors. The UN General Assembly has welcomed their application by "an increasing number of States, United Nations agencies and regional and non-governmental organizations" and encouraged all relevant actors to make use of them in relation to situations of internal displacement (Resolution 62/153, 200), par. 10.

34 The Council of Delegates is the body where the representatives of all the components of the International Red Cross and Red Crescent Movement (the ICRC, IFRC, and National Red Cross and Red Crescent Societies) meet to discuss matters which concern the movement as a whole (Article 12, Statutes). The components of the movement include the ICRC, the IFRC, and the 189 National Red Cross and Red Crescent Societies. All of the components are independent of one another but work toward a common humanitarian mission and shared principles. For more information, see www.icrc.org/eng/who-we-are/movement/overview-the-movement.htm.

35 While the 2001 Council of Delegates refers to "International Disaster Response Law," by the time of the 28th International Conference of the Red Cross and Red Crescent in 2003, this area of focus was referred to as the "laws, rules and principles applicable to international disaster response." "Agenda for Humanitarian Action: Final Goal 3.2," in Resolution 1 of the 28th International Conference of the Red Cross and Red Crescent, Geneva, December 2-6, 2003.

The IDRL Guidelines were formally adopted in 2007 by the 30th International Conference of the Red Cross and Red Crescent. This means that state parties to the 1949 Geneva Conventions (who send representatives to the International Conference), as well as the components of the International Red Cross and Red Crescent Movement, have endorsed the IDRL Guidelines, and agreed to promote their use.[36] Their explicit purpose is to "contribute to national legal preparedness by providing guidance to States interested in improving their domestic, policy and institutional frameworks concerning international disaster relief and initial recovery assistance."[37] The IDRL Guidelines are therefore primarily intended as a tool for strengthening national laws and policies in advance of a disaster, and more generally to ensure the effective exercise of state sovereignty in international disaster response.[38]

The substance of the IDRL Guidelines comprises a series of recommendations aimed at addressing a wide range of common regulatory problems that arise in major disaster operations such as barriers to the entry of relief personnel, goods, and equipment caused by delays with visas, difficult customs procedures, expensive duties and taxes, and bureaucratic licensing requirements.[39] While primarily focused on strengthening the capacity of the affected state to receive relief, the IDRL Guidelines also highlight the difficulties that can be caused by international actors sending unwanted or inappropriate assistance, or by operating without coordinating with national authorities or beyond their competence.[40]

The IDRL Guidelines have received some criticism for being overly technical in nature. However, in the aftermath of a disaster, it has been contended that paying appropriate attention to these factors may mean "the difference between life and death."[41] As past experiences have shown, the period following a major disaster is

[36] Resolution 4 of the 30th International Conference of the Red Cross and Red Crescent, Geneva, November 26-30, 2007. The International Conference of the Red Cross and Red Crescent, held every four years, includes representatives of the components of the International Red Cross and Red Crescent Movement as well as representatives of the state parties to the 1949 Geneva Conventions (as the Geneva Conventions are universally recognized, this is effectively all states): "together, they examine and decide upon humanitarian matters of common interest and any other related matter" (Article 9, Statutes of the Movement).

[37] Guidelines for the Domestic Facilitation and Regulation of International Disaster Relief and Initial Recovery Assistance, par. 1.3

[38] "The Right Aid at the Right Time." *Progress Report on the Guidelines for the Domestic Facilitation and Regulation of International Disaster Relief and Intial Recovery Assistance*, (2009): 2. www.ifrc.org/PageFiles/41203/IDRL-Progress-Report_en.pdf.

[39] Aside from the so-called core responsibilities of affected states and assisting actors, the IDRL Guidelines also cover early warning and preparedness, the initiation and termination of relief, and criteria covering the eligibility for legal facilities of foreign assistance actors to receive a range of exemptions, waivers, and special services to expedite their activities (referred to as "legal facilities" in the guidelines).

[40] Guidelines for the Domestic Facilitation and Regulation of International Disaster Relief and Initial Recovery Assistance, par. 4.

[41] Elizabeth Ferris, "Beyond the Disaster: A Call for Japanese Leadership in International Disaster Response Law," March 22, 2011. www.brookings.edu/research/opinions/2011/03/22-japan-leadership-ferris.

the wrong time to try to develop new rules and policies since the capacity of national authorities may be stretched to the limit and poor decisions may be made in haste. In providing a comprehensive and practical "checklist" against which domestic laws and policies may be measured, the IDRL Guidelines constitute a valuable tool to aid states' legal preparedness for such scenarios. The manner in which the IFRC and National Red Cross and Red Crescent Societies have sought to, and continue to seek to, promote use of the IDRL Guidelines through cooperative efforts with states is explored further in the section that follows.

"TOP-DOWN" APPROACHES VERSUS "BOTTOM-UP" APPROACHES

The development of appropriate disaster laws has remained, by and large, a national and, to a growing extent, regional concern. Top-down approaches, with their focus on creating binding international rules and global mechanisms, have had little concrete success in strengthening disaster frameworks to date, for reasons previously explored. However, these government-led efforts may produce positive outcomes in the future by establishing a global understanding and unified discourse on appropriate international disaster response. On the other hand, bottom-up approaches, namely those primarily aimed at strengthening regulatory disaster relief frameworks at the domestic level such as that promoted in the context of the IDRL Guidelines, may be inherently more suited to addressing the specific technical challenges of international disaster response in a concrete manner. That being said, bottom-up approaches have the potential to be both time-consuming and labor-intensive and may engender an unhelpfully fragmented approach to the development of global laws and policies for disaster response.

At a fundamental level, it is important that the desire to develop international law be a jurisdictionally appropriate response to the actual problem to be addressed. Rather than the need to coordinate an excess of assistance from outside of an affected territory, a more commonly experienced challenge in disaster response is the fact that too little international aid is either available or offered.[42] This point is frequently overlooked in scholarship on international disaster response. Most often, a state will need to rely upon its own resources when combating disasters occurring on its territory, not as an assertion of sovereign rights, but as a matter of simple practicality. This does not mean that there is no place for any international laws in disaster response, but rather that the need for measures to strengthen states' domestic preparedness should be prioritized.

The overwhelming global focus on catastrophic emergencies that transcend national borders – such as the 2004 Indian Ocean tsunami – or that by their unique nature attract a large international response – such as the 2010 earthquake in Haiti – belies the fact that most disasters are local in nature and involve domestic responses

[42] Harvey and Harmer, "Building Trust," 36.

only. In spite of continued efforts to rationalize the system, the IFRC noted in its desk study undertaken in 2007 that the somewhat arbitrary nature of the availability of international relief has persisted.[43] Consequently, although the challenges in international disaster response are both compelling and worthy of attention, it should not be forgotten that the majority of disasters occur within national borders and are dealt with by domestic means.

Compared to the various initiatives aimed at developing international norms in this area, the bottom-up approach seeks to address the problem of international disaster relief first and foremost at its domestic core. The aim of the bottom-up approach is to reinforce the primary role of the affected state in disaster response, in order to ensure that it is reasonably equipped to manage and implement its legal responsibilities to coordinate and deliver aid within its territory. Favoring such an approach would therefore appear to be commensurate with the inherently domestic nature of most disaster assistance. The practical process of reviewing domestic laws and policies to determine states' capacity to receive and coordinate incoming relief can also play a valuable role in both clarifying and mapping domestic disaster arrangements as well as potentially identifying gaps and areas for improvement. The practical experience of the United Kingdom and various other countries in this regard are illustrative of this point.

The susceptibility of different states to natural disasters varies greatly, as do their existing domestic capacities to respond effectively and to manage potential incoming assistance. The structure and style of different countries' response frameworks will be dictated by a variety of legal, economic, cultural, and social factors. Consequently, any attempt to impose uniform international rules in this area will have limited success, unless imbued with sufficient flexibility to allow for consideration of national requirements. An important advantage of the bottom-up approach in the context of the IDRL Guidelines is that domestic disaster responses may be tailored to the specific needs of a state. This means that existing principles, guidance, and best practices are able to be utilized and incorporated only to the extent necessary and appropriate for the particular national context.

A common feature of many domestic disaster frameworks is that they are highly devolved and involve responses primarily at the local level. This means that relevant legislation, policies, and guidance may be found at all levels (local, regional, and national) within a state. Consequently, it is possible that local, regional, and national actors within a territory may all play potential roles in the receipt and coordination of international disaster relief. Arguably, a bottom-up approach may allow for, and even emphasize, greater examination of these domestic complexities.

Finally, the process of working directly with states to build their domestic capacity to receive incoming disaster assistance can help to promote greater

[43] Fisher, *Law and Legal Issues*, 32.

understanding and trust between national governments and aid actors, both domestic and international. For example, the IDRL Guidelines set out clear responsibilities for assisting actors, including the requirements to abide by the laws of the affected state and to coordinate with domestic authorities.[44] States may be reticent to engage in dialogue on such matters at the international level since acknowledging a potential future requirement for outside assistance may be perceived as an indication of sovereign weakness or sovereign incompetence. In this particular context, direct, and where necessary, private dialogue with states that adopts a supportive approach and emphasizes the primacy of their role may arguably achieve more concrete outcomes than seeking to curb state behaviour or to shame them into action at the global level.

BOTTOM-UP APPROACH CASE STUDY: THE UNITED KINGDOM

As just discussed, a bottom-up approach entails the development or amendment of appropriate rules, policies, and guidance for incoming disaster assistance, initially and primarily at the domestic level. In many countries, rather than having a single instrument for disaster response, relevant provisions will be spread across a wide variety of legislation that spans many different sectors and potentially involves different responsible ministries and other bodies. A comprehensive review of these provisions will most likely require significant resources and specialized expertise.

The organizations of the International Red Cross and Red Crescent Movement are uniquely placed to assist states in this regard, where necessary. There are formally recognized National Red Cross and Red Crescent Societies in 189 countries, supported at the international level by the IFRC, lending the movement an extensive global presence. As auxiliaries to their respective public authorities in the humanitarian field, National Red Cross and Red Crescent Societies have a duty to provide appropriate support to their governments in relation to matters of common humanitarian concern.[45] Moreover, in many countries national societies play a significant operational role in disaster response activities in support of the public authorities,[46] and a large number also contribute to the Movement's international disaster relief activities. Following the development of the IDRL Guidelines by the IFRC and their subsequent adoption in 2007,

[44] British Red Cross, "Analysis of Law in the United Kingdom Pertaining to Cross-Border Disaster Relief," 2010, par. 4.1.

[45] Article 4(3) of the Statutes of the Movement requires that national societies be duly recognized by the government of their respective countries on the basis of the 1949 Geneva Conventions and of national legislation as voluntary aid societies, "auxiliary to the public authorities in the humanitarian field."

[46] Article 3(2) of the Statutes of the Movement states that national societies organize, in liaison with the public authorities, emergency relief operations to assist the victims of armed conflicts and of natural disasters. In a number of states the roles of the national society in emergency response is formalized in national legislation, policies or plans.

considerable expertise in legal preparedness issues relating to cross-border disaster relief has been developed both within the IFRC and within a growing number of national societies in different regions.[47]

In 2009, the British Red Cross Society commissioned an eighteen-month national study to examine the extent to which the legal, administrative, and operational framework for disaster response within the United Kingdom was sufficiently equipped to facilitate potential international relief into its territory.[48] The study was undertaken as part of a wider project involving similar national studies by National Red Cross Societies in six European Union (EU) countries, and supported by the IFRC. The UK government agreed to the development of the study, and provided valuable input in order to shape the report's recommendations.

The United Kingdom is in the fortunate position of being less susceptible to large-scale disasters than many other countries. Its lack of land borders also decreases the likelihood of it being affected by disasters occurring in neighboring countries.[49] Still, the United Kingdom is not entirely immune from the effects of all hazards. National risk assessments show that the country is considered to be most at risk from the spread of pandemic diseases and the effects of severe weather and flooding, both coastal and inland.[50] Taking into account these circumstances and using the IDRL Guidelines as a guide, the study aimed to provide realistic and practical recommendations.

The nature of the study necessitated a detailed analysis of the United Kingdom's domestic disaster response framework at all levels from central government to local authorities. The study also examined the extent of the involvement of civil society organizations in domestic disaster response as well as the role of other key institutions such as the British Red Cross Society. One important outcome of this comprehensive approach was to highlight possible capacity and coordination issues at and between local, regional, and national levels. Given the highly devolved and extensive nature of the civil contingencies framework in the United Kingdom, such practical challenges may not have been so readily identified in the absence of a detailed domestic study.

Another clear benefit in undertaking the national study was the fostering of positive relations between the British Red Cross Society and the UK Cabinet

47 For example, most recently, representatives from twenty-two national societies from across the Americas have established the movement's first regional network focused on matters relating to legal preparedness for disasters. IDRL matters may also be discussed in the context of informal regional legal advisers' networks established within the movement (in Europe, Africa, and the Asia-Pacific).

48 British Red Cross, "Analysis of Law in the United Kingdom Pertaining to Cross-Border Disaster Relief." The study was undertaken in association with the British Institute of International and Comparative Law.

49 Ibid., 16.

50 The UK National Risk Register identifies those hazards considered most likely to occur within the United Kingdom, normally covering a five-year period. British Red Cross, "Analysis of Law in the United Kingdom Pertaining to Cross-Border Disaster Relief" 22.

Office. While the British Red Cross Society has a long-established and strong relationship with the UK government in areas such as international humanitarian law and the provision of humanitarian assistance outside of the United Kingdom, there had been little previous sustained contact with relevant areas of government on matters of strengthening legal preparedness for possible incoming disaster response. The British Red Cross was able to assure relevant officials that, rather than seeking to encourage unnecessary sweeping changes to existing legislation, policies, and plans, the purpose of the study would be to take a pragmatic view of the risks to the United Kingdom and its current capacities. The forging of stronger ties with relevant officials through the process of the national study and the continuing dialogue and exchanges since that time have brought benefits both for the British Red Cross Society and the UK government.

In general, the study concluded that the United Kingdom is, on the whole, well equipped to respond to disasters that occur within its territory and would be unlikely to rely on outside assistance. Perhaps for this reason, it does not have an explicit set of operational rules governing possible incoming relief, and it appears that most decisions in this regard would be taken on an ad hoc basis, using, for example, relevant emergency powers provisions of existing civil contingencies legislation.[51] In fact, the study found only one instance where international assistance had been formally requested by the United Kingdom, as well as a small number of requests made informally, primarily at the local level. The formal request was made in 2010, after unusually severe winter weather, when the United Kingdom experienced a road salt shortage. A request was sent through the EU Civil Protection Mechanism seeking any surplus supplies of road salt for purchase from participating states.[52]

Even so, the study identified a number of practical ways in which the United Kingdom's domestic ability to manage cross-border assistance could be enhanced in the event that such assistance was required in the future. For example, it was recommended that the UK government may wish to consider developing a specific policy on requesting and accepting offers of international assistance, specifying clearly the department responsible for coordination, and disseminating the policy to all government ministries and emergency responders.[53] Introducing

[51] The UK's Civil Contingencies Act of 2004 allows for the making of temporary emergency regulations. Other emergency powers may be invoked under specific legislation and by specified authorities or agencies, for example, the Local Government Act 1972. Non-statutory emergency powers may also be invoked by virtue of, for example, the Royal Prerogative, however such use would be rare. British Red Cross, "Analysis of Law in the United Kingdom Pertaining to Cross-Border Disaster Relief," (2010): 30–32.

[52] Conversely, the United Kingdom declined offers of international assistance (on behalf of the relevant Overseas Territories) when Hurricane Ivan struck the Caribbean Islands, after the 2005 London bombings and following widespread flooding in northern parts of the United Kingdom in 2007. Ibid., 26.

[53] In its most recent Concept of Operations (CONOPS), the UK government recognized that central government will work with international partners to "request assistance if necessary." It

a standard format and procedure for such matters could help to avoid any difficulties caused by unsolicited assistance from abroad.[54]

Importantly, it was acknowledged by relevant UK officials that the study had made a useful contribution by grouping together, for the first time, the extensive and somewhat disparate laws, regulations, policies, and plans related to domestic disaster response in one resource. As explained previously, part of the reluctance of states to support binding international rules on technical issues in cross-border disaster response may be due to the practical difficulties in assessing the many different elements of their national disaster framework spread across a variety of sectors and levels of government.

Although the recommendations of the UK study were suitably modest, it is perhaps useful in demonstrating that undertaking such a process can nevertheless generate positive effects even if the perceived disaster risks to a country are relatively low and domestic capacities considered to be robust. Positive externalities associated with the process include the provision of greater clarity and understanding for all stakeholders in relation to an often complex legal and regulatory domestic framework as well as the facilitation of more frequent contact and stronger relations between responsible authorities and civil society organizations involved in domestic disaster response. National authorities may also usefully draw on these experiences in order to provide legal and technical assistance to other states in this area by assuming a leadership role on such matters at regional and international levels.

PRACTICAL BENEFITS OF THE BOTTOM-UP APPROACH

While this chapter argues in favor of a bottom-up approach, it is important not to overstate the progress achieved to date. As reported to the 31st International Conference of the Red Cross and Red Crescent in 2011, there still remains much work to be done in encouraging the use of the IDRL Guidelines to strengthen domestic disaster response capacities. However, given the relatively recent inception of the IDRL Guidelines, and the fact that facilitating their use in practical terms is an often lengthy and complex process, it is reasonable to conclude that the IDRL initiative has yielded a number of notable achievements over the past five years at both the domestic and regional levels.[55]

also confirmed that the UK cabinet office (in particular, its civil contingencies secretariat) is responsible for requesting international civil protection assistance through EU mechanisms and NATO mutual aid arrangements. Ibid., 35.

54 Ibid., 11. A small number of examples of unsolicited disaster assistance to the United Kingdom came to light during the course of the study. These included the delivery of hovercraft by Rome's Civil Protection Agency to Gloucestershire in 2007 to assist with widespread flooding. Ibid., 25

55 Global developments in the area of IDRL are continuously updated at: www.ifrc.org/idrl

Developments at the National Level

Having adopted the IDRL Guidelines during its 30th session in 2007, the International Conference of the Red Cross and Red Crescent requested the submission of intermittent reports on the progress made regarding their utilization and implementation. In 2009, in response to a survey on progress completed by 110 National Red Cross and Red Crescent Societies and governments, the IFRC was able to report that:

> Encouragingly, the large majority of respondents to the survey indicated that they have begun to take steps to follow-up on the IDRL Guidelines. In particular, over two dozen national review processes using the IDRL Guidelines to examine strengths and weaknesses in national laws are either in process or under advanced discussion.[56]

These national review processes involved countries spanning Europe (including the United Kingdom study discussed earlier), the Asia-Pacific region, the Americas, and Africa.[57]

In 2011, a report made to the 31st International Conference of the Red Cross and Red Crescent noted "tangible progress at the national level in several dozen countries."[58] Importantly, nine countries were reported to have already adopted new laws, regulations, or procedures consistent with the IDRL Guidelines in addition to those that had undertaken national reviews of their disaster response frameworks. Examples of such concrete developments include the following:[59]

- The adoption of a new regulation by Indonesia in 2008 on the "Participation of International Institutions and Foreign Non-Governmental Institutions in Disaster Management," which drew heavily on the IDRL Guidelines;
- The development of a set of standard operating procedures by New Zealand in 2009 for an "International Assistance Cell" tasked with coordinating and facilitating international relief;
- The development of a new migration regulation in Norway in 2010, establishing an "emergency visa" category for relief personnel as part of the wider examination of a variety of national laws in the light of the IDRL Guidelines; and

[56] "The Right Aid at the Right Time." *Progress Report on the Guidelines for the Domestic Facilitation and Regulation of International Disaster Relief and Intial Recovery Assistance*, (2009): 4. www.ifrc.org/PageFiles/41203/IDRL-Progress-Report_en.pdf

[57] The completed national reports may be accessed at: www.ifrc.org/idrl

[58] IFRC, "Progress in the Implementation of Guidelines for the Domestic Facilitation and Regulation of International Disaster Relief and Initial Recovery Assistance," *Background Report for the 31st International Conference of the Red Cross and Red Crescent* (October 2011): 1.

[59] These and other examples are set out in the relevant background report submitted to the 31st International Conference, ibid.

- The adoption of a landmark new Disaster Risk Reduction and Management Act by the Philippines in 2010 for which the "Implementing Rules and Regulations" provide that "foreign donations and importations of humanitarian assistance shall also be guided with the International Disaster Response Law (IDRL) and other related guidelines as may be deemed appropriate."

While the primary focus of the IDRL Guidelines is on effecting changes at the domestic level, they appear increasingly to influence the development of guidance, policies, and mechanisms at the regional level. While the IFRC has scaled up its efforts to influence such developments, a number of governments have also reported on their efforts to promote the IDRL Guidelines in the context of regional initiatives and dialogue.[60]

Developments at the Regional Level

In general, states have become increasingly concerned with exploring regional cooperation mechanisms in disaster response. From the Civil Protection Mechanism established by the EU to the African Union's recently adopted Convention on the Protection and Assistance of Internally Displaced Persons in Africa,[61] such initiatives demonstrate that states will often look first and foremost to their immediate neighbors for assistance if circumstances exceed national capacities.

There are a number of examples of such mechanisms or initiatives that draw upon the recommendations of the IDRL Guidelines.[62] For instance, in June 2012, the General Assembly of the Organization of American States (OAS) adopted a new plan of action on disaster management, highlighting the importance of legal preparedness for international disaster cooperation.[63] Nonbinding in nature, the plan emphasizes some of the recommendations of the IDRL Guidelines. As another example,

60 For example, the governments of Belgium, South Africa, and Thailand reported such activities to the IFRC. IFRC, "Progress in the Implementation of Guidelines for the Domestic Facilitation and Regulation of International Disaster Relief and Initial Recovery Assistance," *Background Report for the 31st International Conference of the Red Cross and Red Crescent*, (October 2011): 7.

61 Adopted in 2009, the so-called Kampala Convention places a positive duty on states parties to seek assistance from domestic and international actors when their own resources are inadequate to "protect and assist" internally displaced persons. While note referring directly to the IDRL Guidelines, its provisions on incoming disaster relief are consistent with them.

62 The IFRC has reported that "on every continent, at least one regional or sub-regional organization has taken up the issue [of legal preparedness for international disaster response] and has started to make use of the IDRL Guidelines." IFRC, "Progress in the Implementation of Guidelines for the Domestic Facilitation and Regulation of International Disaster Relief and Initial Recovery Assistance," *Background Report for the 31st International Conference of the Red Cross and Red Crescent*, (October 2011): 15.

63 Inter-American Plan for Disaster Prevention and Response and the Coordination of Humanitarian Assistance.

the Association of Southeast Asian Nations (ASEAN) developed a set of operating procedures that drew heavily on the IDRL Guidelines to help operationalize its Agreement on Disaster Management and Emergency Response (which entered into force in 2009).[64] In Europe, references to the IDRL Guidelines have been included in the 2007 European Consensus on Humanitarian Aid[65] and more recently in the EU Host Nation Support Guidelines, adopted in January 2012.[66]

The IDRL Guidelines also appear to have influenced positively a number of decisions and recommendations taken at the international level. Since 2008, the UN General Assembly has recognized the IDRL Guidelines or the subject of IDRL more generally in ten resolutions, in many cases calling on member states to take the guidelines into account when strengthening their operational and legal frameworks for international disaster relief.[67] The IDRL Guidelines have also been utilized extensively in relation to the technical provisions of the ILC's work on the "Protection of Persons in the Event of Disasters."[68]

While the overall progress in the implementation of the IDRL Guidelines is not overly extensive, in relative terms it would seem more promising than various top-down efforts have been to date. The preceding examples present a limited view of the intensive work being undertaken in particular by the IFRC and a number of National Red Cross and Red Crescent Societies to encourage states to engage in the IDRL endeavor. The increasing references to and endorsement of the IDRL Guidelines at the regional and global levels are welcome. However, the real value lies in the progress achieved at the domestic level. While only a relatively small

[64] ASEAN has requested ongoing assistance from the IFRC in relation to IDRL matters, including participation in annual simulation exercises and relevant ASEAN disaster cooperation meetings, and the development of a questionnaire for member states on their implementation of the Agreement on Disaster Management and Emergency Response. IFRC, "Progress in the Implementation of Guidelines for the Domestic Facilitation and Regulation of International Disaster Relief and Initial Recovery Assistance," *Background Report for the 31st International Conference of the Red Cross and Red Crescent*, (October 2011): 9.

[65] The EU Consensus on Humanitarian Aid states at paragraph 40 that: "The EU also notes positively the work on International Disaster Response Laws, Rules and Principles (IDRL)." Full text available at: http://ec.europa.eu/echo/policies/consensus_en.htm

[66] The IDRL Guidelines are expressly mentioned as one of the documents relevant to the EU Host Nation Support Guidelines. See: www.ifrc.org/en/what-we-do/idrl/latest-news/disaster-law-newsletter-february-2012/host-nation-support-guidelines-approved-by-eu/

[67] IFRC, "Progress in the Implementation of Guidelines for the Domestic Facilitation and Regulation of International Disaster Relief and Initial Recovery Assistance," *Background Report for the 31st International Conference of the Red Cross and Red Crescent*, (October 2011): 11.

These include, most recently, Resolution 67/87 of 7 December 2012 and Resolution 66/227 of 15 March 2012. The IFRC maintains a record of UN General Assembly resolutions making note of IDRL or the IDRL Guidelines at www.ifrc.org/en/what-we-do/idrl/research-tools-and-publications/key-resolutions/un-general-assembly-resolutions-on-the-idrl-guidelines/

[68] IFRC, "Progress in the implementation of the Guidelines for the Domestic Facilitation and Regulation of International Disaster Relief and Initial Recovery Assistance," *Background report for the 31st International Conference of the Red Cross and Red Crescent* (October 2011): 12.

number of countries have so far introduced legislative or regulatory changes to their disaster frameworks (e.g., Indonesia), the implementation of a bottom-up approach has merit as the UK study attests because it involves and engages national authorities in assessing the efficacy of disaster frameworks. Such efforts can lead to improvements at the domestic level where appropriate, can encourage greater engagement with other individual states on such matters, and perhaps facilitate even closer attention to these concerns at the regional and even global levels.

CONCLUSION

The need to address common challenges in the field of international disaster response has received increasing levels of attention in recent times. The extensive global media coverage of major disasters serves as a continual reminder of the widespread suffering and destruction that transverses national borders and exceeds domestic capacities. An increasingly large and sophisticated international aid sector may be mobilized in response to such crises, and assistance may be offered by and received from a plethora of different actors, including governments, international and nongovernmental organizations, national militaries, and private entities. In such complex circumstances, the need for states to be legally prepared to receive and coordinate potential incoming disaster relief is all too apparent.

Efforts to address transboundary disaster response challenges at the domestic, regional, and global levels generally utilize one of two major methods of approach. The first method involves seeking support for rules, principles, guidelines, and other instruments at the international level, with the objective of effecting changes to domestic disaster frameworks through the adoption of a top-down approach. The second method encourages the development of laws, policies, and guidance at the domestic level, on the assumption that such a bottom-up process will facilitate and influence advancements at the regional and international levels.

In the field of international disaster response, legal preparedness activities undertaken by individual states are the key to ensuring the effective facilitation and regulation of cross-border assistance, over and above the potential evolution of global and regional frameworks. The need to observe and to promote the primary role of states in responding to disasters on their own territory plus the reality that disaster response is, on the whole, a national rather than an international undertaking favors the adoption of an approach aimed primarily at strengthening domestic capacities. The inherent challenges in developing binding rules at the international level, as demonstrated by a number of previous and ongoing initiatives, also lend credence to the adoption of a bottom-up domestic driven, rather than a top-down internationally led approach.

The efforts spearheaded by the IFRC to encourage states to examine their legal preparedness to receive international disaster assistance have demonstrated the tangible benefits of utilizing a bottom-up approach. The IDRL Guidelines, developed to address the common technical and regulatory barriers to receiving incoming assistance, have proven to be of practical value in providing guidance to states seeking to develop and strengthen relevant laws, policies, and procedures. The experience of the British Red Cross in overseeing a national study on cross-border disaster relief based on the IDRL Guidelines is indicative of the fact that working directly with individual states on such matters and taking into account each state's relative capacities and national contexts may produce the most appropriate and practical disaster response outcomes.

6

Release of Radioactive Substances into the Sea and International Law: The Japanese Experience in the Course of Nuclear Disaster

Yukari Takamura

On March 11, 2011, the Great East Japan Earthquake and subsequent tsunami triggered an extremely severe nuclear accident at the Fukushima Daiichi nuclear power plant (Fukushima Daiichi NPP), owned and operated by the Tokyo Electric Power Company (TEPCO). The accident led to massive leaks of radioactive substances into the environment, especially the release of radioactive water into the ocean as an emergency response taken by TEPCO. Presently, no international nuclear treaty contains provisions specifically prohibiting or regulating the release of radioactive materials into the sea. The release of radioactive water into the sea, categorized as marine pollution from land-based sources, falls outside of the scope of the 1996 Protocol to the 1972 London Dumping Convention. While the Global Programme of Action for the Protection of the Marine Environment from Land-based Activities (GPA) exists, the lack of clear legally binding obligations for states to address pollution from land-based sources leaves wide discretion to states. While legally binding instruments exist in some regions for this purpose, there is no legally binding framework for the regional seas surrounding Japan.

The principal question emerging from this experience is whether the release of radioactive water would have caused or may still cause a risk to the marine environment at the level that triggers an obligation to prevent transboundary damage to the environment and the obligation to notify others. Japanese practice has revealed the necessity of making the triggering conditions for these obligations more objective and to clarify and develop international rules to this effect. Given the number of nuclear power plants operating within the earthquake-prone Ring of Fire, it is critically important for states in the region to negotiate in advance of future disasters a regional legal framework to protect human health and environment from radiation.

INTRODUCTION

On March 11, 2011, a 9.0 magnitude earthquake and ensuing tsunami disabled the reactor cooling systems of the Fukushima Daiichi NPP, leading to massive leaks of radioactive substances into the environment. The report of the Fukushima Nuclear Accident Independent Investigation Commission,[1] established under the Japanese Diet, estimated that while the radioactive substances that were emitted and the surface area that was contaminated are relatively small compared to the Chernobyl accident in the former Soviet Union, there were still impacts over 1,800 square kilometers with radioactive levels equalling a potentially cumulative dose of 5mSv/year or more. Radioactive substances were not confined to Japanese territory. Small amounts of radioactive material were detected on the West Coast of the United States on March 18, 2011.

According to the report of the French government's nuclear research institute, Institut de Radioprotection et de Sûreté Nucléaire (IRSN), it is estimated that 27.1 thousand terabecquerels of radioactive cesium-137 had leaked into the ocean by July 2011, causing significant marine pollution.[2] Even in July 2013, TEPCO announced that contaminated ground water in the area of Fukushima Daiichi NPP had been leaked into the plant port.[3] There is ongoing concern that some of the radioactive material may be bio-accumulating in fish and marine animals.

Preventing nuclear accidents is undeniably crucial, but once an accident happens, the focus must be on reducing the risk of a nuclear disaster and mitigating damage by radioactive substances. By examining Japanese response measures to the nuclear disaster, especially focusing on release of radioactive substances into the sea, this chapter evaluates whether these domestic measures were compatible with

[1] The Act regarding Fukushima Nuclear Accident Independent Investigation Commission (Act No. 112 of 2011) was enacted on October 7, 2011 and came into effect on October 30, 2011. The Act created an independent commission to investigate the Fukushima accident. This Investigation Commission was established on December 8, 2011, and issued its report on July 5, 2012. The full report (in Japanese) http://warp.da.ndl.go.jp/info:ndljp/pid/3856371/naiic.go.jp/blog/reports/main-report/. The English summary of the report http://warp.da.ndl.go.jp/info:ndljp/pid/3856371/naiic.go.jp/wp-content/uploads/2012/09/NAIIC_report_lo_res10.pdf. On the other hand, the Investigation Committee of the government was established on May 24, 2011 by a Cabinet decision. The Committee published its interim report on December 26, 2011 and its final report on July 23, 2012. The interim report and final report (in Japanese) www.cas.go.jp/jp/seisaku/icanps/. The English version of the interim and final report www.cas.go.jp/jp/seisaku/icanps/eng/. Both reports concluded that all causes of the accident were foreseeable, and that although the accident was triggered by the earthquake and the subsequent tsunami, the nuclear radiation accident was caused by human error.

[2] Institute de radioprotection et de sûreté nucléaire, Government of France, *Updated summary of knowledge concerning the impact on the marine environment of radioactive release from the damaged nuclear site of Fukushima Dai-ichi*, July 13, 2012.

[3] Agency for Natural Resources and Energy, METI of Japan 2nd Fact sheet: Overview of Contaminated Water Issue at TEPCO's Fukushima Daiichi NPS, September 3, 2013, www.meti.go.jp/english/earthquake/nuclear/decommissioning/pdf/20130904_01b.pdf.

relevant international rules. The chapter then explores how effective the existing international legal system was in reducing the risk of marine pollution from the Fukushima accident and what lacuna still need to be addressed as part of an emerging international law of disaster.

FUKUSHIMA NUCLEAR ACCIDENT AND ITS CONSEQUENCES

The Fukushima Daiichi NPP is located on the coast of the island of Honshu, in the Fukushima Prefecture, 200 km northeast of Tokyo. On March 11, 2011, the Great East Japan Earthquake triggered an extremely severe nuclear accident at the Fukushima Daiichi NPP, owned and operated by TEPCO. This devastating accident was ultimately declared a Level 7 (Major Accident) by the assessment scale of International Nuclear Event Scale (INES).[4]

When the earthquake occurred, Units 1, 2, and 3 of the Fukushima Daiichi NPP were operating normally while Units 4, 5, and 6 were undergoing periodic inspections. The emergency shutdown feature went into operation at Units 1, 2, and 3 immediately after the commencement of the seismic activity. The seismic tremors damaged electricity transmission facilities resulting in a total loss of offsite electricity. The tsunami caused by the earthquake flooded and totally destroyed the emergency diesel generators, the seawater cooling pumps, the electric wiring system, and the power supply, eventually resulting in loss of all power. The loss of electricity made it very difficult to effectively cool down the reactors. Cooling the reactors had been attempted by water injection using fire trucks and helicopters while the plant worked to recover its electricity supply and normal cooling systems.

The Japanese Diet's Investigation Commission recognized that reactor core damage within Unit 1 had already started on the very day of the earthquake, March 11, 2011, and eventually led to a hydrogen explosion at its reactor building. For Unit 2, reactor core damage started on March 14, 2011; Unit 3's reactor core was damaged on March 13, 2011 leading to a hydrogen explosion as well.[5] These explosions made recovery operations for the units even more difficult due to high-level radiation and difficulties with accessing the units. In retrospect, the

[4] International Nuclear Event Scale (INES) is a tool "for promptly and consistently communicating to the public the safety significance of events associated with sources of radiation," originally developed in 1990 by international experts convened by the International Atomic Energy Agency (IAEA) and OECD Nuclear Energy Agency (OECD/NEA) and updated by the INES member states in the biennial meeting. Events are classified on the scale at seven levels in order of increasing severity. Level 7 (major accident) which includes events such as the meltdown at the Chernobyl nuclear power plan is defined: "an event resulting in an environmental release corresponding to a quantity of radioactivity radiologically equivalent to a release to the atmosphere of more than several tens of thousands of TBq of Iodine-131." IAEA, *INES: The International Nuclear and Radiological Event Scale User's Manual, 2008 Edition*, amended in 2013 (Vienna, 2013), 1–2, 17.

[5] Independent Investigation Commission, National Diet of Japan, *The official report of The Fukushima Nuclear Accident Independent Investigation Commission*, 2012, 12–14.

hydrogen explosions from these units are considered one of the main causes of discharge of radioactive substances into the atmosphere.

Emergency cooling of reactors by water injection added another problem for response measures; the cooling produced a large quantity of contaminated water, which needed to be stored until normal cooling systems would be recovered. From April 4 to April 10, TEPCO, after having informed the Minister of Economy, Trade and Industry and having received advice from the Nuclear Safety Commission (NSC) of Japan, released more than 10,000 tons of "low-level" radioactive water into the sea in order to ensure a storage space for high-level radioactive water.[6] This level is well above the standard of discharge under the national law.[7] This chapter focuses on examining whether the release of low-level radioactive water as a disaster response conformed with international law.

In the course of the Fukushima Daiichi NPP shutdown, an enormous amount of radioactive substances were emitted into the environment. The accident resulted in the release of radioactive substances into the atmosphere, in the form of radioactive gases or very fine radioactive aerosol particles dispersed into the air, of which a portion settled on the surface of the soil; and into the marine environment, directly in the form of water releases into the sea and indirectly due to fallout on the sea's surface from radioactive aerosols dispersed over the ocean.[8]

As for the release into the atmosphere, IRSN estimates that most of these releases occurred between March 12 and March 25, with lower level atmospheric releases over the course of the following months that were not easily detectable in the environment. All land surfaces were contaminated, which resulted in an increase in radiation levels across Japan and the contamination of agricultural products.[9] Comparing the surface areas contaminated after the Fukushima and Chernobyl accidents, relatively smaller surface areas were affected in Japan than in Ukraine. This is explained in particular by the fact that a major portion of the releases from the Fukushima accident drifted over the Pacific Ocean.

6 *Act on the Regulation of Nuclear Source Material, Nuclear Fuel Material and Reactors: Article 64(1)*, (Act No. 166 of June 10, 1957), provides that when an earthquake, fire, or any other disaster occurs and threatens to cause or causes a disaster involving nuclear fuel material, material contaminated by nuclear fuel material, or reactors that a licensee of nuclear energy activity possesses, the licensee of nuclear energy activity shall take emergency measures immediately, pursuant to the provisions of the ordinance of the competent ministry. Based on that provision, TEPCO did the release as emergency measures.

7 Japanese Ministry of Economy, Trade and Industry, *Public notice regarding dose limits pursuant to the Regulation on installment, operation etc. of commercial power reactors*, No. 187 of March 21, 2001, (as amended by Public notices of Ministry of Economy, Trade and Industry No. 275 of October 26, 2005 and No. 295 of November 22, 2005).

8 Institute de radioprotection et de sûreté nucléaire (IRSN), *Fukushima, one year later: Initial analyses of the accident and its consequences*, Report IRSN/DG/2012-003, March 12, 2012, 43.

9 Ibid., 83–98.

Compared to the 13,000 square kilometers of land around Chernobyl, only around 600 square kilometers of land in Japan is contaminated by more than 600,000 Bq/m2 of cesium-137. However, as IRSN indicated, in addition to the 80,000 residents evacuated as an emergency measure in the 20-km zone, a population of 70,000 still lives in these areas, which in total represents more than half the number of people who were living in the most heavily contaminated areas around Chernobyl (270,000 people for cesium-137 fallout measurements greater than 555,000 Bq/m2).[10] More than 80,000 people are still not allowed to return to their homes because of high levels of radioactive concentration.[11]

As for the releases into the ocean, the marine environment was affected by fallout of aerosols emitted into the atmosphere above the ocean as well as by the release of radioactive water directly into the sea. The high concentrations measured in the seawater in the immediate vicinity of the Fukushima Daiichi NPP were the result of radioactive water releases from several origins including the water used to cool the damaged reactors, "intentional" release of radioactive water, and leak of heavily contaminated water from a crack in the pit adjacent to the reactor 2 turbine hall. In the future, the leaching of contaminated soils will be the main source of long-term contamination of the marine environment. IRSN estimated that a total of 271,015 Bq of cesium-137 was released into the sea up to mid July 2011, with most being released before April 8.[12]

Unlike in the terrestrial environment, where a residual deposit will persist for several years, the acute contamination period of the marine environment was contained within a period of about six months. This duration is the result of especially favorable hydrodynamic conditions related to the dynamics of the currents that are directed toward the open sea, and the size of the receiving medium (the Pacific Ocean). If this event had taken place in an enclosed sea (for instance, to the west of Japan), the consequences could have been ten times worse.[13]

Some research demonstrates the existence of a potential ecological risk, especially to the terrestrial ecosystem in the evacuation zone, particularly for reproduction. The knowledge about a long-term radiological risk to the ecosystem is still very limited in part because of scientific uncertainty. Continued environmental monitoring and increased knowledge is a key that will allow us to evaluate the ecological impact of the accident on the environment.[14]

[10] Ibid., 68.

[11] Japanese Ministry of Economy, Trade and Industry, "Overview of areas to which evacuation orders have been issued, as well as population and the number of households living in each of these areas" (as of August 30, 2013) (in Japanese), (September 13, 2013). www.meti.go.jp/earthquake/nuclear/pdf/130830a.pdf.

[12] IRSN, *Fukushima, one year later*, 107. About the updates, see also IRSN *Updated summary of knowledge*.

[13] IRSN *Fukushima, one year later*, 105. [14] Ibid., 122.

RELEVANT INTERNATIONAL NUCLEAR LAW REGULATING NUCLEAR ACTIVITIES

The International Atomic Energy Agency (IAEA) was established in 1957 by the Statute of International Atomic Energy Agency adopted in 1956.[15] Article 2 of the Statute states that: "[T]he Agency shall seek to accelerate and enlarge the contribution of atomic energy to peace, health and prosperity throughout the world" and thus refers to an enhancement of peaceful use of nuclear energy as its objective. Article III.A.6 of the Statute provides that the IAEA is authorized:

> To establish or adopt . . . standards of safety for protection of health and minimization of danger to life and property (including such standards for labour conditions), and to provide for the application of these standards to its own operations as well as to the operations making use of materials, services, equipment, facilities, and information made available by the Agency or at its request or under its control or supervision; and to provide for the application of these standards, at the request of the parties, to operations under any bilateral or multilateral arrangements, or, at the request of a State, to any of that State's activities in the field of atomic energy.

The IAEA thus has a broad power to establish standards of safety for protection of health and minimization of danger to life and property including standards for radiation protection, transport and handling of radioactive materials, disposal of radioactive waste, and safety of nuclear facilities.[16] These standards usually only apply to the operations of the IAEA as well as to the operations making use of materials, services, equipment, facilities, and information made available by the IAEA or at request or under control or supervision of the IAEA. They may also apply to a state's activities based on the consent of that state. In the case of Japan, while the standards established by the IAEA are not legally binding, some of them have been integrated into its domestic law.

Additionally, international treaties relating to nuclear activities were developed in the 1960s under the IAEA to regulate nuclear activities for civil use. The first category of treaties involves nuclear accidents or emergency. The April 26, 1986 Chernobyl nuclear accident led to the adoption of two additional international treaties: the Convention on Assistance in Case of a Nuclear Accident or Radiological Emergency[17] and the Convention on Early Notification of a Nuclear

[15] The statute was approved on October 23, 1956, put into force on July 29,1957, and ratified by Japan on July 16, 1957.

[16] A comprehensive body of safety standards has been elaborated and regularly updated under the IAEA, which is composed of fundamental safety principles, safety requirements, and safety guides.

[17] International Atomic Energy Agency, "Convention on Assistance in Case of a Nuclear Accident or Radiological Emergency," Adopted on September 26, 1986 and entered into force on February 26, 1987. 457 UNTS 133, 25 ILM 1377 (1986). Japan deposited its instrument of acceptance on June 9, 1987 and it entered into force on July 10, 1987.

Accident (Early Notification Convention).[18] The former convention obliges parties to strengthen cooperation between state parties and with the IAEA in order to facilitate prompt assistance in the event of a nuclear accident or radiological emergency to minimize its consequences and to protect life, property, and the environment from the effects of radioactive releases.[19] The latter convention, as a whole, establishes a notification system for nuclear accidents that have the potential for international transboundary release that could be of radiological safety significance for another state.

The second category involves nuclear safety and includes the Convention on Nuclear Safety which, adopted in 1994, requires countries operating nuclear installation within their jurisdiction to achieve and maintain a high level of nuclear safety through the enhancement of national measures and international cooperation including setting international benchmarks.[20] The obligations of the parties are based to a large extent on the principles contained in the IAEA Safety Fundamentals document, such as establishing a legislative and regulatory framework, which should define the responsibilities of the government, the regulatory body, and the operators, and securing safe operation and maintenance of the facilities.[21] The 1997 Joint Convention on the Safety of Spent Fuel Management and on the Safety of Radioactive Waste Management[22] applies to spent fuel and radioactive waste resulting from civilian nuclear reactors, other civilian applications, and certain military programs. The Joint Convention requires parties to take the appropriate steps to ensure that at all stages of spent fuel management as well as of radioactive waste management, individuals, society, and the environment are adequately protected against radiological hazards.[23]

18 International Atomic Energy Agency, "Convention on Early Notification of a Nuclear Accident," Adopted on 26 September 26, 1986 and entered into force on October 27, 1986. 1457 UNTS 133, 25 ILM 1369 (1986). Japan deposited its instrument of acceptance on June 9, 1987 and it entered into force on July 10, 1987.

19 "Convention on Assistance in Case of a Nuclear Accident or Radiological Emergency," Article 1.1.

20 Adopted on June 17, 1994 and it came into effect on October 24, 1996. 33 ILM 1514(1994). Japan deposited its instrument of acceptance on May 12, 1995 and it entered into force on October 24, 1996.

21 The Convention had been developed based on two drafts elaborated by the IAEA at that time, namely, Safety Fundamentals: The Safety of Nuclear Installations (1991) and Draft Safety Fundamentals: The Principles of Radioactive Waste Management, a publication within the RADWASS (1991). The former document was later published in the Safety Series, No. 50, The Safety of Nuclear Installations, December, 5 1993. Odette Jankowitsch-Prevor, "The Convention on Nuclear Safety," *International Nuclear Law in the Post-Chernobyl Period* (OECD Nuclear Energy Agency and International Atomic Energy Agency, 2006) 157. The most recently updated standards is contained in IAEA, *Fundamental safety principles: safety fundamentals* (Vienna : IAEA, 2006).

22 Adopted on September 5, 1997 and it came into effect on June 18, 2001. 36 ILM 1431 (1997). Japan deposited its instrument of accession on August 25, 2003 and it entered into force on November 24, 2003 for Japan.

23 Ibid., Articles 4 and 11.

A series of conventions related to civil liability for nuclear damage have also been developed, starting with the 1963 Vienna Convention on Civil Liability for Nuclear Damage and its 1996 Protocol to Amend the Vienna Convention on Civil Liability for Nuclear Damage.[24] Another group of civil liability conventions including the Paris Convention on Nuclear Third Party Liability and its 2004 Protocol as well as the 1963 Brussels Supplementary Convention on Nuclear Third Party Liability and its 2004 Protocol, was adopted by the Organisation for Economic Co-operation and Development (OECD) representing the interests of many of the largest nuclear power states.[25]

The first generation of these international treaties relating to nuclear activities focused on radiation protection for human health and property. The protection of the environment was not specifically addressed. However, the second generation of these treaties adopted after the Chernobyl accident started to include provisions referring to the environment. For instance, the Convention on Assistance in Case of a Nuclear Accident or Radiological Emergency includes language to "protect the environment from the effects of radioactive releases" as an objective. Another example is the 1996 Protocol to Amend the Vienna Convention on Civil Liability for Nuclear Damage that covers some damage to the environment by including recovery for the costs of measures of reinstatement of impaired environment[26] as well as for loss of income derived from an economic interest in any use or enjoyment of the environment. Presently, no international nuclear treaty has provisions specifically prohibiting and regulating the release of radioactive materials into the sea as was experienced by the Fukushima Daiichi NPP case.

INTERNATIONAL ENVIRONMENTAL LAW REGULATING RADIOACTIVE RELEASES

International environmental law has a different origin than international nuclear law and each of the fields has evolved independently from the other. Generally

[24] IAEA, *Information Circular*, INFCIRC/500, March 20, 1996. 2 ILM 727 (1963); IAEA, *Information Circular*, INFCIRC/566, July 22, 1998. 36 ILM 1462 (1997). Japan has not ratified any of the civil liability conventions under the IAEA and the OECD.

[25] OECD Nuclear Energy Agency, "Convention on Third Party Liability in the Field of Nuclear Energy of 29th July 1960, as amended by the Additional Protocol of 28th January 1964 and by the Protocol of 16th November 1982," July 29, 1960. 1041 UNTS 358 (1960); OECD Nuclear Energy Agency, "2004 Protocol to Amend Convention on Third Party Liability in the Field of Nuclear Energy, as amended by the Additional Protocol of 28th January 1964 and by the Protocol of 16th November 1982," February 12, 2004; OECD Nuclear Energy Agency, "Convention of 31st January 1963 Supplementary to the Paris Convention of 29th July 1960, as amended by the additional Protocol of 28th January 1964 and by the Protocol of 16th November 1982 ("Brussels Supplementary Convention")," January 31, 1963. 956 UNTS 264 (1963); OECD Nuclear Energy Agency, "2004 Protocol to Amend the Brussels Supplementary Convention on Nuclear Third Party Liability," (2004). See www.oecd-nea.org/law/legal-documents.html#agreements.

[26] Article 1 of the 1963 Vienna Convention amended by its 1996 Protocol that inserted (iv) and (v) in that article. IAEA, *Information Circular*, INFCIRC/566.

speaking, most international environmental law does not specifically attempt to regulate nuclear activities or to explicitly exclude nuclear activities from its scope of application. Only the 1958 Convention on the High Seas[27] and the 1972 London Convention[28] explicitly prohibit dumping of radioactive waste into the oceans. Other relevant treaties such as the United Nations Convention on the Law of the Sea (UNCLOS)[29] can also be implicitly interpreted to apply to nuclear activities and their impact on the environment.[30]

In the context of the Fukushima accident, the first treaty that attracted attention of Japanese citizens and neighboring states to Japan was the 1996 Protocol to the 1972 London Convention. Just after the release into the sea from Fukushima Daiichi NPP, claims for violation of the 1996 Protocol were raised both within and outside of Japan.[31] Dumping of radioactive waste into the sea has been one of the key concerns from the early period of the development of international environmental law. Since 1946, a number of countries, including the United Kingdom, the United States, the Netherlands, Japan, and Korea had disposed of low-level radioactive wastes into the sea since there was no international regulation on the matter.[32]

In 1958, the Convention on the High Seas was adopted and it obliges parties to "take measures to prevent pollution of the seas from the dumping of radioactive waste, taking into account any standards and regulations which may be formulated by the competent international organizations"[33] (and to "cooperate with the competent international organizations in taking measures for the prevention of pollution of the seas or air space above, resulting from any activities with radioactive materials or other harmful agents").[34] The 1972 London Convention specifically prohibits dumping of high-level radioactive wastes into the sea by listing them in Annex I.[35] Under the 1972 London Convention, parties adopted a nonbinding

[27] Adopted on April 29, 1958 and it entered into force on September 30, 1962. 450 UNTS 11 (1964). Japan ratified it on June 10, 1968.

[28] Adopted on November 13, 1972 and it entered into force on August 30, 1975. 11 ILM, 1291 (1972). Japan ratified it on October 15, 1980 and it entered into force for Japan on November 14, 1980.

[29] Adopted on December 10, 1982, it came into force on November 16, 1994. 1833 UNTS 3, 21 ILM 1261(1982). Japan ratified it on June 20, 1996 and it entered into force on July 20, 1996 for Japan.

[30] See for instance Sam Emmerechts, "Environmental Protection under Nuclear Law: Still a Long Way to Go," *International Nuclear Law: History, Evolution and Outlook* (Paris: OECD, 2010): 121–56.

[31] "Release of polluted water: 'Lack of transparency' – Expanding dissatisfaction in neighboring countries," *Asahi Shimbun*, April 6, 2011 (in Japanese), reporting concerns expressed by the government and parliament of Korea as well as non-governmental organizations, such as FoE (Friends of the Earth), indicating that the release might violate the 1996 Protocol. www.asahi.com/special/10005/TKY201104050556.html.

[32] Lasse Ringius, *Radioactive Waste Disposal at Sea: Public Ideas, Transnational Policy Entrepreneurs, and Environmental Regimes.* (MIT Press, 2001), 24–8.

[33] "Convention on the High Seas," Article 25.1.

[34] Ibid., Article 25.2.

[35] "1972 London Convention," Article 4.1.

resolution calling for a moratorium on dumping of low-level radioactive waste pending completion of a two-year technical and scientific review of such dumping by experts.[36]

In the spring of 1985, the experts issued a report in which no scientific consensus emerged to support the decision of a total ban on dumping of radioactive waste.[37] In 1985, parties adopted another resolution calling for extending suspension of such dumping until more detailed review and assessments including impacts on marine environment could be completed.[38] In 1993, based on results of the review and assessments, parties decided to prohibit all dumping of radioactive wastes by amending Annex I and Annex II of the 1972 London Convention.[39] This amendment prescribing overall prohibition of dumping of radioactive waste was entered into force on February 20, 1994.

In November 1996, parties to the 1972 London Convention adopted a new protocol that introduced a clear concept of precaution. The 1996 Protocol to the 1972 London Convention came into force on March 24, 2006, and as of 2014, had forty-four parties including Japan.[40] The 1996 Protocol prohibits the dumping of all radioactive waste and other materials into the sea by adopting a so-called reverse list approach. The protocol allows dumping of very limited materials that are listed in Annex I, such as dredged material, sewage sludge, and vessels and platforms or other man-made structures at sea, under permit system of each contracting party. Article 4 of the protocol provides:

> 1. Contracting Parties shall prohibit the dumping of any wastes or other matter with the exception of those listed in Annex 1.
> 2. The dumping of wastes or other matter listed in Annex 1 shall require a permit. Contracting Parties shall adopt administrative or legislative measures to ensure that issuance of permits and permit conditions comply with provisions of Annex 2. Particular attention shall be paid to opportunities to avoid dumping in favour of environmentally preferable alternatives.[41]

Dumping of radioactive waste and other materials is similarly prohibited under the 1996 Protocol, since these materials are not listed in Annex 1.

[36] International Maritime Organization, "London Convention Resolution LDC.14(7): Disposal of Radio-Active Wastes and Other Radio-Active Matter at Sea," (1983).

[37] See *Radioactive Waste Disposal*, 144.

[38] International Maritime Organization, "London Convention Resolution LDC.21(9) Dumping of Radioactive Wastes at Sea," (1985).

[39] International Maritime Organization, "London Convention Resolution LC.51(16) Amendments to the annexes to the Convention on the Prevention of Marine Pollution by Dumping of Wastes and Other Matter, 1972, concerning disposal at sea of radioactive wastes and other radioactive matter," 12 November 1993.

[40] Adopted on November 7, 1996, and it entered into force on Mar 24, 2006. 2006 ATS 11. Japan deposited its instrument of accession on October 2, 2007 and it entered into force for Japan on November 1, 2007.

[41] Ibid., Article 4.

Is radioactive water from the Fukushima Daiichi NPP considered a "waste or other matter" that is regulated under the 1996 Protocol? In the protocol, "wastes or other matter" is defined as "material and substance of any kind, form or description."[42] This comprehensive definition naturally covers radioactive water, of which dumping at sea is prohibited. Is the release of radioactive water from Fukushima Daiichi NPP considered "dumping" as defined in the 1996 Protocol? Article 1.4 of the protocol defines "dumping" as follows:

1. Any deliberate disposal into the sea of wastes or other matter from vessels, aircraft, platforms or other man-made structures at sea;
2. Any deliberate disposal into the sea of vessels, aircraft, platforms or other man-made structures at sea;
3. Any storage of wastes or other matter in the seabed and the subsoil thereof from vessels, aircraft, platforms or other man-made structures at sea; and
4. Any abandonment or toppling at site of platforms or other man-made structures at sea, for the sole purpose of deliberate disposal.[43]

In the case of Fukushima Daiichi NPP, IRSN estimates that the radioactive pollution is from two main sources: liquid radioactive discharges from the site and atmospheric fallout on the surface of the sea. The latter is estimated to represent about 0.3 percent of the overall radioactivity into the ocean (a total of 271,015 Bq of cesium-137 up to mid July 2011 as referred to previously)[44], which implies that most of the discharge comes from land-based sources. Neither release from land-based sources nor fallout from the atmosphere would be classified within the definition of dumping under the 1996 Protocol. There has been no deliberate disposal into or storage in the sea of radioactive wastes or other matter originating from vessels, aircraft, platforms, or other man-made structures at sea. If a vessel had been loaded with radioactive water and then discharged into the sea, such discharge would fall within the scope of the 1996 Protocol. It would be the same case if radioactive water was discharged from a platform at sea.[45] However, the release from Fukushima Daiichi NPP falls outside of the scope of the 1996 Protocol that countries agreed to.

INTERNATIONAL RULES ADDRESSING POLLUTION FROM LAND-BASED SOURCES

Obligation to Protect the Marine Environment

Although the 1996 Protocol does not cover the release of radioactive water from land-based sources, the 1982 UNCLOS contains provisions related to such release.

[42] Ibid., Article 1.8. [43] Ibid., Article 1.1. [44] *Fukushima, one year later*, 110.

[45] In these cases falling within the scope of the 1996 Protocol, these dumpings would be prohibited under the Protocol unless they fell within one of the Article 8 exceptions.

UNCLOS stipulates in Article 192 that "States have the obligation to protect and preserve the marine environment." While "States have the sovereign right to exploit their natural resources pursuant to their environmental policies and in accordance with their duty to protect and preserve the marine environment,"[46] states shall take all measures "necessary to prevent, reduce and control pollution of the marine environment from any source, using for this purpose the best practicable means at their disposal and in accordance with their capabilities."[47] In addition, UNCLOS has a specific provision on pollution from land-based sources. Article 207 of UNCLOS requires states to "adopt laws and regulations to prevent, reduce and control pollution of the marine environment from land-based sources, including rivers, estuaries, pipelines and outfall structures, taking into account internationally agreed rules, standards and recommended practices and procedures"[48] and "take other measures as may be necessary to prevent, reduce and control such pollution."[49] These laws, regulations, and other measures include "those designed to minimize, to the fullest extent possible, the release of toxic, harmful or noxious substances, especially those which are persistent, into the marine environment."[50]

Naturally, "States are responsible for the fulfillment of their international obligations concerning the protection and preservation of the marine environment" including these obligations.[51] However, states enjoy some degree of discretion in how and to what extent they implement these obligations. For instance, Article 194.1 obliges states to take all measures but only to the extent of "the best practicable means at their disposal and in accordance with their capabilities." For pollution from land-based sources, even a greater degree of discretion has been left to states because states are only obliged to take into account internationally agreed upon rules and standards in adopting laws, regulations, and other measures. The choice of measures and the standards is largely left to the discretion of states since there are few internationally agreed upon rules and standards governing land-based pollution.

Global Programme of Action for the Protection of the Marine Environment from Land-based Activities (GPA)

Article 207.4 of UNCLOS obliges states to "endeavor to establish global and regional rules, standards and recommended practices and procedures to prevent, reduce and control pollution of the marine environment from land-based sources," which should be taken into account when states adopt law, regulations, and other measures. While it is estimated that about 80 percent of marine pollution is from land-based sources, there is no legally binding treaty on this matter at the global level and only general regional agreements.[52] In 1985, the United Nations

[46] "UNCLOS," Article 193. [47] Ibid., Article 194.1. [48] Ibid., Article 207.1.
[49] Ibid., Article 207.2. [50] Ibid., Article 207.5. [51] Ibid., Article 235.1.
[52] United Nations Environment Programme, *UNEP in 2006* (Nairobi: UNEP, 2007), 56. For a comprehensive study on legal framework to control pollution from land-based sources, see

Environment Programme (UNEP) adopted a nonlegally binding document entitled the Montreal Guidelines for the Protection of the Marine Environment Against Pollution from Land-based Sources.[53] While most of the guidelines originate from relevant provisions of the UNCLOS, the guidelines classify "radioactive materials" as part of a blacklist of pollution that states should undertake to eliminate.[54] In 1992, the UN implemented Agenda 21, which stipulates objectives and actions to address land-based pollution issues in its Chapter 17. In order to implement Agenda 21, states adopted the Global Programme of Action for the Protection of the Marine Environment from Land-based Activities (GPA),[55] which is nonlegally binding but the only universal instrument to address pollution from land-based sources.

The GPA aims at "preventing the degradation of the marine environment from land-based activities by facilitating the realization of the duty of States to preserve and protect the marine environment."[56] The GPA, therefore, is designed to be "a source of conceptual and practical guidance to be drawn upon by national and/or regional authorities in devising and implementing action to prevent, reduce, control and/or eliminate marine degradation from land-based activities."[57] The GPA lists eight source categories on which action will be needed at the global level, either to address global effects or to facilitate action at the national or regional levels. One of the source categories is radioactive substances. The objective or proposed target for radioactive substances is to "reduce and/or eliminate emissions and discharges" in order to prevent, reduce, and eliminate pollution of the marine and coastal environment.[58] The GPA also provides guidance for actions to be taken by individual states. For example, states should "not promote or allow the storage or disposal of high-level, intermediate-level and low-level radioactive wastes near the marine and coastal environment" unless they determine on the basis of scientific evidence and the appropriate application of the precautionary approach that the storage or disposal poses no unacceptable risk to people and the marine and coastal environment or does not interfere with other legitimate uses of the sea.[59]

Regional Framework to Address Pollution from Land-Based Sources

Together with the GPA, the Regional Seas Programme launched in 1974 by the UNEP has provided an important forum to tackle pollution from land-based

Daud Hassan, *Protecting the Marine Environment from Land-Based Sources of Pollution: Towards Effective International Cooperation* (Ashgate Publishing, 2006); Yoshifumi Tanaka, "Regulation of Land-Based Marine Pollution in International Law: A Comparative Analysis Between Global and Regional Legal Frameworks," *Zeitschrift für ausländisches öffentliches Recht und Völkerrecht* 66 (2006): 535–574.

[53] Governing Council of UNEP, Decision 13/18/II, 24 May 1985.

[54] Annex II Classification of Substances of the Montreal Guidelines.

[55] UNEP(OCA)/LBA/IG.2/7, 5 December 1995. [56] Ibid., 8, par. 3.

[57] Ibid., 10, par. 14. [58] Ibid., 42, par. 109. [59] Ibid., 43, par. 111.

sources. The Regional Seas Programme covers eighteen regions including the East Asian Seas and the Northwest Pacific. For many of these regions, there are protocols that specifically address land-based sources activities

In the Northwest regional sea, the region most likely to be affected by radioactive substances released due to Fukushima Daiichi NPP, there is no binding instrument related to land-based pollution. Instead, there is the Action Plan for the Protection, Management and Development of the Marine and Coastal Environment of the Northwest Pacific Region (NOWPAP).[60] NOWPAP is a nonbinding framework where the four coastal states of the region, namely China, Korea, Japan, and Russia, cooperate on protecting marine resources. While the program calls for "cooperation to control and reduce the total discharges from land-based sources of domestic wastes, industrial effluents and those from agricultural non-point sources,"[61] there are no clear standards of discharge from land-based sources nor explicit prohibition on releases of radioactive substances.

In the neighboring marine areas of the geographical scope of NOWPAP, there is an East Asian Seas Action Plan steered by the Coordinating Body on the Seas of East Asia (COBSEA), consisting of ten member countries.[62] However, Japan is not member of COBSEA, and although COBSEA selected marine- and land-based pollution as one of the three priority areas in the New Strategic Direction (2008–12) it adopted, few concrete measures have been prescribed, including any provisions related to radioactive substances.[63]

As is apparent in the vague obligations under UNCLOS, the general language of the GPA,[64] and the lack of binding legal frameworks in the region, states could not easily agree on more stringent obligation to abate pollution from land-based sources, since they fear that such obligations might affect economic activities and limit exercise of national sovereignty.[65] In the case of release of radioactive water by Fukushima Daiichi NPP into the sea, since there are neither internationally nor regionally agreed legally binding obligations applicable to the countries involved and because a large degree of discretion has been left to states, it would be difficult

[60] Adopted in September 1994. www.nowpap.org/. [61] NOWPAP, 14, par. 21.

[62] The Action Plan for the Protection and Development of the Marine Environment and Coastal Areas of the East Asian Seas Region (the East Asian Seas Action Plan) was approved in 1981 and was revised in 1993. UNEP(OCA)/EAS IG5/6 Annex IV (1994).

[63] UNEP, *New Strategic Direction for COBSEA (2008–2012)* (United Nations Environment Programme, 2008).

[64] David VanderZwaag and Anne Powers, "The Protection of the Marine Environment from Land-Based Pollution and Activities," *The International Journal of Marine and Coastal Law* (2008): 423–52. Hassan, *Protecting the Marine Environment*, 179–206. (Some states were willing to have legally binding obligations while others were not.)

[65] In the course of negotiations toward the UNCLOS, countries wished to keep as much flexibility as possible in implementing obligations. Patricia Birnie, Alan Boyle, and Catherine Redgwell, *International Law of the Environment* 3rd ed. (Oxford University Press, 2009), 452–4; M. H. Nordquist et al., *United Nations Convention on the Law of the Sea 1982: A Commentary* (Brill, 1995), 132–4.

for a state to sustain a claim of a violation of state responsibility for such a land-based pollution under the existing international rules to protect the marine environment, including UNCLOS.

General Obligation of States to Prevent Transboundary Damage to the Environment

Article 194.2 of UNCLOS provides that "States shall take all measures necessary to ensure that activities under their jurisdiction or control are so conducted as not to cause damage by pollution to other States and their environment, and that pollution arising from incidents or activities under their jurisdiction or control does not spread beyond the areas where they exercise sovereign rights in accordance with this Convention."[66] This provision reaffirms the obligation of states to prevent transboundary damage to the environment as declared in Principle 21 of the Stockholm Declaration in the context of protection of marine environment.[67] After the adoption of the UNCLOS, this principle embodied in Article 194.2 has evolved so as to acquire the status of customary law, which has been confirmed by the Advisory Opinion of the International Court of Justice (ICJ) in the Legality of the Threat or Use of Nuclear Weapons[68] when the ICJ opined that: "The existence of the general obligation of states to ensure that activities within their jurisdiction and control respect the environment of other states or of areas beyond national control is now part of the corpus of international law relating to the environment."

The obligation of states to prevent transboundary damage to the environment is one of due diligence. According to the 2001 draft articles on Prevention of Transboundary Harm from Hazardous Activities adopted by the ILC, the obligation of due diligence means that states are under an obligation to take unilateral measures to prevent significant transboundary harm or at any event to minimize the risk thereof arising out of activities.[69] The standard of due diligence depends on what is generally considered to be appropriate and proportional in relation to the degree of risk of transboundary harm in the particular case. The required degree of care is proportional to the degree of hazard involved.

While the degree of care in question is that expected of a good government, the expected degree of care would change according to capabilities and resources that are available to that state. The harm itself should be foreseeable and the state must know or should have known that the given activity has the risk of significant harm.

66 UNCLOS, Article 194.2.

67 " Report of the United Nations Conference on the Human Environment," UN Doc. A/Conf.48/14/Rev. 1 and Corr. 1 (United Nations: Stockholm, 1973); 11 I.L.M. 1416 (1972).

68 Reports of Judgments, Advisory Opinions and Orders, Legality of the Threat or Use of Nuclear Weapons, Advisory Opinion of 8 July 1996, 19, par. 29.

69 "Draft articles on Prevention of Transboundary Harm from Hazardous Activities 2001," *Report of the International Law Commission on the work of its fifty-third session*, 154–155.

Whether the release of radioactive substances into the sea from Fukushima Daiichi NPP triggers an obligation for Japan to prevent transboundary damage to the environment should be judged by considering whether Japan satisfied the degree of due diligence by taking appropriate and reasonable actions based on its capabilities in light of the level of hazard of the release and foreseeability of the harm that would eventually occur by the release, presupposing that occurrence of damage could be proved.[70]

Obligation to Provide Timely Notification

The 2001 ILC Draft Articles provide some procedural obligations for states in order to prevent transboundary damage to the environment. These procedural obligations include notification of and consultation with countries likely to be affected prior to authorization of activities based on risk assessment.[71] The obligation to notify together with the obligation to consult with countries likely to be affected is well-established customary law especially in the context of shared resources such as international watercourses.[72] In the *Corfu Channel Case*, the ICJ also confirmed that states have obligations to notify other states of the existence of danger in their jurisdiction and to warn them particularly of imminent danger.[73]

Ambiguity remains in terms of the timing and the mode of such notification. However, since notification is to allow the other state to take necessary measures to avoid or mitigate the adverse effects of an imminent danger, it should be done as soon as the notifying state becomes aware of the danger. The 2001 ILC Draft Articles state that the state shall, without delay and by the most expeditious means at its disposal, notify the state likely to be affected of an emergency.[74]

Similar obligations to notify other states of the risk of significant harm are already included in several international treaties that Japan has already ratified. For instance, the Early Notification Convention applies in "the event of any accident involving nuclear facilities or nuclear activities of a State Party or of persons or legal entities under its jurisdiction or control, from which a release of radioactive material occurs or is likely to occur and which has resulted or may result in an international transboundary release that could be of radiological safety significance

[70] This paper limits itself to examining compatibility of the release of radioactive substances into the sea undertaken by Japan with the general obligation to prevent transboundary damage to the environment; however, whether Japan had satisfied its obligation to prevent transboundary damage due to Fukushima Daiichi NPP accident would also be examined in a similar way, by considering whether Japan had taken appropriate and reasonable actions based on its capabilities in light of the level of hazard and foreseeability of the harm that would eventually occur.

[71] "ILC 2001 Draft articles," 159–61.

[72] United Nations Reports of International Arbitral Awards, "Affaire du lac Lanoux," vol. XII, 310, par. 17.

[73] *Corfu Channel Case*, Judgment of April 9th, 1949: I.C.J. Reports 1949, 22.

[74] "ILC 2001 Draft articles," 169.

for another State."[75] In the event of such a nuclear accident, the state party shall notify those states that are or may be physically affected and the IAEA of the nuclear accident, which must include any available information needed to minimize the radiological consequences of the accident.[76]

Another example of a notification requirement is found in Article 198 of UNCLOS, which stipulates that when a state becomes aware of cases in which the marine environment is in imminent danger of being damaged or has been damaged by pollution, it shall immediately notify other states likely to be affected by such damage as well as the competent international organizations.

Does Japan's release of radioactive water into the sea violate this obligation to notify? According to the fact-finding by the Japanese Diet's Investigation Commission, the option of release of radioactive water had been considered by TEPCO since the end of March 2011, but had been rejected given that the level of radioactivity of the water to be released was well above the dose limit under domestic law.[77] On the morning of April 4, 2011, the release option was seriously considered for fear of an imminent risk of leaks of high-level radioactive water and of possible loss of power that would lead to stopping operation of the cooling systems for Units 5 and 6.[78]

The government and TEPCO started and completed the necessary procedures for release of radioactive water, including seeking advice from the NSC of Japan around 3:20 PM.[79] TEPCO and the government announced this plan in press briefings held at 4:00 PM and at 4:03 PM, respectively.[80] For notification to other countries and relevant international organizations, at 4:00 PM, the Ministry of Foreign Affairs gave a briefing on the matter in the regular meeting with foreign embassies in Tokyo[81] although only fifty-one countries attended the briefing and members of the Russian and Korean embassies were not in attendance.[82] At 7:03 PM, TEPCO started the release, and the Ministry of Foreign Affairs sent a notice to all diplomatic corps in Japan by fax and by e-mail at 7:05 PM.[83] On the next day, April 5, 2011, the Ministry of Foreign Affairs and the former Nuclear and Industrial Safety Agency (NISA) organized a briefing to explain relevant information to other countries.[84] The investigation report noted that NISA, which was in charge of procedures for the release, had not been aware of the existence of the international obligation to notify neighboring countries prior to the release;[85] one official of NISA happened to watch the briefing of Chief Cabinet Secretary Yukio Edano on TV, and, at 5:46 PM, sent a notification on this matter to the IAEA by e-mail.[86]

75 "Early Notification Convention," Article 1. 76 Ibid., Article 2.

77 Investigation Committee of the Government, *Interim Report* (English translation): 340-341. On this point, the Final Report referred to this part of the Interim Report.

78 Ibid., 328–4. 79 Ibid., 334–5. 80 Ibid., 335–6. 81 Ibid., 358.

82 "Korea and Russia are Absent in the Briefing. Information Not Reached Before the Release," *Kyodo*, April 18, 2011 (in Japanese). www.47news.jp/CN/201104/CN2011041701000327.html.

83 Investigation Committee of the government, *Interim Report* (in Japanese), 358

84 Ibid. 85 Ibid., 357. 86 Ibid., 357–8.

The report noted further that the Ministry of Foreign Affairs, which is responsible for matters involving international notification, had not been originally informed of the planned release until an official from the ministry, who just happened to be in TEPCO headquarters, informed the ministry around 3:30 PM.[87] Eventually, while recognizing that response measures to control four reactors had been urgently needed at that time, Japan did not promptly notify some of its neighboring countries, including Russia and Korea, who were most likely to be affected. This failure to promptly notify other states might be incompatible with Japan's obligations under both general international law as well as the Early Notification Convention.

Japan argues that it did not violate these obligations to notify because the release in question had not caused and was not likely to cause transboundary adverse effects to the environment of other states or to the marine environment of areas beyond the limits of national jurisdiction. Former Chief Cabinet Secretary Edano, in the press briefing on April 5, 2011, stated that Japan recognized its obligation to prevent marine pollution under the UNCLOS; however, the release had not or was not likely to cause imminent risk of adverse effects of pollution to neighboring countries.[88] On the other hand, former Minister of Foreign Affairs Takeaki Matsumoto, in his press briefing on the same day, explained that the release did not entail obligation to provide the states and the IAEA under the Early Notification Convention and that Japan notified the release to the IAEA on a voluntary basis under Article 3 of the Convention.[89] The Japanese government considered that it was in compliance with its obligations both under the Early Notification Convention and UNCLOS.[90]

When they learned of the release, a couple of neighboring states, including China and Korea, expressed a deep concern about impacts of the release on the marine environment and especially the lack of prompt notification of the planned release.[91] None of the neighboring states pursued international legal action against

87 Ibid., 358.

88 "Press briefing by Chief Cabinet Secretary, April 5, 2011 (in Japanese)." http://nettv.gov-online.go.jp/prg/prg4640.html.

89 Article 3 of the Convention provides: With a view to minimizing the radiological consequences, States Parties may notify in the event of nuclear accidents other than those specified in article 1. Press briefing by Minister of Foreign Affairs, April 5, 2011 (in Japanese). www.mofa.go.jp/mofaj/press/kaiken/gaisho/g_1104.html#2. See also Investigation Committee, *Interim Report*, 358–9.

90 Negotiation indicates that the Convention leaves it to the state where a nuclear incident occurs to decide whether the incident is likely to result in "an international transboundary release that could be of radiological safety significance for another State." Nonetheless, Article 3 permits states to notify the incident that they consider outside of the scope. In fact, Japanese notification was made under this article. Peter Cameron, "The Vienna Conventions on Early Notification and Assistance," in *Nuclear Energy Law After Chernobyl*, ed. Peter Cameron et al. (Graham & Trotman and International Bar Association, 1988), 23–5.

91 Eben Harrell, "Fukushima Dumping: A Violation of International Law?" *TIME.com*, April 6, 2011, (introducing concern expressed through diplomatic channel by South Korea Foreign

Japan, but what is interesting is that Japan, while confirming the existence of relevant obligations under the UNCLOS and the Early Notification Convention, argued that the release did not trigger any of these obligations since the release in question neither caused transboundary adverse effects to the environment nor implied an international transboundary release of radiological safety significance for another state, and never invoked any circumstances precluding the potential wrongfulness of the release, such as necessity.[92]

In this context, one of the most important issues from the perspective of international law is whether the release of radioactive water would reach the level that triggers obligation to notify, and who has the authority to decide it. Negotiation toward the Early Notification Convention indicates that these questions had been left unsolved; the Convention lacks an objective definition on when a release would reach "radiological safety significance for another State,"[93] and leaves much discretion to states where accidents occur.[94]

The Japanese practice relating to the obligation to notify has realized these concerns. Even when countries recognize a duty to notify neighboring states in the case of an imminent harm, they may avoid doing so for fear that such notification might be considered an admission of causing a release resulting in "radiological safety significance for another State," which might lead to a violation of a primary legal obligation to prevent transboundary damage to the environment. States need to have clearer guidance regarding what objective conditions trigger an obligation to notify. Given the lack of guidance, in June 2011, after the Fukushima accident Russia submitted to the IAEA a formal proposal to amend the Early Notification Convention in order to provide more clarity.[95]

Affairs Minister) http://science.time.com/2011/04/06/fukushima-dumping-a-violation-of-international-law/; "Pacific waters off Japan's Fukushima clearly affected by nuclear crisis: China's oceanic administration," *Xinhua*, July 31, 2011, http://news.xinhuanet.com/english2010/china/2011–07/31/c_131021046.htm.

92 "Draft articles on Responsibility of States for Internationally Wrongful Acts, with commentaries," *Report of the International Law Commission on the work of its fifty-third session*, 71.

93 "Early Notification Convention," Article 1.

94 A. O. Adede, *The IAEA Notification and Assistance Conventions in Case of a Nuclear Accident: Landmarks in the Multilateral Treaty-Making Process* (Martinus Nijhoff, 1987); Cameron et al., "The Vienna Convention," 23–5; Alan E. Boyle, "Nuclear Energy and International Law: An Environmental Perspective," *British Yearbook of International Law* 1989, 60(1): 283–284 (1989).

95 S.V. Kirienko, (statement, The IAEA Ministerial Conference on Nuclear Safety, Vienna, June 20, 2011), www-pub.iaea.org/MTCD/meetings/PDFplus/2011/cn200/plenary/p_d1_russia.pdf. The proposal was subsequently circulated to all contracting parties on July 12, 2011. As of August 2012, the required majority of states parties in order for the depositary to convene a diplomatic conference to consider the proposed amendments is 56. IAEA, Progress in the Implementation of the IAEA Action Plan on Nuclear Safety Supplementary Information, GOV/INF/2012/11-GC(56)/INF/5, August 15, 2012

Duty to Cooperate to Manage Potential Adverse Impacts

In the context of the release of low-level radioactive material from the Fukushima Daiichi NPP, there is a question as to whether Japan appropriately engaged the precautionary approach. The MOX Plant case, a dispute between Ireland and the United Kingdom over alleged marine pollution caused by discharges of radioactive materials, provides a useful parallel for examining the Fukushima radioactive water release. The International Tribunal for the Law of the Sea (ITLOS) issued an order to prescribe provisional measures requiring parties to cooperate and to enter into consultations in order to "(a) exchange further information with regard to possible consequences for the Irish Sea arising out of the commissioning of the MOX plant; (b) monitor risks or the effects of the operation of the MOX plant for the Irish Sea; and (c) devise, as appropriate, measures to prevent pollution of the marine environment."[96]

Likewise, in a subsequent case between Malaysia and Indonesia,[97] the ITLOS referred to the provisional measures from the MOX plant case, and then prescribed an order for provisional measures directing the States to cooperate in order to 1) establish promptly a group of independent experts with the mandate to conduct a study to determine the effects of Singapore's land reclamation, 2) to propose, as appropriate, measures to deal with any adverse effects of such land reclamation; and 3) to prepare, as soon as possible, an interim report; and to exchange, on a regular basis, information on, and assess risks or effects of, Singapore's land reclamation works.[98]

Both orders for provisional measures by the ITLOS indicate that even in the middle of dispute, states have an obligation to cooperate and consult in order to exchange relevant information, to undertake monitoring and risk assessment even when states in a dispute have a different assessment about the existence and gravity of potential risk of marine pollution. The decision by the ITLOS is very much in line with the concept of precaution, which requires states to continually reevaluate potential risk of an activity in light of scientific developments in order to implement the obligation to prevent transboundary damage to the environment.

[96] The MOX Plant Case (Ireland v. UK), Request for provisional measures, ITLOS, Order, December 3, 2001. www.itlos.org/fileadmin/itlos/documents/cases/case_no_10/Order.03.12.01.E.pdf. The arbitral tribunal established under Annex VII of the UNCLOS affirms the provisional measure prescribed by ITLOS in its Order of 3 December 2001 and provisional measures were kept effective until the final order on termination of proceedings on June 6, 2008. The MOX Plant Case, Ireland v. United Kingdom, Arbitral Tribunal, Order N° 3, June 24, 2003. The MOX Plant Case, Ireland v. United Kingdom, Arbitral Tribunal, Order No. 6, June 6, 2008. www.pca-cpa.org/showpage.asp?pag_id=1148.

[97] *Case Concerning Land Reclamation by Singapore in and around the Straits of Johor (Malaysia v. Singapore)*, Provisional measures, ITLOS, Order, October 8, 2003. www.itlos.org/fileadmin/itlos/documents/cases/case_no_12/Order.08.10.03.E.pdf.

[98] Ibid., 21.

CONCLUSION

International rules have evolved toward more effective regulation of nuclear activities that impact the ocean,[99] but there remain lacunae in international rules relating to nuclear pollution both within international nuclear law and international environmental law. Legally binding instruments to address land-based pollution exist at the regional level, but there was and still is no legally binding framework for the regional seas surrounding Japan.

The principal question emerging from this chapter is whether the release of radioactive water would have caused or may still cause a risk to the marine environment at a level that triggers both the obligation to prevent transboundary damage to the environment and the obligation to notify. Japanese practice has revealed once more the necessity to clarify and to make the triggering condition of obligation to notify more objective. Likewise the Russian proposal, using more objective criteria such as the International Nuclear Event Scale (INES), to assess the "radioactive significance" of incidents is worthy of consideration.

Since March 21, 2011, strong levels of radioactive pollution have been observed in the marine environment near the Fukushima Daiichi NPP.[100] However, contaminated waters were carried away to the open seas and diluted in the ocean waters.[101] The initial assessment was that these releases were not problematic for the environment. The acute contamination period of the marine environment was contained within a period of about six months and the release, diluted at a depth of 0 to 100 meters throughout the Pacific Ocean only added a concentration of 0.002 Bq/L to the ocean, which are similar to the levels of fallout from atmospheric nuclear tests (0.002 Bq/L). These concentrations would represent only 1/3000th of the natural potassium-40 concentrations in seawater (12 Bq/L).[102]

However, long-term impact of the release on the marine ecosystem and food chain are not so certain. Even if the cesium contamination in the seawater has decreased sharply in the vicinity of the Fukushima Daiichi NPP and the estimated level of radioactivity seems low, there is a justification for continuing to monitor the marine species in the coastal waters of northeast Japan, and to exchange information with other potentially affected states.

The potential remains in Asia for another fatal nuclear disaster since numerous nuclear power plants continue to operate in the Ring of Fire. While Japan has suspended operation of its nuclear reactors until it can determine compliance with new nuclear safety regulations, Japan has fifty-four reactors (including four reactors of Fukushima Daiichi NPP) and three more under construction, all located on the coast of Japan. Korea has twenty-three reactors and five more under construction on its coast. Both China and Taiwan have several reactors

[99] David D. Caron and Harry N. Scheiber (eds.), *The Oceans in the Nuclear Age: Legacies and Risks* (Martinus Nijhoff Publishers, 2010).

[100] IRSN (2012), 102–6. [101] Ibid., 106. [102] Ibid., 108.

located adjacent to the coast.[103] In light of this situation in Asia, it is critically important that states cooperate and negotiate in advance of future disasters and develop a regional framework to strengthen nuclear safety as well as protection of human health and environment from radiation.

[103] World Nuclear Association, *World Nuclear Power Reactors & Uranium Requirements*. www.world-nuclear.org/info/reactors.html.

7

The International Law of 96 Hours: Urban Search and Rescue Teams and the Current State of International Disaster Response Law

Kirsten Nakjavani Bookmiller

This chapter highlights the contributions made by the International Search and Rescue Advisory Group (INSARAG), a transnational network of emergency management officials and practitioners, to the growth of International Disaster Response Law (IDRL). The soft law guidelines and protocols developed by the group have been credited by the international humanitarian community with strengthening the United Nations' facilitated disaster response not only within the search and rescue domain but for other sectors as well. Yet, INSARAG was ultimately unsuccessful in its quest to secure a binding search and rescue convention, challenged by states' insistence to retain sovereign control over immigration and customs matters. The micro-universe of urban search and rescue therefore represents both great promise and unresolved issues within the larger world of IDRL. After an introduction to the world of international urban search and rescue (USAR) and INSARAG, the chapter will highlight the group's accomplishments, particularly in the area of transnational coordination and standard setting, and suggest future directions for the network, including enhancing national and regional capacity in search and rescue response.

INTRODUCTION

> The United Nations should come up with an international protocol of how the local authorities should cooperate with urban search and rescue teams that all countries would have to abide with.[1]
>
> Every damaging earthquake is a sort of library, a storehouse of lessons for those in search of them.[2]

[1] An anonymous USAR team member quoted in Brendon Morris, *Coordinated Use of International Urban Search and Rescue Assistance By Earthquake Affected Countries*, (November 2006): 139. http://afetkoordinasyon.org/pdf/BrendonMorris.pdf.

[2] Larry Collins, "Earthquakes: Lessons from the Past, Part I," *Fire Engineering* 153, 6 (June 2000): 67.

After structures collapse due to a natural disaster such as an earthquake or a human-caused tragedy such as a terrorist attack, the clock begins for those involved in urban search and rescue (USAR).[3] A critical window of approximately ninety-six hours makes up the "survival interval,"[4] which begins with the disaster's onset and lasts to the maximum threshold when people will most likely be located alive. While this period will vary by individual and disaster context, the challenge is still the same. Every hour is critical.

Does international law provide mechanisms by which transnational responders can make the most of this interval? The answer is mixed. Within the USAR universe itself, a collection of soft law guidelines and protocols crafted by the INSARAG has been lauded by the wider relief community for literally shaving off lost hours in the mobilization, deployment, and coordination stages. INSARAG is a transgovernmental network[5] of national and subnational emergency management officials and practitioners (and to a much lesser degree, non-governmental participants), that emerged in the wake of several high profile, poorly organized transborder USAR responses in the 1980s, including to the 1985 Mexico City and 1988 Soviet Armenia earthquakes. The INSARAG Guidelines, developed in tandem with the United Nations Office for the Coordination of Humanitarian Affairs (OCHA), contain a sophisticated set of pre-departure and post-arrival protocols that were unanimously endorsed in 2002 by the UN General Assembly. One of the protocols, the Disaster Assessment and Coordination (UNDAC) system, has been adopted far outside the tight knit world of USAR and is now standard operating procedure for large scale disaster responses facilitated by the UN.

Yet for all of the progress made within the international USAR realm, search and rescue teams when deployed abroad experience many of the same frustrations that other relief providers do, particularly at the point of entry into another state. While

[3] USAR is the most commonly used acronym for this activity globally. In the United States – which is a leader in the field – it is also common to see the acronym US&R so as not to be confused with the United States Army Reserve.

[4] The length of the survival interval after an earthquake is at the center of a highly contested debate. While the ninety-six-hour reference is frequently cited, studies are ongoing. See Anthony G. Macintyre et al., "Survival Interval in Earthquake Entrapments: Research Findings Reinforced During the 2010 Haiti Earthquake Response," *Disaster Medicine and Public Health Preparedness*, (2011): 13–22.

[5] INSARAG's heterogeneous membership of national, state/provincial and local officials plus non-governmental USAR experts does not fit neatly into the "like kind" membership of transgovernmental networks advanced most prominently in Anne-Marie Slaughter, *A New World Order* (Princeton University Press, 2004). Structurally speaking, however, INSARAG is primarily a peer-to-peer network of participants largely operating in an official capacity on behalf of national governments, but not directly representing the highest levels of officialdom or home foreign ministries. As members are also contributing to international problem solving through the application of highly specialized knowledge and technical expertise, they represent an "epistemic community" within the international system, as conceptualized by Peter Haas, "Introduction: Epistemic Communities and International Policy Coordination," *International Organization*, (Winter 1992): 1–35.

mobilization at home is now finely tuned for many of the major international USAR teams and UNDAC support is available following entry into the disaster affected state, the recipient country's border controls in many cases still cost vital, lifesaving time. Deploying with potentially up to 100 members per team, tons of highly specialized rescue equipment, tons of emergency supplies, and sometimes search dogs, a single USAR team's arrival can prompt numerous bureaucratic headaches related to immigration, customs, and animal quarantine. Addressing these myriad border challenges has been at the heart of the International Disaster Response Laws, Rules and Principles (IDRL) Programme promulgated by the International Federation of the Red Cross and Red Crescent Societies (IFRC) and reflective of larger questions of state responsibility as framed by the International Law Commission's (ILC) work on the Protection of Persons in the Event of Disasters.

This chapter will shed light on INSARAG's contributions to the recently emerging field of IDRL as well as advance the following: that the micro-universe of USAR and its continually enterprising transgovernmental advisory group, represent both great promise and unresolved issues within the larger world of IDRL. After an introduction to the field of USAR, this chapter will trace the emergence and growth of INSARAG while highlighting the group's significant impact upon both operational standards inside the field and other international humanitarian responses in the late twentieth century. As the Guidelines have recently marked their twentieth year in operation, the chapter closes by reflecting upon the following: has INSARAG reached the outer limits of what its soft-law based initiatives can achieve in an international relief landscape still dominated by sovereign states, and if so, what happens next?

USAR'S ROLE IN DISASTER RESPONSE

Within the constellation of international relief actors, USAR teams fill a very specific response niche. Search and rescue (SAR) activities assume many forms depending on the environment (maritime, wild lands, etc.) in which they take place. As the name implies, *urban* search and rescue teams are deployed to areas where victims have been entrapped in collapsed structures. While earthquakes are the most common catalyst, USAR teams may also be activated as the result of other natural events, including hurricanes, tsunamis, and tornadoes. Industrial-related accidents as well as acts of terrorism, such as the 1998 bombings of the United States' embassies in Kenya and Tanzania and the 2001 World Trade Center attacks, will also require victim location and extraction from a structural cave-in. After such a tragedy occurs, USAR is the very first stage in the emergency response cycle.

While search and rescue within a metropolitan environment is not a new disaster response mechanism (the World War II London Blitz, as one example), the 1980s were a key turning point for the emergence of the first significantly organized urban search and rescue units, with the earliest technical teams being organized in the United States and Western Europe. By this time, global urbanization was in full swing and coupled with a massive building boom, the trend led to the appearance of higher occupancy dwellings like high-rise apartments and commercial complexes, typically made of concrete and steel. Many such structures are comprised of precast concrete, where walls, floors, and other building components are fabricated off-site and then later connected on-site. These structures are particularly vulnerable to seismic and other events.[6] Further, by the late twentieth century, urban growth patterns resulted in the emergence of "megacities" particularly within the Global South, where construction standards – often undermined by cheap building materials, weak government regulation, and/or corruptive practices – were even more problematic than in some of their northern counterparts.[7] The latter factor in particular may help explain why, in spite of progress in earthquake engineering in the modern urban era, death tolls due to earthquakes have continued to rise, according to a groundbreaking study by Nicholas Ambraseys and Roger Bilham. As the authors reported in their 2011 work linking corruption and levels of earthquake deaths: "Of all earthquake fatalities attributable to building collapse in the past three decades, 8.26% occur in societies that are anomalously corrupt. . ."[8] Or as Ambraseys stated in conference remarks he made more than thirty years ago: "Today's 'Act of God' is likely to become tomorrow's 'Act of criminal negligence'."[9]

These intertwined developments generated a demand for trained professionals, heavy equipment, and technically intensive forms of rescue on an unprecedented scale to respond to the inevitable building collapses. In the immediate minutes and hours after an event, local inhabitants and response entities (if available) are the first to pull out survivors near the surface of collapsed buildings. More sophisticated SAR assistance will be required to free victims that are more deeply entrapped under the rubble, have not initially succumbed to traumatic injuries (typically crush-related) and are situated in such a way that they have access to voids and spaces where they can breathe.[10] When such conditions exist, multifaceted action is required, involving advanced skill sets (among both personnel and search dogs), technology for locating victims (including thermal, visual, and acoustic), and specialized tools (including drills, cutters, and rigging and shoring equipment).

6 Robert Rhea and Brian Rousseau, *Fundamentals of Technical Rescue* (Sudbury: Jones and Bartlett Publishers, 2009).

7 Andrew Revkin, "Disaster Awaits Cities in Earthquake Zones," *The New York Times*, February 24, 2010. www.nytimes.com/2010/02/25/science/earth/25quake.html?pagewanted=all

8 Nicholas Ambraseys and Roger Bilham, "Corruption Kills," *Nature*, (2011): 155.

9 *The Guardian*. "Third World Report: Acts of God Increasingly Amount to Acts of Criminal Negligence," December 30, 1988.

10 Rhea and Rousseau, *Fundamentals of Technical Rescue*. For a survey of a multitude of factors which may contribute to greater survival, refer to Sherif El-Tawil and Aguirre Benigo, "Search and Rescue in Collapsed Structures: Engineering and Social Science Aspects," *Disasters*, (2010): 1084–1085.

A successful mission involves locating the entombed, stabilizing the structure as well as the victims, and safely extracting them and their rescuers out of tons of heavily reinforced concrete and twisted steel. A typical single operation requires dozens of highly trained responders with a variety of specialties ranging from structural engineering to emergency medicine.[11]

Likely due to the fact that the need for what has become regarded as "heavy" USAR capability is not common to every disaster – as opposed to sustenance, shelter and medical care – the UN does not possess its own USAR resources. USAR units only exist within individual countries' respective emergency response infrastructures and are primarily located within the public fire and rescue sector at either the local, state/provincial, or national level. While frequently a civilian endeavor, governments may also have USAR units situated within their military forces. Some non-governmental teams as well as volunteer units also exist. Generally speaking, while lesser economically developed countries may possess such capabilities, their capacity is typically limited to a single team.

THE RISE OF USAR ASSISTANCE: MEXICO CITY AND ARMENIA

The impetus for governments to consider organizing urban-specific search and rescue teams both domestically and internationally was the same catastrophic event – the 1985 Mexico City earthquake.[12] "The Mexico City earthquake . . . was one of the first large-scale urban earthquakes that resulted in the collapse of many modern reinforced concrete buildings, trapping victims under heavy and difficult to dismantle rubble piles."[13] The reverberations of the 1985 quake response were felt far beyond the SAR domain. An increasingly interconnected global community began to realize its transborder relief capability was in dire need of policy attention. Further, if the 1985 disaster served as a wake-up call, the 1988 Soviet Armenia earthquake spurred international relief stakeholders into action.

1985 Mexico City Earthquake

On September 19, 1985, Mexico City and several central Mexican states were struck by an 8.1 magnitude event. A strong aftershock, almost as severe as the initial quake, struck a day later. While final numbers of fatalities have never been confirmed, the quake killed at least 9,500 (and perhaps more than 30,000) while 30,000 were injured and between 100,000 and 250,000 people were left homeless. While the epicenter was located 250 miles away in Michoacán state, the most impacted section of Mexico's capital was situated on an ancient lake bed formed of sediment

[11] Rhea and Rousseau, *Fundamentals of Technical Rescue*, 125–8.

[12] El-Tawil and Aguirre, "Search and Rescue," 1090. [13] Ibid.

which further amplified the ground motion. As a result, more than 400 buildings collapsed and more than 3,000 were considerably damaged. A significant number of destroyed or impaired buildings were office complexes. Because the city was struck at 7:19 AM local time, when most inhabitants were commuting and not yet at their desks, the death toll was prevented from reaching much higher numbers.[14]

While initial media coverage praised the extraordinary worldwide outpouring of relief supplies and aid workers, it soon became apparent that the transnational response, from an organizational perspective, was in complete disarray. At first slow to realize the full impact of the quake and therefore delaying a formal request for international assistance, the Mexican government was then overwhelmed by more than 250 offers of help proffered by governments, intergovernmental organizations (IGOs), and non-governmental organizations (NGOs).[15] Ultimately fifty-two countries and four international organizations sent relief.[16] Mexican authorities registered 177 international flights arriving from thirty-one countries with more than 1,000 tons of materials in the first ten days following the crisis.[17] "Food, medical supplies, heavy equipment, clothing and other goods began arriving by the ton at Mexico City International Airport, much of it unrequested; most of it untargeted, with no designated recipient organization or group."[18]

At the time of the disaster, there were neither organized USAR groups nor formal registration procedures for arriving international responders. However, rescuers from nine countries ultimately deployed – including six Western European countries, Israel, the United States, and Venezuela. They represented a mix of personnel from local fire services, national military units, and citizen search dog groups as well as specialists from the private sector, each with their own variations of training, equipment, and technologies.[19] The United States, for instance, sent regular firefighters and paramedics, but also U.S. Forest Service fire response units, mining search and rescue personnel, heavy demolition workers, and private search dog teams.[20]

[14] W. C. Stone et al., *Engineering Aspects of the September 19, 1985 Mexico Earthquake (NBS BSS 165)*. (May 29, 1987). www.nist.gov/manuscript-publication-search.cfm?pub_id=908821.

[15] E. L. Quarantelli, "Organizational Response to the Mexico City Earthqauke of 1985," *Natural Hazards*, (1993): 24.

[16] Louise Comfort, "International Disaster Assistance in the Mexico City Earthquake," *FMHI Publications, Paper 24*, (1986): 1. http://scholarcommons.usf.edu/fmhi_pub/24, p. 1.

[17] Pan American Health Organization. "Emergency Preparedness and Disaster Relief Coordination Program," *The Earthquake in Mexico, Disaster Report Series* No. 3, 1985: 31.

[18] Comfort, "International Disaster Assistance," 1.

[19] Ibid., 15. See also *Washington Post*. "World Ready to Help if Mexico Makes Request," September 21, 1985; *Washington Post*. "Rescue Workers Dig for Quake Survivors," September, 22, 1985; *The Globe and Mail*. "Mexican President Faces Criticism as Rescue Teams Race Against Time," December 23, 1985.

[20] *Washington Post*, "Donations to Agencies Favored in Quake Relief," September 22, 1985. See also U.S. Embassy-Mexico City, *Final Disaster Summary Report Mexico—1985 Earthquake*," (January 1986): 27.

The trials experienced in Mexico by international search and rescue responders have since become a part of international USAR lore. Among the more infamous documented incidents included clashes between American and French search and rescue personnel regarding the best way to access victims trapped in a hospital. As the author of one of the most definitive studies on the USAR response to the Mexico City disaster explains:

> The U.S. team, mindful that each hour of delay reduced the chances of getting the victims out alive, advocated the use of heavy equipment to clear access to the victims more quickly. The French team, mindful of the risk of the unsafe building collapsing altogether and crushing the victims in the process, favored a slower, tunneling approach that might delay the rescue but be less likely to precipitate further structural collapse of the building. Unable to resolve the dispute, the French team left the site and left Mexico City the next day.[21]

Further disagreement erupted over whether British military helicopters, located in neighboring Belize, would be useful for lifting heavy rubble or would undermine location efforts with their loud engines. The British offer was initially turned down by the Mexican authorities, but they later reversed their decision. Ultimately the helicopters remained grounded outside of the capital.[22]

Partly due to the Mexican government's initial hesitation to request international aid, and to some extent owing to the rudimentary nature of international SAR efforts, many of the major teams did not arrive until almost fifty-six hours after the quake had struck. Total confusion as to which emergency sites had already been searched and which sites still needed to be searched may have sacrificed numerous additional potentially lifesaving hours.[23] One newspaper report indicated that the West German team "appeared to be dispatching their scores of workers around the city themselves, using police radio frequencies."[24] In another account, a loudspeaker broadcast the need for a team right after a California contingent had just left the exact same area.[25]

All media and scholarly accounts of the SAR response to the 1985 Mexico City earthquake pointed in the end to the same critical problems – coordination, communication, and standardization among the multitude of national responders were nonexistent. Recognizing that structural collapse emergencies would only continue to intensify in the coming years with full-tilt urbanization, several countries began to take initial steps to formally organize USAR units

[21] Comfort, "International Disaster Assistance," 30. [22] Ibid., 30–1.

[23] E. L. Quarantelli, "Preliminary Paper #187: Organizational Response to the Mexico City Earthqauke of 1985: Characteristics and Implications." *University of Delaware Disaster Research Center*, (1992): 10. See also Comfort, "International Disaster Assistance," 4.

[24] *The Globe and Mail*. "Canadian Rescuers Thwarted." September 25, 1985.

[25] *The Globe and Mail*. "Hopes Fade for Survivors in Mexico's Quake Chaos." September 25, 1985.

for both domestic and international service. As USAR expert Louise Comfort explains:

> Pride and commitment to specific technologies by separate national teams tended to obstruct the genuine exploration of cooperative solutions to the extremely difficult problems of search and rescue amidst the urban mountains of collapsed concrete and steel. The criterion of fit performance, nonetheless, remained the successful rescue of live victims. The low ratio of victims found to effort expended compelled all participating teams to reflect soberly upon their collective performance.[26]

A 1985 Pan American Health Organization report on the disaster, in its summary findings, also called for the international community to "develop techniques and methodologies for the rescue of trapped persons in collapsed buildings" as well as enhance the effectiveness of international assistance for recipient countries.[27]

The 1988 Soviet Armenia Earthquake

On December 7, 1988, at 11:41 AM local time, a 6.8 magnitude earthquake struck northern Armenia, then a part of the Soviet Union. The epicenter was located north of the city of Leninakan; the region in general is known as one of the most seismically active areas in the world and has had a history of devastating earthquakes.[28] Unlike the early morning quake that shook Mexico City, the Armenian event (often known as the "Spitak" earthquake after the town that is situated closest to the fault line) occurred while buildings were filled with local residents at work or in school. Moreover, a large majority of buildings were fabricated either with unreinforced stone masonry or precast concrete, both of which were more vulnerable to giving way.[29] The results were catastrophic. In total, 40,000 inhabitants were trapped under the rubble of collapsed structures, 25,000 of who perished before they could be extricated.[30]

At the time of the disaster, Soviet leader Mikhail Gorbachev rushed home from the United States, where he was visiting the UN. It was sixteen hours before he received initial briefings about the enormity of the devastation caused to some of the larger population centers such as Leninakan, and it was more than a week before reports were updated to include massive devastation to twenty more remote villages.[31]

[26] Comfort, "International Disaster Assistance," 32–3.
[27] Pan American Health Organization, 32.
[28] U.S. Geological Survey. "Historic Earthquakes: Notes About the Armenia Earthquake." *Earthquake Hazards Program*, 1988. http://earthquake.usgs.gov/earthquakes/world/events/1988_12_07_ev.php
[29] Ibid., and Eric Noji et al., "The 1988 Earthquake in Soviet Armenia: A Case Study," *Annals of Emergency Medicine*, (1990): 894.
[30] Noji, "The 1988 Earthquake," 891.
[31] "Tardy Quake Response Belies Perestroika." *The Globe and* Mail, December 20, 1988.

Partly attributable to a sluggish Soviet central government bureaucracy, delays in organizing relief were also caused by the significant damage the disaster caused to the local Armenian public sector and communication infrastructure as well as the wide swath of geographic destruction.[32]

Two days after the quake hit, the Soviet Union – a country which historically did not request or welcome outside offers of humanitarian assistance and had not accepted aid from the United States since World War II – made its first appeals for international emergency relief. In addition to medical supplies, the early requests focused on the specific need for search and rescue teams to assist in locating the tens of thousands of victims trapped in the rubble, who by this time had already been exposed to freezing winter temperatures for forty-eight hours.[33]

The Soviet authorities waived traditionally cumbersome visa requirements and opened national airspace for incoming civilian and military planes flying to Moscow or directly to Yerevan, Armenia. Ultimately sixty-seven countries, including the United States, provided some form of assistance.[34] In many respects the international humanitarian response to Armenia was a frenzied replay of Mexico City, but the opening days of relief disarray were made exponentially worse due to the greater magnitude of the damage and the correlating need for disaster relief in the Soviet case. A UN official assessing the scene referred to it as "chaotic and disorganized,"[35] while a representative with US NGO AmeriCares referred to it as "chaos, absolute chaos."[36] The influx of relief was so great that within a two-day period, not one but two aircrafts carrying supplies – a Soviet military plane and a Yugoslavian Air Force plane – both crashed, the first killing seventy-eight people. Both crashes were blamed on high levels of air traffic congestion caused by the massive response.[37]

While many international USAR teams were poised to deploy upon a Soviet appeal, Moscow's delay in making its request meant that the first team to arrive, a several hundred member corps sent by France, did not reach the area until fifty-three hours after the event, shortly followed by full or partial units from Switzerland, the United Kingdom, Italy, and Germany.[38] With a greater distance to travel, the newly created "International Search and Rescue Team" from the United States did not arrive and deploy in Leninakan until five days after the quake.[39] Again, with

[32] Ibid., and Charles Krimgold, "Search and Rescue," *Earthquake Spectra*, (1989): 140.

[33] John Goshko, *The Washington Post* "400,000 Homeless in Armenian Quake; U.S. Responding to Requset for Help," December 10, 1988; Irvin Molotsky, *The New York Times* "Soviets Accept U.S. Aid for the First time Since 40's," December 10, 1988.

[34] Krimgold, "Search and Rescue," 141.

[35] Reuters. *The New York Times*. "In Ravaged City, theThe Voices in theThe Rubble Begin to Fade.." December 12, 1980.

[36] *The Washington Post*. "Relief Workers Find Chaos, Cooperation." December 14, 1988.

[37] *The Globe and Mail*. "Mountains Pose Danger to Quake Air Relief." December 14, 1988.

[38] Krimgold, "Search and Rescue," 142.

[39] Ibid., 137 and Steve McConaughy and Carol McConaughy, *The Washington Post* "Many Lights in Armenia, Two American Members of a Rescue Team Tell Their Story," January 1, 1989.

no central registration in existence for incoming or departing USAR teams, there is no official count of the responders who participated. However, Charles Krimgold, who was in Armenia and who has studied the disaster extensively, holds that eighteen countries – double the number of Mexico City – sent government teams totaling more than 1,000 personnel altogether. Non-governmental teams also came from four countries.[40]

In the three years after Mexico City, many national governments had begun to systematize their USAR response for domestic as well as transborder purposes, and several of the incoming groups possessed a greater cohesion within their national teams than in Mexico City. What was missing though was any central international mechanism to coordinate the global USAR response, to facilitate standardization among the myriad rescuers, and to calibrate the response to the host country's actual needs. This weakness reflected the larger decentralization and fragmentation of the humanitarian response infrastructure in the global community overall. International teams kept arriving a week after the quake itself – falling far outside the window of survivability of the victims. Even if the teams had arrived earlier, the groups were wildly uneven in quality and levels of training.[41] With no operating standards across national teams, responders once again were unable to determine which buildings had been previously searched as each country possessed their own marking system.

On the recipient end, the Soviets – unused to such an extraordinary degree of external involvement on their sovereign territory – had neither an apparatus in place to receive nor process arriving teams nor ways to deploy them to best effect. USAR personnel had difficulties securing transportation and few had sufficient translation assistance. More deliberately, Soviet authorities actively impeded the response with customs officers confiscating communications equipment from the U.S. team, the KGB harassing British rescuers as they attempted to use their satellite equipment, and other government officials attempting to split up and reconfigure various national teams whose members were then unable to communicate with each other in a common language.[42]

In the end, despite the widespread global media focus on the worldwide outpouring of response to the disaster and the extraordinary efforts made to deploy rescue personnel, USAR teams were responsible for less than 1 percent of live rescues (with Soviet SAR specialists accounting for 2.5 percent and the remainder of the live rescues by local residents).[43] The efficacy of international USAR

[40] Krimgold, "Search and Rescue," 143. [41] Ibid., 142.

[42] David Webster and Peter Zimmerman, *The New York Times*, "Fatal Delays in Aremenian Relief," December 17, 1988; Esther Fein, *The New York Times*, "Soviets Say Rescue Won't Be Halted," December 17, 1988; Bill Keller, *The New York Times*, "From Soviet Quake, Echoes Widen," December 17, 1988.

[43] Eric Noji, Haroutune Aremnian, and Ashot Oganessian, "Issues of Rescue and Medical Care Following the 1988 Armenian Earthquake," *International Journal of Epidemiology* (1993): 1070.

assistance was beginning to be questioned by direct participants, relief analysts, and policymakers. Eric Noji, the lead author of several medical studies on the Armenian earthquake wrote: "The relatively few people rescued by specialized foreign rescue teams in Armenia indicates that the benefits of deploying international rescue teams to distant earthquake sites may not be worth the cost unless the teams arrive much more quickly than they did in Armenia."[44]

Krimgold discussed how much members of the American team might learn from their counterparts from abroad and asserted: "There is clearly a great need to coordinate international search and rescue efforts," including "to establish some acceptable international standard for prequalification of rescue personnel. . ."[45] In a pivotal March 1991 position paper for the National Red Cross and Red Crescent Societies on international search and rescue teams, Peter Walker, in a review of earthquake responses to date, contended that the teams were drawing a disproportionate share of global media attention while they were "under skilled and under equipped," did not "interface well with the stricken community" and arrived "too late to provide any meaningful assistance in terms of saving lives."[46] In his summary recommendations, Walker also called for a preregistration process that would establish screening standards and elevate the overall quality of responders.[47]

EMERGENCY RESPONDERS AND INTERNATIONAL LAW: THE INTERNATIONAL SEARCH AND RESCUE ADVISORY GROUP (INSARAG)

The 1985 Mexico City earthquake spurred many national governments, particularly the more affluent countries, to assess their own urban search and rescue capacity. The 1988 Spitak disaster placed the issue of transborder USAR assistance squarely on the international humanitarian policy agenda. Less than a year after Armenia, the United Nations Disaster Relief Organization (UNDRO) convened a meeting of national emergency managers and diplomats in Geneva to examine nine problematic areas of international disaster relief, with search and rescue activities included among them. The conference recommended a preregistration and screening process to enhance the quality of participating teams, urged SAR capacity building for disaster vulnerable countries, and called for the assembling of a new international USAR directory. It also proposed that UNDRO hold a workshop for USAR officials to share best practices and formulate transnational standards.[48]

44 Ibid., 1074. 45 Krimgold, "Search and Rescue," 142.

46 Peter Walker, "International Search and Rescue Teams: A League Discussion Paper," (1991): 26.

47 Ibid., 28.

48 Office of the United Nations Disaster Relief Coordinator (UNDRO), "Third Meeting of Officials in Charge of National Emergency Relief Services (NERS III) Final Report," Geneva, October 1989, 11.

The First International Search and Rescue Workshop

In April 1991, UNDRO sponsored and the Austrian government hosted the first international USAR-specific conference ever to be held. Delegations attending the "International Search and Rescue Workshop" in Wattener Lizum were a hybrid mix of search and rescue professionals from seventeen countries – many of whom had responded to the Mexico City and Spitak quakes.[49] as well as high level diplomatic representatives concerned with strengthening their home countries' efforts at "disaster diplomacy." While the "disaster diplomacy" concept, most prominently advanced by Ilan Kelman, primarily focuses upon humanitarian responses between "enemy" states, the term is used here to cover relief outreach in general, regardless of the nature of the relationships between governments.[50]

The meeting was also taking place against a larger backdrop of initial post-Cold War attention to disaster mitigation and response, as reflected in the adoption of UN General Assembly Resolution 44/236 in December 1989 launching the International Decade for Natural Disaster Reduction (IDNDR)[51] as well as reforms focusing on greater coordination of the global humanitarian response system, most notably, negotiations leading up to the passage of Resolution 46/182 in December 1991. The latter resolution led to the creation of the UN Department of Humanitarian Affairs (later to become OCHA) and also served as the impetus for the establishment of the Inter-Agency Standing Committee (IASC) relief coordinating body.[52] The critical evaluation of USAR just recently completed by the IFRC was also foremost on the delegates' minds.[53]

Recalling their recent experiences, participants at the historic Austria gathering agreed upon eight working areas of focus for future responses by international USAR teams. The proposals were largely dedicated to establishing a framework of greater coordination, operational standardization, and network building within the USAR community of responders who typically served only within a single metropolitan region of their home countries, possessed a dizzying array of training backgrounds, protocols, and approaches and only sporadic exposure to disasters outside their own borders. The eight areas of future focus coming out of the workshop included:

1) capability assessment of UNDRO-registered teams;
2) common operating procedures on-site;
3) ongoing information exchanges among teams outside the disaster response phase;

[49] Disaster Relief: If You Want Help, Stay Away. (April 14, 1991). In *The Observer (London)*.

[50] Ilan Kelman, *Disaster Diplomacy* (London: Routledge, 2011)

[51] UNGA Res. 44/236, UN Doc. A/RES/44/236 (December 22, 1989).

[52] UNGA Res. 46/182, UN Doc. A/RES/46/182, (December 19, 1991).

[53] "The Story of INSARAG Twenty Years On." *United Nations Office for the Coordination of Humanitarian Affairs*. (2010). www.insarag.org/images/stories/INSARAG_booklet_FINAL_VERSION_V3_screen_view.pdf.

4) a forum to address medical treatment on-site;
5) responsibilities of both responding and disaster-affected countries;
6) establishment of a new On-site Operations Coordination Center;
7) a forum for USAR teams to discuss response logistics; and
8) protocols regarding communications.[54]

Working groups were organized to further develop some of the key recommendations. With the exceptions of items 5 and 6, the objectives outlined could be largely accomplished "internally" within the group, while the responsibilities of sending and disaster affected states as well as the successful setup of an on-site coordination center post-disaster would require a genuinely intergovernmental level of discussion.

The December 1991 Birth of INSARAG

Just a few days before the successful passage of UN Resolution 46/182 in New York, the inaugural meeting of the International Search and Rescue Advisory Committee was held in Beuggen, Germany, again under the auspices of UNDRO. In addition to USAR specialists, international relief officials, and diplomats from fifteen countries, the forum also included several intergovernmental representatives from UNDRO, the International Civil Defense Organization, and the Pan American Health Organization, as well non-governmental participation from the IFRC, the Asian Disaster Preparedness Center, Central American Coordination Center for Natural Disaster Prevention, and Partners of the Americas.[55]

The Beuggen meeting set in motion many of the post-disaster response innovations associated with INSARAG today. Working from a draft prepared by the Swedish National Rescue Services Board, the delegates endorsed a plan for the group to organize themselves along regional lines with an executive body and working or ad hoc bodies as needed; the new group would be housed under the auspices of UNDRO. The participants also agreed to progress the "On-site Operations Coordination Center" concept, working off a proposal submitted by the U.S. Office of Foreign Disaster Assistance (OFDA). The German government outlined parameters for operations standardization, including initial response times, information sharing, and sensitivity to local customs and laws, while UNDRO's working paper focused on quality control for rescuers, including recommendations for a new classification system.[56]

[54] Office of the United Nations Disaster Relief Coordinator (UNDRO), Inaugural Meeting of the International Search and Rescue Advisory Committee (INSARAC), (December 11-13, 1991). Beuggen, Federal Republic of Germany, 4.
[55] INSARAG "Twenty Years On," 6–7. [56] UNDRO, "The Inaugural Meeting."

OVER TWENTY YEARS OF INNOVATION: INSARAG, 1991-PRESENT

An International USAR Network

International USAR assistance today is nearly unrecognizable from that of the 1980s when it was every team for itself. At the heart of this transformation is the transnational expertise-sharing and peer to peer professional network institutionalized by INSARAG. Membership is open to either countries or organizations, and notably comprises not only countries that provide international USAR assistance but also highly encourages participation by states more vulnerable to disasters, in particular earthquakes. Of the fifty-five INSARAG member countries participating as of September 2013, half have experienced earthquakes themselves, including the two that inspired the entire INSARAG movement, Mexico and Armenia. Team participation has grown from an initial fifteen to ninety.[57] The secretariat is housed within UNOCHA's Emergency Services Branch in Geneva, specifically within the Field Coordination Support Section. The executive function is carried out by the Steering Group which meets yearly and consists not only of USAR officials but also representatives from the INSARAG Secretariat, UNOCHA, and the IFRC. The leadership is supported by annual meetings of three regional groupings and the International USAR Team Leaders forum, a global knowledge exchange between search and rescue officials on lessons learned from recent responses. In addition, ad hoc "Working Groups" are formed and disbanded as individual issues are addressed.[58]

While the formal INSARAG policymaking bodies are credited with developing innovative response strategies in a variety of areas to be highlighted later, USAR officials frequently reiterate that given that no such networks existed before 1991 the mere act of teams connecting across borders has been in and of itself highly invaluable. As one consulting report indicated about the response to the 1999 Marmara, Turkey earthquake:

> Many of the lead officers within the international SAR teams knew each other, were familiar with and expected a coordinating platform and had previous contact with UNDAC members, which assisted in no small way in the exchange of information and trouble-free operations on-site. This reflects positively on the important role played by INSARAG through meetings, seminars, workshops, training and exercises since its inception in 1991.[59]

The group's overriding concern from the outset has been to improve response coordination, from mobilization through deployment, so that lost hours are

[57] "INSARAG-USAR Directory." http://vosocc.unocha,org/USAR_Directory/USARTeamsByRegion.asp

[58] "Structure of INSARAG." www.insarag.org/en/about/structure.html.

[59] Joseph Bishop, "Rapid Study Report on the International Search and Rescue Response to the Izmit Earthquake: Turkey," 1999.

reduced, and the likelihood of successful rescues is increased. Over time, INSARAG has tackled the issue of coordination from four principal directions. They include the development of the On-Site Operations Coordination Center concept; the drafting of the INSARAG Guidelines and Methodology; the creation of the External Classification System; and the push for a protocol of state responsibilities related to USAR. In December 2002, the UN General Assembly recognized INSARAG's efforts in these areas by its passage of Resolution 57/150, "*Strengthening the effectiveness and coordination of international urban search and rescue assistance.*"[60]

OSOCC: Coordination on the Ground

One of the group's major priorities from the outset was to address the challenge of managing the external USAR response more effectively prior to and immediately upon deployment. In a related concern, INSARAG sought to better calibrate international responses based on the actual needs of the distressed country, rather than flooding it with superfluous resources. Mexico City and Armenia had shown that a disaster-impacted state's Local Emergency Management Authority may struggle to take the lead in conducting a needs assessment or furnish support of various kinds to incoming USAR teams. A solution refined over time but advocated from the outset during the 1991 meetings, was the On-Site Operations Coordination Center (OSOCC), which provided a central rendezvous point for incoming international USAR teams and would assist search and rescue personnel in deploying to appropriate sites. The model was further expanded to include a Reception and Departure Center (RDC).

Typically located at designated entry points in the disaster-stricken country, the RDC is responsible for proactively addressing issues that may impede the successful deployment of USAR personnel, including customs and immigration, transportation logistics, and interpretation support. It is also tasked with registering incoming teams and conveying their status to the OSOCC. As personnel exit the country, the RDC conducts a debriefing, logs their departure, and works to resolve any outstanding matters. Both the RDC and the OSOCC are set up either by the first USAR team to arrive in country, or by representatives from the UNDAC team. A concept which also emerged from INSARAG discussions in the early 1990s, UNDAC is a corresponding rapid response mechanism, involving a roster of national and IGO-affiliated emergency management officials who mobilize quickly and assist the affected country in articulating its needs and coordinating the influx of global relief when appropriate.[61]

[60] UNGA Res. 57/150, UN Doc. A/RES/57/150A/RES/57/150 (Dec. 16, 2002).
[61] "Achievements of INSARAG", www.insarag.org/en/about/achievements.html.

The first mechanism of its kind to be actually implemented within the international relief system in any sector, the INSARAG-proposed OSOCC concept has needed time to take root. The strategy assumes that incoming teams will automatically report to the RDC and to OSOCC, which has not always been the case. During the 1999 Taiwan earthquake, an Australian USAR member stationed within the UNDAC mission describes efforts to track down teams that had not participated in the registration process.[62] The operations center during the Turkish earthquake the same year also experienced similar challenges.[63] Successful implementation of the OSOCC strategy is also predicated upon a single designated entry point or a crisis that is confined within a more limited geographic area. While certainly unprecedented in its dimensions, the multinational scale of destruction caused by the 2004 Indian Ocean earthquake and subsequent tsunami placed significant stress on the more centralized model.[64] Even in a "smaller" event like the 2005 South Asian quake, the issue of multiple ports of entry posed challenges according to Yosuke Okita, who participated in the response on behalf of the Japanese government and noted that the:

> Japan Disaster Relief Rescue Team arrived in Lahore and headed for the site of activity at Baragram without stopping at Muzaffarabad where the OSOCC was located. For that reason, although the Pakistan government was aware of the activities of Japan Disaster Rescue Team, the team was not registered initially with the OSOCC. This occurred as a result of reception centers being located only in Islamabad, and using Pakistan as an example, it is conceivable that, teams may have also entered the country via Lahore (as Japan did), Peshawar and Karachi, in addition to Islamabad.[65]

Nonetheless, for countries willing to accept UN-facilitated external aid in the wake of a disaster, the establishment of an OSOCC is now standard operating procedure during a large-scale international relief operation, and the efforts have paid off. As Okita asserts: "As a result of the growing awareness of this system's existence among donor countries and organizations, its effects are beginning to show in the reduction of organizations that begin activities without registering at reception centers and the thorough implementation of information sharing at OSOCC."[66] In the 2005 Humanitarian Response Review, commissioned by the UN, INSARAG and OSOCC received favorable reviews: "The overall impression is that the search and rescue teams, under the coordination of the International Search and Rescue Advisory Group (INSARAG) and the On-Site Operations Coordination Centre (OSOCC), are working well, with disciplined coordination

[62] Mark O'Connor, "Taiwan Earthquake," *Australian Journal of Emergency Management*, (2000): 32–33.
[63] Bishop, "Rapid Study Report," 3.
[64] Yosuke Okita, "UNOCHA's Coordination in International Disaster Response and its Contribution from Japan," *Technology and Development*, (2007): 66.
[65] Ibid. [66] Ibid.

and actively functioning groups."[67] The structure has expanded in tandem with new technologies, and a "Virtual OSOCC" now allows for nearly instantaneous knowledge sharing and coordination efforts over the Internet, pre-response. All of these international relief mechanisms, originally devised by INSARAG, are now regarded as standard tools to be used for a variety of sudden-onset events beyond that of earthquakes and types of relief other than search and rescue.[68]

Coordination Through Common Language

A different but equally important track to improving coordination has been the drafting and ongoing revision of the INSARAG Guidelines and Methodology. First endorsed in June 1999 as the "International Search and Rescue Guidelines" (ISARR)[69], the document is a living handbook for the transnational USAR community and sending and recipient governments, and has been continually updated at the various INSARAG meetings as the result of lessons learned in recent disasters and other developments. The majority of guidelines are dedicated to establishing a common frame of reference for international responders, whose primary job is to work in search and rescue within their home countries, and therefore have highly diverse training backgrounds and operating procedures.

The latest revision was undertaken in 2011, and the document has grown to 205 pages in length. The comprehensive document includes everything from the first steps to be taken upon notification of a disaster to outlining professional ethics, establishing common marking systems, and providing mission summary report forms. UNGA Resolution 57/150 encourages both sending and receiving states to adhere to the Guidelines and Methodology thereby giving the document an elevated legitimacy within international relief circles. As UN Under-Secretary-General for Humanitarian Affairs Valerie Amos asserted at INSARAG's Global Meeting held in Kobe, Japan in September 2010: "INSARAG has shown how countries and organizations, responders and beneficiaries, can work together to develop professional standards in their field of humanitarian response. It is a model of international cooperation, from which we can all learn."[70]

67 Costanza Adinolfi, *Humanitarian Response Review, Commissioned by the United Nations Emergency Relief Coordinator and Under-Secretary-Gernal for Humanitarian Affairs*. (Geneva and New York: United Nations, 2005).

68 "Achievements of INSARAG."

69 United Nations Office for the Coordination of Humanitarian Affairs. "International Search and Rescue Response: Meeting of International Search and Rescue Team Leaders on Lessons Learnt and Follow Up to the 1999 Turkey and Taiwan Earthquakes." Neuhasen, Germany, (December 9-12).

70 N. Erdogan, "INSARAG Global Meeting: A Global Commitment to Saving Lives." December 2010. www.unspecial.org/UNS701/UnSpecial_Decembre_2010_WEB.pdf.

Coordination Through Classification: The International External Classification (IEC) System

By the early 1990s, it also became increasingly clear to the global relief community that not every arriving search and rescue team was adding value to an international response. In the late 1980s and early 1990s, study after study revealed a truth that still remains a search and rescue maxim today: The greatest proportion of live rescues will happen in the first twenty-four hours by the local inhabitants.[71] Those extricated during this period are typically closer to the surface, and do not require sophisticated skill sets and equipment (although there is always a danger with unstable rubble that more injuries may be sustained in the process).

After the wave of earthquakes in the late 1980s and 1990s, search and rescue teams of every skill level and preparedness were deploying and putting pressure on local infrastructure. In fact, what the recipient country really needed by the time USAR personnel were arriving from abroad were the advanced teams trained in highly technical rescues and possessing specialized equipment. This has proven to be an ongoing problem since groups may deploy with the more self-serving motive of "disaster tourism" to "say they were there."[72] At the time of the 2005 South Asia earthquake, a professional British USAR specialist recounted that "One such team turned up at the Margala Towers without liaison. They posed for pictures, carried out television interviews and then left. This sort of action by a theoretically specialized rescue team does nothing to gain the local people's confidence."[73] Moreover, many teams did not come sufficiently supplied to be self-sufficient, thereby placing further burdens on an already stressed local supply chain.

At the 1989 UNDRO meeting examining the Armenian quake response, it was clear that teams needed "to be fully self-sufficient, well qualified and properly equipped."[74] The 1991 International Red Cross and Red Crescent Report also concluded that "quality screening" had to become a major priority[75] while the 2002 UNGA Resolution 57/150 urged assisting states "to take the necessary measures

[71] See for example, Michael Durkin, "The Survival of People in Collapsed Buildings/Earthquake in Mexico, September 19 and 20, 1985," *Disaster Chroniclse*, n.d.: 32–43; Eric Noji, et al., "The 1988 Earthquake in Soviet Armenia: A Case Study," *Annals of Emergency Medicine* (1990): 891–897 at 894; Osamu Kunii, Masumi Akagi, and Etusko Kita, "The Medical and Public Health Response to the Great Hanshin-Awaji Earthquake in Japan: A Case Study in Disaster Planning," *Medicine and Global Surivival*, (1995): 214-226 at 223; Anthony G. Macintyre et al., "Survival Interval in Earthquake Entrapments: Research Findings Reinforced During the 2010 Haiti Earthquake Response," *Disaster Medicine and Public Health Preparedness*, 2011: 13–22.

[72] Morris, "Coordinated Use," 141.

[73] Arju Katoch, "The Responders' Cauldron: The Uniqueness of International Disaster Response," *Journal of International Affairs*, (Spring/Summer 2006): 167.

[74] Office of the United Nations Disaster Relief Coordinator, UNDRO/89/20 Third Meeting of Officials in Charge of National Emergency Relief Services, Final Report (1989).

[75] Walker, "International Search and Rescue Teams," 28.

to ensure that international urban search and rescue teams under their responsibility are deployed and operate in accordance with internationally developed standards as specified in the Guidelines of the International Search and Rescue Advisory Group, particularly concerning timely deployment, self-sufficiency, training, operating procedures and equipment…"[76]

Upon a unanimous recommendation by USAR team leaders at their annual meeting in 2005, which was later unanimously passed at all three INSARAG Regional Groups and the Steering Committee the same year, INSARAG launched an international peer-based evaluation process referred to as the International External Classification (IEC).[77] Previously, teams "self-classified" themselves when submitting their materials to be included in the international USAR directory relied upon by OCHA.[78] The Guidelines identify three classifications: "Light"; "Medium"; and "Heavy." The categories vary depending on a number of elements, including the location of victims requiring USAR, building material involved, level of medical expertise required, ability to mobilize quickly, required length of continuing operations, self-sufficiency, scope and sophistication of equipment, and depth of personnel expertise.

The "Light" designation applies to first responders who have the ability to assist victims lightly trapped or close to the surface and who are typically reached in the hours right after the event takes place. These rescuers on average are either from the country impacted or from neighboring states. USAR teams who meet the requirements of the "Medium" or "Heavy" classification must demonstrate best practices in five areas: management, logistics, search, rescue, and medical. Personnel in these categories are trained to respond to highly complex and technical rescues. The highest classification of "Heavy" means that the team already contains expertise in locating the "deeply entombed" in structures with reinforced steel; possesses the ability to search at two sites simultaneously; has built-in capacity to locate victims with both search dogs and technology; has proficiency in both heavy rigging and lifting; and is able to conduct continuous twenty-four-hour operations and be self-sufficient for up to ten days at two different sites. In addition, they must have the capability to be on-site and operational within forty-eight hours.[79]

The 2005 decision to implement the peer-review classification process was partially an attempt to sift out search and rescue teams who would simply replicate local capacity. As long-time UN humanitarian affairs official Arjun Katoch explains, "It simultaneously established a process for mutual assessment of international USAR team operational capabilities and made it clear that light USAR teams should not respond internationally as they lack adequate operational capability."[80] The Classification Guidelines also mandate that those national teams

76 UNGA Res. 57/150, par. 5.
77 "INSARAG External Classification/Reclassification Guidelines." (2011). http://ochanet.unocha.org/p/Documents/INSARAG%20Guidelines%202011-Latest.pdf.
78 Ibid. 79 Ibid., 8–10. 80 Katoch, "The Responders' Cauldron," 167.

that have achieved the medium or heavy classification will also participate in the UN OCHA coordination mechanisms, including the OSOCC, the RDC processes, and be a part of, or collaborate with, the UNDAC teams. Further, the teams take on the responsibility to respect the authority of the Local Emergency Management Authority.[81]

As of 2013, more than forty teams from twenty-seven countries had voluntarily participated in the rigorous international peer review process, with the first "heavy" classification achieved by a Hungarian team in 2005.[82] The IEC system is again considered an innovative process within the larger relief field, a domain where no actual training or expertise is required for a government or group to respond to an international disaster. In other words, literally anyone can show up and identify themselves as a disaster responder. As INSARAG Chairman Toni Frisch wrote in an introduction to the latest edition of the IEC guidelines, "To this very day it remains a truly unique process that establishes verifiable operational standards and a humanitarian example of how independent peer review can be a great added value in response preparedness and at the times of response."[83]

Achieving a medium or heavy INSARAG classification is a significant point of pride for USAR teams around the globe that has elevated national, regional, and international capacity for urban search and rescue assistance in a remarkable period of time. Yet the system has caused unease within some corridors because smaller, highly professional non-governmental USAR assistance has been nearly completely shut out in favor of large, government-affiliated USAR teams.[84] Another challenge is that at this time there is still simply no way to prevent an ill-equipped and ill-prepared group from arriving and potentially deploying in a disaster-impacted country. During the 2010 Haiti earthquake, as an example, two private South African groups arrived with inadequate vaccinations, food, clothing, management, equipment, and physical fitness, who then also ended up requiring assistance from responders.[85]

In a 2006 interview project with international USAR personnel from more than twenty different countries, mostly government affiliated respondents expressed concern over "a lack of policing and rules" that would instill a greater sense of discipline within transborder USAR assistance.[86] It had been, in fact, the hope of the very first participants at the 1991 Beuggen meeting and later of INSARAG to have "an international Convention on the Principles and Procedures for urban SAR operations."[87] This last major pillar of needed coordination, a formal

[81] 'INSARAG External Classification/Reclassification Guidelines,' 10.
[82] "INSARAG External Classification (IEC)." n.d. www.insarag.org/en/iec/iec-leftmenu.html.
[83] "INSARAG External Classification/Reclassification Guidelines," 3.
[84] Morris, "Coordinated Use," 159; and Okita "UNOCHA's Coordination," 63–65.
[85] Daniel J. van Hoving, "Haiti: The South African Perspective," *The South African Medical Journal*, (2010): 513–515.
[86] Morris, "Coordinated Use," 139.
[87] United Nations Office for the Coordination of Humanitarian Affairs. "International Search and Rescue Response: Meeting of International Search and Rescue Team Leaders on Lesson

commitment by sending and receiving states in the area of international USAR assistance, may be where the exceptionally accomplished INSARAG has met its match.

COORDINATION THROUGH CODIFICATION? INSARAG AND IDRL

At first glance it may seem extraordinary that a highly specialized global web of search and rescue practitioners – a mixture of public emergency response officials that may easily be from a local firefighting unit as well as NGO and volunteer responders – would succeed in having the UN General Assembly unanimously pass a resolution recognizing their efforts, as the group achieved in December 2002 with Resolution 57/150. But the reality is that INSARAG's membership is more complex. Throughout its history, INSARAG successfully secured "buy-in" from both responding and disaster-impacted states related to its activities – the UNGA resolution itself was sponsored by Turkey, as the result of its own experience in the devastating 1999 Marmara earthquake[88] – and official recognition by both the UN and the IFRC, each playing instrumental roles in the push for its passage.[89] As a 2012 study commissioned to evaluate INSARAG's governing structure refers to it, INSARAG has "two souls": "This idea basically consists of recognizing that INSARAG is both a 'group,' as a gathering of member states with an interest in USAR activities, and a 'network,' as a gathering of practitioners and experts from governments, NGOs, voluntary associations and other sources."[90]

As the result of both Resolution 57/150 and the Guidelines and Methodology, INSARAG's efforts have been recognized as part of the diverse corpus of IDRL. In the defining survey conducted by David Fisher at the IFRC in 2007, IDRL possesses no central legal reference point, but rather consists of "a number of relevant instruments, including some multilateral treaties, a multitude of bilateral agreements between states (as well as between states and international organizations), and an important number of resolutions, declarations, guidelines, and models issued by various authoritative bodies."[91] According to Fisher's desk study, INSARAG's soft law contributions through 57/150 and the Guidelines have gone far to enhance the level of comfort among disaster-impacted states related to response

Learnt and Follow Up to the 1999 Turkey and Taiwan Earthquakes." Neuhasen, Germany, 9–12 December.

88 *Official Records*, UNGA, 57th session, 75th Plenary Meeting, December 16, 2002, A/57/PV.75: 7. International Disaster Response Law Project Update. (Archives: January 2003), In *International Disaster Response Laws, Rules and Principles Programme*, http://w3.ifrc.org/what/disasters/idrl/archives/articles/2003/jan2003.asp.

89 Ibid.

90 INSARAG Steering Group Meeting. "Annex H: INSARAG Governance Discussion Study Paper." February 2012. www.insarag.org/en/global-structures/steering-group.html.

91 David Fisher, "Law and Legal Issues in International Disaster Response: A Desk Study." *International Federation of Red Cross and Red Crescent Societies*, (2007): 15.

initiation, the transborder mobility of goods and equipment (particularly in relation to search dogs), and quality and accountability.[92] And though states are not legally obligated to comply with the Guidelines, the global attention brought about by Resolution 57/150's passage may have encouraged a greater pool of states to take the standards seriously, according to the account of an UNDAC team leader following the 2003 Algerian earthquake:

> That process of winning agreement on the UN resolution meant that awareness and understanding of the issues involved in search and rescue were raised from a practitioner's level to that of foreign ministry in many countries. This contributed to a far better readiness to plug into the coordination system set up in Algeria, and made the job of the Algerians in coordinating international assistance easier...[93]

Dug Cubie captures it similarly when examining the debate surrounding soft law (which he reterms "nonbinding normative standards") and hard law and international humanitarian assistance more generally:

> For those attempting to provide international humanitarian assistance and protection to persons affected by disasters, the wider recognition and acceptance that an agreed body of humanitarian principles and performance standards have at an inter-State level, the stronger their persuasive value when appeals for access to victims need to be made.[94]

Despite its measure of success in this regard, INSARAG's original intention since 1991 was to secure an actual convention related to USAR. In 1999, a year that witnessed three devastating earthquakes in Turkey, Greece, and Taiwan, a core group of nearly thirty states sought to establish a legal framework related to search and rescue activities, and met four times between 2000 and 2002 to draft a protocol. However, the quest to codify INSARAG's work in the area of coordination and standardization via a treaty was abandoned in favor of a more realistic push to obtain a General Assembly resolution.[95]

No doubt one of the major contentious pieces was framing a set of responsibilities for both assisting states and impacted states following a sudden-onset disaster. Using Resolution 57/150 as a guide, governments sending USAR assistance are expected to recognize the sovereignty of the disaster-affected state in the provision of assistance, to respect the host government's authority, and to provide

[92] Ibid., 91, 113, and 139.

[93] Quoted in Nick Carter, "Legal Protection for Disaster Victims," *The Magazine of the International Red Cross and Red Crescent Movement* (2003) www.redcross.int/EN/mag/magazine2003_4/14-15.html.

[94] Dug Cubie, "An Anaylsis of Soft Law Applicable to Humanitarian Assistance: Relative Normativity in Action?" *International Humanitarian Legal Studies*, (2011): 214.

[95] Office of the Coordination of Humanitarian Affairs (OCHA), International Search and Rescue Advisory Group. "Report on the Meeting of the INSARAG Regional Group of the Americas," May 2–5, 2002, www.insarag.org/end/regional-groups/americas/meetings-americas/105-regional-group-meeting-2002.html; Fisher, "Law and Legal Issues," 28.

USAR aid that is timely, technically appropriate, and self-sustaining while in country. However the expectations for the host state are far more sweeping. The resolution:

> *Urges* all States, consistent with their applicable measures relating to public safety and national security, to simplify or reduce, as appropriate, the customs and administrative procedures related to the entry, transit, stay and exit of international urban search and rescue teams and their equipment and materials, taking into account the Guidelines of the International Search and Rescue Advisory Group, particularly concerning visas for the rescuers and the quarantining of their animals, the utilization of air space and the import of search and rescue and technical communications equipment, necessary drugs and other relevant materials.[96]

This one paragraph alone touches upon most of the larger international legal lightning rods related to transborder disaster response, particularly as they correspond with facilitation of assistance. The right of entry remains fully within the national domain. A number of sectoral-specific conventions touch upon admission of relief to some degree, including those for customs[97] and transport.[98] More centrally, the Tampere Convention focuses on telecommunications resources during disaster relief operations.[99] However the treaties in question or relevant annexes tend to have very low levels of state participation or have language that is overly general or less compulsory in nature.[100]

INSARAG officials are even reporting a newer form of sovereign push back, despite the group's stellar reputation in the global humanitarian community. Over the past few years, disaster affected countries in need of international USAR aid have begun to increasingly resist INSARAG's "open invitation" model, in which the Geneva-based office determines the ultimate roster of responding teams by matching needs with the specialized expertise available globally. Instead, receiving governments are preferring to handpick which teams (and more pointedly, which countries) will be permitted to participate.[101]

From the perspective of treaty making and international USAR assistance, INSARAG – after two decades of forward momentum in other areas – has in all likelihood reached the outer boundaries of its influence. It had legal room to run as long as the network of field practitioners remained focused on highly technical

[96] UNGA Res. 57/150 at par. 3.

[97] See e.g., The 1973 Kyoto International Convention on the Simplification and Harmonization of Customs Procedures at its 1999 Protocol of Amendment; and the 1990 Istanbul Convention on Temporary Admission. It should be noted that the latter contains sections related to animal import relevant to the entrance of search dogs.

[98] See e.g., The 1944 Chicago Convention on International Civil Aviation and the 1965 Convention on Facilitation of International Maritime Traffic.

[99] 1998 Tampere Convention on the Provision of Telecommunication Resources for Disaster Mitigation and Relief Operations

[100] Fisher, "Law and Legal Issues," 40–45.

[101] Ibid.

problem solving as expressed in both the Guidelines and the External Classification system. This is primarily for two reasons; the first related to the functional nature of negotiations between search and rescue officials. Referring to horizontal networks operating in the financial and telecommunications fields but also applicable in this context, Jenia Iontcheva Turner writes: "...networks are most likely to arise in response to cross-border problems whose solutions depend largely on technical expertise rather than on political judgment...The less politicized the issue, the more likely participants are to agree on the policies to be pursued through transgovernmental cooperation."[102] The second element is one of autonomy. While USAR specialists developed operational guidelines and team classifications for their field, they were generally left alone to do their work as they saw fit.

Anne-Marie Slaughter, writing from the perspective of transgovernmental networks of domestic regulatory agencies, could have been referring to the activities of INSARAG when describing the agencies' use of nonbinding memorandums of understanding: "These agreements embrace principles that can be implemented by the regulators themselves...Widespread use of Memorandums of Understanding (MOUs) and even less formal initiatives has sped the growth of transgovernmental interaction exponentially, in contrast to the lethargic pace at which traditional treaty negotiations proceed."[103] On the other hand, depending on the context, networks – through formation of soft law – may serve as a "stepping stone" to an international treaty because: "Networks...may smooth the negotiation of a treaty by creating convergence in principles, policies, and rules."[104] Yet in INSARAG's case, while a technical convergence has been largely achieved, the larger political consensus regarding the domestic facilitation of international relief has far to go not only in the field of USAR but for all sectors involved in disaster response. More recent developments involving INSARAG may indicate that the group has chosen another path in terms of confronting the wall of state sovereignty, working from within the country rather than standing on the outside.

CONCLUSION

The year 2010 was a significant year for INSARAG on two major fronts. First, the USAR response to the devastating earthquake in Haiti was commended for one of its highest numbers of rescues, with 132 individuals pulled out of the rubble. Eight teams participating were IEC-classified while another eight were partaking in the review process. This was the highest number of IEC teams involved since the

[102] Jenia Iontcheva Turner, "Transnational Networks and International Criminal Justice." *Michigan Law Review*, (2007): 992.

[103] Anne Marie Slaughter, "The Accountability of Government Networks," *Indiana Journal of Global Legal Studies* (2001): 359.

[104] Kal Raustiala, "The Architect of International Cooperation: Transgovernmental Networks and the Future of International Law," *Virginia Journal of International Law* (2002): 86.

system was put into place in 2005.[105] Nine months later, another milestone was achieved when INSARAG held its inaugural "Global Meeting" in Kobe, Japan. It was the first time that all three INSARAG regional groups came together in a single forum, and did so during the fifteenth anniversary year of the Kobe earthquake and ahead of INSARAG's own commemoration of two decades in existence in 2011. Two hundred participants attended from seventy countries and six organizations.[106]

At the conclusion of the historic gathering, the participants adopted the Hyogo Declaration on "Recognition and Strengthening of International Urban Search and Rescue Operational Standards."[107] The Declaration reaffirmed many of the themes important to INSARAG throughout its operational history including urging assisting states to send only the most qualified teams, requesting states to abide by the Guidelines and Methodology and encouraging more countries to join the international external classification process. There were new goals as well, but one of them in particular stands out: the introduction of a national, rather than external, classification system of USAR teams.[108] According to one INSARAG official, this latest initiative "makes the rules of the game from the grassroots rather than a top down treaty, incorporating methodology into national laws."[109] INSARAG has recognized from the outset, that no matter how quickly the most efficient and well-equipped teams can mobilize from abroad, the challenges of globe-spanning distances and recalcitrant governments uneasy about external involvement cannot easily be overcome. In the end it remains a truism that a local community after a disaster will need to take the lead to save itself. While after twenty years of effort the INSARAG is beginning to see the early fruits of its labor in the area of international USAR assistance, it is now time to turn its attention to the extraordinary goal of national capacity building over the next decades, so that the need for international USAR itself might even become a thing of the past.

[105] INSARAG. "2010 Haiti Earthquake Response: An After Action Review of Response." n.d. Available at www.alnap.org/pool/files/printed-haiti-book-low-definition-version.pdf.

[106] INSARAG Global Meeting. "Chairman's Summary." (September 15–16, 2010).www.insarag.ord/images/stories/Documents/Global_Meeting/2010_INSARAG_Global_Meeting_Chairman_Summary_English.pdf.

[107] INSARAG. "Hyogo Declaration from the First Global Meeting of the International Search and Rescue Advisory Group." (September 14–16, 2010) www.advant.se/insarag/images/stories/v2_IGM_Declaration_final_20100916_r1.pdf.

[108] Confidential interview by author. [109] Ibid.

PART III

The Right of Access to International Disaster Relief

FIGURE 6 **Pakistan Disaster Relief Mission during 2010 catastrophic floods in Khyber Pakhtunkhwa Province**. Source: U.S. State Department Image, August 10, 2010.

8

Legal Framework Applicable to Humanitarian Actors Responding to Disasters in Weak and Fragile States

Catherine Gribbin and Ilario Maiolo

Traditionally, international human rights law imposes on states the obligation to protect and care for victims of disasters. Rights like the right to life, dignity, food, and shelter are the cornerstones of a state's obligation to respond to the needs and protect the victims of the disasters from being victimized again by an inadequate, inappropriate, or unjust response. However, more and more, disasters are occurring in areas of the world where a state's authority and capacity to act is weakened or in some cases nonexistent. International humanitarian organizations are often called on to supplement, or in some cases, outright substitute essential public humanitarian services. Recently, the term "republic of NGOs" has been coined to describe situations where humanitarian organizations are effectively running the basic infrastructure necessary to save the lives of victims, thereby assuming the basic elements of a state's responsibility.

In these situations, what obligations does international law place on humanitarian agencies and groups? Standards for humanitarian response have been primarily developed by humanitarian agencies in an effort to self-regulate. Some of the only hard law standards that can be effectively imposed on humanitarian agencies working in such context are laws from their home state that primarily deal with financial management and accountability to certain extraterritorial dispositions of a criminal nature. This chapter examines whether there is a gap in legal protection for victims, whether current efforts at self-regulation are adequate, and whether possible solutions imposing minimum standards are achievable.

INTRODUCTION

The past two decades have seen an exponential increase in the number of nongovernmental organizations (NGOs),[1] with more than 3,500 currently recognized

[1] Maryann Zarnegar Deloffre, "NGO Accountability Clubs in the Humanitarian Sector: Social Dimensions of Club Emergence and Design," in *Voluntary Regulation of NGOs and*

by the UN Economic and Social Council alone.[2] A significant number of these humanitarian actors (HAs) engage in disaster relief work. This increase, combined with increased linkages among HAs around the world, has given the international HA community growing legitimacy, might, and authority in terms of the delivery of humanitarian aid during times of emergency.[3] This chapter argues that the Guidelines for the Domestic Facilitation and Regulation of International Disaster Relief and Initial Recovery Assistance (IDRL Guidelines)[4] provides an excellent starting point for a binding legal framework that can provide more guidance and more accountability for HAs to fulfill their obligations toward their beneficiaries as well as the obligations that they owe to the country in which they are operating.[5]

There is a need for a binding legal framework as HAs have come to increasingly represent the privatization of public interest as more and more services traditionally provided by the state are now being provided by HAs in post-disaster settings.[6] This is especially true during times of overwhelming need, such as in the aftermath of natural disasters or when HAs must coordinate with, and in, weak or failed states particularly where populations have come to rely on the assistance provided by HAs. In some places, HAs now increasingly operate in contractual environments[7] providing more services in more countries today than ever before,[8] including everything from education and healthcare to water and shelter.[9]

In addition, it is often naively said that HAs do not need rules and laws in order to do good work in difficult situations because they are motivated and guided by altruistic principles and staffed with "good people." While the work of HAs is extremely valuable, it has become apparent over the last several years and after thousands of humanitarian relief operations that HAs independently operating in desperate situations, already exacerbated by a lack of government leadership in the prevention of such harm, may unintentionally harm those they intend to help, with few accountability mechanisms available to the state or the beneficiaries who may suffer as a result of this harm.

Nonprofits, eds. Mary Kay Gugerty and Aseem Prakash (New York: Cambridge University Press, 2010), 169.

2 Havrda, Marek, and Petr Kutilek, "Accountability 2.0 – In Search for a New Approach to International Non-Governmental Organistations' Accountability." In *Evalutating Transnational NGOs: Legitimacy, Accountability, Representation*, by Jens Steffek and Kristina Hahn, (New York: Palgrave MacMillan, 2010): 157–59.; UN NGO Branch, Department of Economic and Social Affairs. http://csonet.org/index.php?menu=14.

3 Ibid., 159–60.

4 "International Disaster Response Law Guidelines," *International Federation of Red Cross and Red Crescent Societies* www.ifrc.org/PageFiles/41203/1205600-IDRL%20Guidelines-EN-LR%20(2).pdf.

5 For the purposes of the paper we will focus particularly on HAs originating from Canada.

6 Jem Bendell, *Debating NGO Accountability*. New York and Geneva: United Nations Non-Governmental Liason, 2006; Havrda and Kutilek, "Accountability 2.0," 157–60.

7 See Havdra and Kutelik at 157–159.

8 Bendell, *Debating NGO Accountability*, 9.

9 InterAction, *Haiti Aid Map*.+65 (2011), http://haiti.ngoaidmap.org/.

Due to their presence in countries and regions with fragile governments or governments with very limited resources, HAs have become increasingly relied on around the world. In these evolving conditions, HAs increasingly operate in contractual environments:[10] from education and healthcare to water and shelter, in many parts of the world HAs are providing more services in more countries today than ever before.[11] Haiti, for instance, clearly exemplifies the shift in service provision away from the state and toward the international community of HAs after a large disaster. The majority of HAs operating in Haiti since the 2010 earthquake have been providing traditionally state-supplied services in sectors such as education, health, water sanitation, housing, economic recovery, disaster management, human rights, and democracy development.[12]

When a state provides basic services to its citizens, rules, social norms, and laws are meant to govern their provision. In fragile states, however, such rules, norms, and laws either do not exist or simply cannot be enforced due to a lack of resources and will. Yet this lack of framework affects all service providers operating within a country including HAs. It is not the proposal of this chapter that HAs be subject to the same laws and rules as the state, but rather that there is a fundamental need for an overarching set of laws that will both improve the accountability of HAs and ensure the quality of the provision of humanitarian assistance. And as the state is the only entity in a position to create a legal means by which to hold actors accountable, it is its responsibility to do so.

INTRODUCTION OF "HUMANITARIAN ACCOUNTABILITY"

HA accountability has been a growing topic of debate over the last twenty years. The discussion has ranged from the application of anti-terrorism legislation and anti-corruption laws to HAs, to the professionalization of staff of HAs, and the need for increased transparency within the community. This debate has led to calls from within the HA community for the improvement of regulations and the need to have more meaningful consequences for those who do harm to others during a humanitarian response by improving all aspects of accountability within the sector.

Yet few accountability frameworks (e.g., the Sphere Project, Humanitarian Accountability Partnership [HAP]) actually result in either legal rights for the beneficiaries, or legal obligations or substantive standards of care on the part of the agencies or professionals involved in a response, as these frameworks are voluntary, not legally binding, and have no real oversight or enforcement mechanisms. Due to the history and functionality of HAs, there have been few binding

[10] Zarnegar Deloffre, Maryam, ", NGO Accountability Clubs in the Humanitarian Sector: Social Dimensions sector: social dimensions of Club Emergence and Design, in Mary Kay Gugerty and Aseem Prakash (eds.), *Voluntary Regulation of NGOs and Nonprofits* (New York: Cambridge University Press, 2010), 169.

[11] Bendell, *Debating NGO Accountability*, 9.

[12] Interaction, *Haiti Aid Map*.

regulatory regimes governing the delivery of humanitarian services. Simultaneously, the majority of states have often not created regulatory frameworks that address the needs of populations in times of emergency, relief, and recovery. Due to this gap, in the majority of states, HAs often are left to identify and implement best practices on their own.[13] This chapter will look at what laws currently make up the regulatory framework of HAs and point out the gaps in the current system.

It was within this vacuum that the idea of International Disaster Response Law (IDRL) was formed. The goal of the advancement of IDRL is to create comprehensive regulatory legal frameworks relating to the provision of emergency relief by both governments and non-state actors in domestic and international settings.[14]

It is with this in mind that this chapter proposes the advancement of IDRL in order to provide a framework for the creation of legal rights and obligations vis-à-vis states and HAs that, in turn, create better regulatory frameworks and provide better guidance, oversight, and accountability during disaster relief. A positive step in this direction would involve implementing the IDRL Guidelines because this would also mean that the HA community is no longer solely responsible for monitoring its own actions; state actors with this regulatory framework will be able to hold HAs to more account.[15]

EXISTING LEGAL FRAMEWORKS AFFECTING HAS

Currently, most of the legal guidance an HA receives from states relates to the regulation of their finances. States usually classify HAs as charities, which makes them subject to domestic laws that can regulate their expenses, funding allocations, provision of services through partners, and accounting practices. Additionally, many international regulations relating to charities have also been focused on finances and the provision and spending of money. With that said, let us first look at the overarching legal frameworks that apply during times of humanitarian response.

International Law

The most obvious and relevant legal framework in the international domain governing disaster relief and recovery assistance is international human rights law (IHRL) and, when applicable, international humanitarian law (IHL). Humanitarian agencies have long used IHRL to inform their efforts. This approach has evolved into a rights-based approach to the delivery of humanitarian assistance. For example, the International Federation of Red Cross and Red Crescent

[13] For further discussion on national and international frameworks, see the International Federation of Red Cross and Red Crescent Societies, "Summary and conclusions of existing frameworks," in *Law and legal issues in international disaster response: a desk study*, (2007): 85.

[14] "International Disaster Response Law Guidelines."

[15] A discussion on the effect of the IDRL Guidelines on states is outside the scope of this paper.

Societies (IFRC) in its comments regarding the International Law Commission's (ILC) progress on its project concerning the Protection of Persons in the Event of Disasters, the International Red Cross and Red Crescent Movement's (the Movement)[16] tradition is "adherence to an approach to disaster response that is based on needs but informed by rights."[17] Much work has been done recently to develop clear sets of guidelines for HAs based on human rights. The International Development Law Organization in *International Law and Standards Applicable in Natural Disaster Situations*[18] developed what is probably the most comprehensive overview of rights, which can in turn be used as guidelines to assist HAs in responding to disasters.

These guidelines cover all aspects of IHRL in disasters; for example, the right to a legal personality and the role documentation can play in the relief process and access to services.[19] Nonetheless, they remain an overview and are not binding but rather are a guide for programmatic response.[20] Although scholars like Andrew Clapham[21] argue that in a certain respect non-state actors including HAs may be obliged to respect IHRL in armed conflict and in some respects can be held to account, the Westphalian reality remains that as a matter of law human rights obligations are still primarily binding on states.

IHRL can affect the provision of relief services provided by a state if the state has adopted IHRL conventions or if the conventions have become customary international law. However, IHRL does not comprehensively affect the work of HAs as most rights recognized under IHRL relate to the actions of the state[22] but not to the actions of individual, private actors such as HAs. Although there have been some developments in the law (for example the resurgence of the use of the *Alien Tort Claims Act* in the United States), the relationship between private actors and IHRL is still determined on a case-by-case and jurisdiction-by-jurisdiction basis. Law that binds all HAs to an international human rights standard in their provision of services is by no means universal.

This is because most HAs, unlike states, do not have an international legal personality and can only be held liable in jurisdictions where the domestic law

16 The International Red Cross and Red Crescent Movement refers to the collective thoughts and actions of the International Federation of Red Cross and Red Crescent Societies, the International Committee of the Red Cross and the 187 Red Cross and Red Crescent National Societies.

17 United Nations General Assembly, IDRL: Protection of *Persons in the Event os Disasters (Statement by Marwan Jilani)*. October 30, 2009. www.ifrc.org/en/news-and-media/opinions-and-positions/speeches/2009/idrl-protections-of-persons-in-the-event-of-disasters/

18 Erica Harper, *International Law and Standards Applicable in Natural Disaster Situations* (Rome: International Development Law Organization, 2009).

19 Ibid., 27.

20 Ibid., 19.

21 Andrew Clapham, *Human Rights Obligations of Non-State Actors* (New York: Oxford University Press, 2006).

22 Henry J. Steiner, Philip Alston, and Ryan Goodman, *International Human Rights in Context*, 3rd. ed. (New York: Oxford University Press, 2007).

captures their offending activities and where the state has the ability to enforce the law.[23] Although standards like those set out by Sphere or the *International Law and Standards Applicable in Natural Disaster Situations*[24] may mention that HAs have an obligation to respect IHRL, the obligation remains more of an ethical or moral nature rather than a clear legal one. In the same vein, in a recent report from the International Council on Human Rights Policy, which looked at the responsibilities of Human Rights NGOs under IHRL, the language used to refer to the obligation was primarily that of principles and moral responsibility rather than law.[25] As was stated at the beginning of the chapter, it is not enough to presume that HAs because of their altruistic mandate will always do "what is right;" a moral or ethical obligation is not enough to hold HAs to an adequate standard that will protect the interests and rights of beneficiaries. Rather, it is with the creation of a binding set of rules that the international community can then in turn expect a certain level of care to be delivered.

IHL applies to all situations of armed conflict and only applies during an armed conflict. IHL lays out the limited legal requirements for those involved in the conflict regarding the provision of humanitarian assistance during times of armed conflict. It is recognized under customary IHL that the use of starvation of the civilian population as a method of warfare is prohibited.[26] Although IHL lists what specific allowances exist related to agencies providing relief,[27] IHL includes only the very minimum of what provisions should be permitted.[28] What IHL does not specifically mention are laws governing what happens to an agency if it does not meet this basic criteria, nor what happens to the agency if it intentionally

[23] *Rights and Responsibilities of Human Rights Organisations*, 2009, par. 88, in Archives of the International Council on Human Rights Policy. www.ichrp.org/files/reports/67/119_report.pdf. United Nations General Assembly. *IDRL: Protection of Persons in the Event os Disasters (Statement by Marwan Jilani, Permanent Observer of the IFRC to the Sixth Committee of the United Nations)*. October 30, 2009. www.ifrc.org/en/news-and-media/opinions-and-positions/speeches/2009/idrl-protections-of-persons-in-the-event-of-disasters/.

[24] Harper, *International Law and Standards*.

[25] Steiner, Alston, and Goodman, *Rights and Responsibilities of Human Rights Organisations*, 153 par. 84.

[26] Customary International Humanitarian Law; edited by: Jean-Marie Henckaerts & Louise Doswald-Beck, International Committee of the Red Cross (Cambridge University Press: Cambridge, 2005), Rules 53–55.

[27] For an international armed conflict, for example, see article 30 of the Fourth Geneva Convention.

[28] The right to receive assistance during an armed conflict and the obligation on a party to a conflict to allow aid the reach civilians finds its roots in the customary rule prohibiting the use of starvation as a method of warfare. Rules 53–55 of the Customary Law Study found that when read in totality, the limitation on the use of starvation as a method of warfare entailed the obligation to allow assistance to reach the civilian population. It is from this prohibition that the right to receive aid and the obligation on parties to a conflict to allow assistance to reach the effected civilian population arises.

violates or fails to act on any of the specific allowances.[29] Moreover, IHL does not apply during times of peace and so is of limited application.[30] Although IHL does provide some legal guidance, it is very thin in regards to the expectations and obligations that are required by the agencies involved in the provision of humanitarian relief. Between the application of IHRL and IHL to the work of HAs, there remain many gaps due to this lack of specificity and clarity.

In addition, under international law, several other treaties directly and indirectly affect the work of HAs. An important example is the recent UN International Convention for the Suppression of the Financing of Terrorism[31] that came about as a result of a growing international dialogue regarding the need to cut off financial support to terrorist groups. The convention includes specific reference to the role played by organizations that claim charitable status and support terrorist groups[32] and has resulted in increased scrutiny of charities, their work, and finances. While this source of international law deals with certain issues that may implicate how HAs finance themselves, the Convention does not include a comprehensive and clear legal framework for HAs finances. HAs are therefore once again left with vague definitions and unclear obligations while states are left to organize domestic means by which to regulate and enforce somewhat vague and uninstructive language of the Convention.[33]

Domestic Law of Beneficiary Country

The second legal framework that applies is the domestic law of the jurisdiction in which the HA is operating. Domestic laws vary from country to country and largely depend on a country's ability and propensity to create and then regulate legal frameworks. In fragile states, both the existence of actual laws as well as any means of legally enforcing them is often lacking. Oftentimes, due to a range of issues,

29 Agencies who undertake to provide relief under these provisions could be held responsible for the provision of their services via contractual terms (i.e. if there is a remedy for breaches and the contract enunciated the expected level of the provision of the services). But there is no obligation laid out in IHL on the contractor to create terms in the contract to meet a particular level.

30 It is important to note that IHL only applies during times of armed conflict as per articles 2 and 3 of the Four Geneva Conventions. Therefore if a conflict is over the rules relating to the provision of assistance during times of armed conflict no longer apply and IHRL (to the exclusion of IHL) applies.

31 UN *International Convention for the Suppression of the Financing of Terrorism*. www.un.org/law/cod/finterr.htm.

32 The sixth paragraph of the Preamble of the Convention recalls the General Assembly Resolution which recognizes the role played by groups who have or claim to have charitable, social or cultural goals who through direct or indirect funding finance terrorists and terrorist organizations. Article 2 of the Convention lists the offences which due to the nature of charitable organizations are directly affected by the creation of this crime.

33 Article 18, *Convention*. www.un.org/law/cod/finterr.htm.

these legal frameworks are not the equivalent to Canadian (or other HAs' countries of origin) legal frameworks; again, particularly in fragile states.

But what is consistent from preliminary research done by the IFRC is that most states have not yet started to analyze their domestic legal frameworks from a disaster response needs lens. For example, although the Canadian Income Tax Act (ITA) does not require that charities comply with laws in foreign jurisdictions, being registered in Canada does not exempt a charity from the laws in the jurisdiction in which they operate.[34] Because of this explicit reference to other countries' jurisdictions, there is an assumption that domestic laws are broad enough to offer some form of protection. It is oftentimes this basic assumption that leads to the misunderstanding that the safety and protection that basic laws should provide actually exists. Therefore, this lack of analysis regarding the law leaves both potential beneficiaries and HAs without comprehensive coverage or guidance.

Domestic Law of Country of Origin

The third legal framework applicable to the work of HAs is the domestic law of the HA's country of origin. This, too, varies by country. Some countries actively support the idea that they have extraterritorial jurisdiction to prosecute crimes committed overseas by their own citizens while other states are less inclined.[35] In Canada, most domestic laws have not been created with the intention to apply extraterritorially. As will be discussed later, due to the nature of the work and the difficult environments in which HAs often work, it is unreasonable to think that a country of origin's regulatory bodies will have the ability, resources, and time to follow up on allegations stemming from work being conducted outside of its borders. In what follows we will look at some specific laws and regulations as they apply to Canadian HAs and their staff working abroad and demonstrate that the current domestic approach is a piecemeal one.

Finances: the Income Tax Act (the ITA)

In Canada, the ITA creates several rules for HAs when they operate as a charity.[36] The ITA applies regardless of where a Canadian registered charity operates and includes several laws regarding charities' activities. Most activities deemed charitable in Canada are, as a matter of Canadian law, considered charitable abroad. Canadian courts have noted that there are exceptions to this general rule. One such

[34] Canada Revenue Agency, *Canadian Registered Charities Carrying Out Activities Outside Canada: Guidance (Reference Number: CG-002)*, July 8, 2012. www.cra-arc.gc.ca/chrts-gvng/chrts/plcy/cgd/tsd-cnd-eng.html

[35] A recent Canadian development of this trend is the ability for the Canadian police and prosecution services to charge and try Canadians involved in child prostitution overseas: *Criminal Code*, R.S.C., 1985, c. C-46, section 7(4.1).)

[36] *Income Tax Act*, R.S.C., 1985, c. 1, 5th Supp.

exception is that increasing the effectiveness and efficiency of Canada's Armed Forces is considered charitable while any support to another country's armed forces is not.[37]

A charity's purpose and activities must not violate officially declared and implemented Canadian policy.[38] Canadian policy is subject to change according to the government in office and, as such, Canadian HAs must ensure that the purpose and activities they engage in are not in violation of officially declared policy. In addition, all charities must meet the public benefit test, which requires that the public benefit of a charity's activities substantially outweighs any likely harm to the charity's staff, the beneficiaries of its program, or any other person.[39] Factors considered in the assessment of the net public benefit include the likelihood and nature of harm, the urgency of need for assistance, the HA's experience, and its proposed risk mitigation measures.[40] Although the public benefit test is a seemingly powerful tool, it is important to note that the most frequent use of the text is to determine to legitimacy of the organization's mandate; not to analyze the standard to which the charity is performing. And again, this text only relates to whether or not the charity receives certain benefits on their taxes.

The ITA also creates rules regarding the use of resources including funds, personnel, and property.[41] A registered charity can only use its resources in two ways, whether inside or outside Canada, by either carrying out its own charitable activities or through an intermediary that is subject to certain criteria while the Canadian charity must still direct and control the activities carried out on its behalf and the use of its resources.[42]

The consequences of not meeting these rules is that the HA can be subject to sanctions including financial penalties and/or the revocation of its registered status. Since few foreign organizations are qualified donees (agencies that meet the ITA criteria), most Canadian charities operating outside of Canada do so in the form of actively delivering their own programming.[43]

37 Canada Revenue Agency. *Canadian Registered Charities Carrying Out Activities Outside Canada: Guidance*.

38 Canada Revenue Agency. *Summary Policy CSP-P13: Public Policy*. www.cra-arc.gc.ca/chrts-gvng/chrts/plc/csp/csp-p13-eng.html.; *Canadian Magen David Adom for Israel v. Canada (Minister of National Revenue)*, 2002 FCA 323 (Federal Court of Appeal, September 13, 2002). http://decisions.fca-caf.gc.ca/en/2002/2002fca323/2002fca323.html.

39 Canada Revenue Agency. *Guidelines for Registering a Charity: Meeting the Public Benefit Test (Reference Number: CPS-024)*. August 28, 2012. www.cra-arc.gc.ca/chrts-gvng/chrts/plcy/cps/cps-024-eng.html.

40 Income Tax Act, R.S.C., 1985, c. 1, 5th Supp.

41 Income Tax Act, s. 149.1(1).

42 *The Canadian Committee for the Tel Aviv Foundation v. Canada*. 2002 FCA 72; *Canadian Magen David Adom for Israel v. Canada (Minister of National Revenue)*. 2002 FCA 323 (Federal Court of Appeal, September 13, 2002)HYPERLINK "http://decisions.fca-caf.gc.ca/en/2002/2002fca323/2002fca; *Bavit Lepletot v. Canada (Minister of National Revenue)*. 2006 FCA 128.

43 Canada Revenue Agency. *Charities in the Interantional Context*. www.cra-arc.gc.ca/chrts-gvnv/chrts/ntrntnl-eng.html.

Criminality: Criminal Code (the Code)

Another overarching set of legal obligations can be found in criminal law. Criminal law (but for a limited number of exceptions) applies within the jurisdiction of the state and typically does not apply to acts that take place in other countries. For example, Canadians who commit offenses while in another country will not be charged in Canada for committing the crime unless it is one of the offenses listed as an exception to this general rule.[44] It should be noted that the exceptions were not written with the accountability of HAs during times of natural disasters in mind. So according to Canadian criminal law, Canada will not charge Canadians with offenses that were committed overseas (so long as they are not one of the exceptions listed in the Code as mentioned previously), which leaves only the beneficiary country with the ability to press charges for criminal acts committed on their territory.

The most robust laws in Canada in regards to the nexus between HAs and criminal law relate primarily to financial accountability and monetary crimes. The Charities Registration (Security Information) Act enacted under the Anti-terrorism Act (not the Code) states that a charity's status may be revoked if its resources are made available to a listed terrorist group or a group that is known to be involved in terrorist activities. It should be noted that information not otherwise publically available can be used for this determination and not in an open court.[45] This act created a test for charities to confirm their non-involvement in contributing to funding terrorist groups which is: are there reasonable grounds to believe, based on criminal and security intelligence reports, that an organization makes its resources available to these groups? If it is determined that a charity is believed to make its resources available, then it could lose its status and face criminal charges. The Canada Revenue Agency has produced a checklist to help Canadian charities identify these types of vulnerabilities which could make HAs vulnerable to terrorist abuse.[46]

Professional Oversight: Disciplinary Actions

The deployment of humanitarian relief is accompanied by the professionals who know how to make the best use of these supplies. HAs often deploy professionals from a range of professions (from electricians to water and sanitation specialists to doctors) to do the work required in a relief effort. There are a range of these professions who are authorized, under law, to self-regulate and it is these regulations that are the last source of authority this chapter will address. Various

[44] C. C-46, ss 6(2) & 7.

[45] *Anti-terrorism Act*, S.C. 2001, c. 41 and the *Charities Registration (Security Information) Act*, S.C. 2001, c. 41, s. 113, in *Department of Justice Canada*. September 15, 2013 http://laws-lois.justice.gc.ca/eng/acts/A-11.7/page-1.html#s-1.

[46] Canada Revenue Agency. *Checklist for Charities on Avoiding Terrorist Abuse*. July 20, 2010. www.cra-arc.gc.ca/chrts-gvng/chrts/chcklsts/vtb-eng.html.

professional bodies have been given statutory authority to regulate their own members. In Canada, provincial legislation most often deals with the regulation of various professions and grants authority to the individual associations to deal with disciplinary measures related to breaches of the professional codes of conduct. As a result, the means, methods, and scope of each regulatory body is different.

This regulation of professions in Canada therefore grants various regulatory bodies jurisdiction. In some cases this jurisdiction is quite far reaching and includes the legal right to consider an individual's conduct, to proceed with hearings, and render decisions.[47] The jurisdiction of each body is dependent on the relevant legislation but it is generally conceded that most professional regulatory bodies have implicit or explicit jurisdiction over the conduct of members when they act or work in other provinces and therefore one might suggest that the regulatory bodies operate without general territorial limitations.[48] The Saskatchewan Court of Appeal held that a legislative provision giving the engineering profession the ability to review conduct that takes place both within and outside the province was within the purview of the provincial legislature. In that case, the provision conferred the power to deal with an individual's conduct outside of Saskatchewan in order to determine fitness for licensing within the province.[49] Though related to external activities, the case indicated the court's willingness to grant jurisdiction to the regulatory body to look at activities that physically take place outside of its geographical base. The Supreme Court of Canada has also weighed in on the subject and has confirmed that the jurisdiction of the regulatory body is personal and can be extra-territorial in some situations.[50]

ENGINEERS Engineers in Canada are regulated by individual provincial boards. The Professional Engineers Ontario (PEO) has jurisdiction over the acts of its members no matter where they are; members are bound by the Professional Engineers Act,[51] and regulations,[52] and a Code of Ethics wherever they are working.

[47] James Casey, "Jurisdiction of the Tribunal Over the Professiona," in *The Regulation of Professions in Canada* (Toronto: Carswell, 1994), 4.

[48] "In the opinion of this Court, the jurisdiction of the Law Society over its members is a personal one, which extends to the member's conduct without territorial limitation." *Legault v. LSUC*. 8 OR (2d) 585 (1975).

[49] *Underwood, McLellan, and Associates Ltd v. Saskatchewan Association of Professional Engineers, et al.* SJ No. 551 (1979).

[50] *Black v. Law Society of Alberta*. 1 S.C.R. 591 "The Law Society can define the appropriate ethical requirements for its members and discipline them, wherever they may reside. Sanctions can be effective even outside provincial boundaries." This statement by the SCC although in a different context, confirms the jurisdictional authority of a profession over its members can be a personal one and can have extra-provincial reach.

[51] Ontario e-Laws. *Professional Engineers Act, R.S.O. 1990, CHAPTER*. www.e-laws.gov.on.ca/html/statutes/english/elaws_statutes_90p28_e.htm.

[52] Ontario e-Laws. *Professional Engineers Act*, R.R.O. 1990, REGULATION 941. www.e-laws.gov.on.ca/html/statutes/english/elwas_statutes_90p28_e.htm.

Although not always easy to investigate complaints arising from acts in other jurisdictions, the PEO is able to do so.

Members of the Association of Professional Engineers and Geoscientists of British Columbia (APEGBC) are also bound by a Code of Ethics[53] and the provincial Engineers and Geoscientists Act[54] wherever they go. As the regulatory body, the APEGBC also has the ability to investigate complaints arising from conduct occurring abroad. The Professional Engineers and Geoscientists of Newfoundland and Labrador (PEGNL) members are also bound by a provincial act and regulations[55] and a code of ethics when working abroad while the PEGNL Board can investigate complaints arising from instances outside of Canada.[56]

PHYSICIANS Doctors in Canada are also subject to the jurisdiction of their profession's provincial regulatory bodies, each of which have authority over actions that take place outside of the province or territory regardless of whether they are related to the individual's practice.[57] This authority includes the ability to investigate and discipline members for incidents that occur outside of their territorial jurisdiction.[58]

Traditionally, licensing authorities prosecute the matter within their own judicial setting although, in some provinces, the case is turned over to the provincial Crown.[59] The consequences of a criminal conviction in most cases allows for the regulatory body to revoke the license of the member[60] while the actual facts surrounding a complaint to a professional licensing authority can also be raised in a civil action where damages can be awarded.[61]

In Ontario, the College of Physicians and Surgeons can investigate complaints from foreign jurisdictions concerning Ontario-registered doctors. Under the provincial Regulated Health Professions Act, the College may take actions it decides are necessary in order to protect the public interest as a result of complaints, criminal, disciplinary, or any other proceeding, against a member from any jurisdiction whether inside or outside of Canada, relating to the applicant's competency, conduct, or character.[62] This includes the ability to find that a member has committed an act of

53 Association of Professional Engineers and Geoscientists of BC. Engineers and Geoscientists Code of Ethics. http://www.apeg.bc.ca/resource/publications/actbylawscode.html.

54 BC Laws. Engineers and Geoscientists Act, RSBC 1996, CHAPTER 116. http://www.bclaws.ca/EPLibraries/bclaws_new/document/ID/freeside/00_96116_01.

55 House of Assembly of Newfoundland and Labrador, *An ACT Respecting the Practice of Engineering and Geoscience, CHAPTER E-12.1.* 2008. http://www.assembly.nl.ca/legislation/sr/annualstatutes/2008/e12-1.c08.htm.; *Engineers and Geoscientists Regulations, Regulation 43/11.* 2011. www.assembly.nl.ca/legistlation/sr/annualregs/2011/nr110043.htm.

56 Professional Engineers and Geoscientists Newfoundland and Labrador, *Code of Ethics By-law*, 2008. www.pegnl.ca/documents/bylaw3_final%20of%20Ethics%20May%2013%202010.pdf.

57 *Halsbury's Laws of Canada: Regulation of Health Professionals.* 1st ed. Vol. 22. (Markham, Ontario: LexisNexis Canada, 2008).

58 Ibid., 1. 59 Ibid., 4. 60 Ibid., 11. 61 Ibid., 13.

62 Regulated Health Professions Act, S.O. 1991, Chapter 18. www.e-laws.gov.on.ca/html/statutes/english/elaws_statutes_91r18_e.htm.

professional misconduct if a governing body of the health profession in another jurisdiction finds that the member committed an act of professional misconduct.[63]

In British Columbia, doctors are bound by the Canadian Medical Association's Code of Ethics, which applies no matter where members are located.[64] Under the provincial Health Professions Act,[65] the College of Physicians and Surgeons of British Columbia's disciplinary committee may take action against a member if a licensing body in another province or jurisdiction finds that the member committed an act that, in the opinion of the discipline committee, constituted unprofessional conduct under the provincial act.[66]

All of these individual regulatory boards, statutes, and regulations leave us somewhat unsatisfied, as not all the regulations are complementary and the drafters of the legislation were not necessarily thinking of regulating situations of overseas disaster response when they were writing the legislation. In addition to the legal gaps that exist, the regulatory bodies have limited financial means and human resources to collect and hear evidence related to breaches of professional misconduct abroad. This leaves the regulation of professionals overseas in a less than robust position. It is with the advancement of IDL that this weakness in the accountability gap could be addressed.

Self-regulation and Voluntary Standards

As mentioned earlier, in the 1990s, HAs started working on a number of initiatives aimed at enhancing their accountability and improving the consistency of responses from the HA community. This need to address accountability within the HA sector arose due to a number of reasons: a failure to respond adequately in the wake of the Rwandan genocide; the exponential increase in the number of HAs; the surge in amount and types of funding for humanitarian activities; the participation of non-traditional actors such as militaries and private contracting firms in aid delivery; and a rise in media attention on HA activities.[67] The increase in the number of people voicing concerns regarding HA accountability has also been linked to the increase of the financial and political impact of the sector.[68] There was a simultaneous realization within the HA community that it was no longer acceptable to act without addressing the need for accountability within its own sector.[69] As a result, initiatives

[63] Ibid., Chapter 18, s. 51 (1)(b).)

[64] Canadian Medical Assocation, *CMA Code of Ethics*, 2004. http://policybase.cma.ca/dbtw-wpd/PolicyPDF/PD04-06.pdf..

[65] British Columbia, *Health Professions Act, RSBC 1996, Chapter 183*, www.bclaws.ca/EPLibraries/bclaws_new/document/ID/freeside/00_96183_01.

[66] Ibid., Chapter 183, s. 39.1(1)(a).)

[67] Maryam Zarnegar Deloffre, "NGO Accountability Clubs in the Humanitarian Sector: Social Dimensions of Club Emergence and Design" in *Voluntary Regulation of NGOs and Nonprofits* (Cambridge University Press: 2010).

[68] Balam Singh Aswal, *An Introduction to NGO Accountability: Issues and Challenges* (New Dehli: Cyber Tech Publications, 2011), 189–208.

[69] Ibid.

such as Sphere and HAP were developed and are now thought of as the most well-respected attempts to improve HA accountability. These initiatives are relied on today to provide HAs with the basics concerning what should be expected from them in their aid delivery.

In spite of these initiatives, it is well known within the HA community that there is no overarching organizing body that regulates, oversees, and polices the activities of HAs. Rather, there exists a set of voluntary codes (many listed below) which set out the best practices, goals, and rules that HAs *should* abide by. It is widely acknowledged within the HA community that not all HAs abide by these best practices and there is no way to ensure that HAs are doing what standards they should be meeting according to these best practices. Yet in spite of these failings, the Sphere Project, HAP, and the *Good Enough Guide*, for example, do provide tremendous leadership in the field. Following are the notable initiatives that provide guidance meaningful guidance to HAs.

Name of initiative	Goal and governing framework
Code of Conduct for the International Red Cross and Red Crescent Movement and NGOs in Disaster Relief	A code of conduct for individual agencies to sign and agree to abide by.[70] HAs are under no obligation to sign the code, as it is voluntary, but several hundred have. HAs commit themselves to a certain standard of conduct to which no other global document alludes.
SPHERE Project and Humanitarian Charter	A document that includes specific (qualitative) expectations of HAs in different types of relief operations, such as minimum sizes for homes in shelter operations and minimum food distributions in food security settings.[71] Only document of its kind for the HA community. Without it as a reference, HAs are at complete liberty to define minimum standards of a humanitarian response. The Charter calls on HAs to abide by four basic protection principles: • avoid exposing people to further harm as a result of your actions; • ensure people's access to impartial assistance; • protect people from physical and psychological harm due to violence or coercion; and, • assist with rights claims, access to remedies and recovery from abuse

(continued)

[70] "The Code of Conduct for the International Red Cross and Red Crescent Movement and NGOs," in *Disaster Relief* www.ifrc.org/Docs/idrl/I259EN.pdf.

[71] *The Sphere Project*, www.sphereproject.org/.

(*continued*)

Name of initiative	Goal and governing framework
Good Humanitarian Donorship (GHD)	An initiative driven by an informal donor forum and network which works to facilitate the collective advancement of GHD principles and good practices. The group works to advance the Principles and Good Practice of Good Humanitarian Donorship which aim to enhance the coherence and effectiveness of donor action, as well as their accountability to beneficiaries, implementing organizations and domestic constituencies, with regard to the funding, coordination, follow-up and evaluation of humanitarian responses.
The Financial Action Task Force (FATF)	An intergovernmental body responsible for developing and promoting policies to combat money laundering in an attempt to limit financing to terrorist organizations.[72] FATF includes thirty-four countries, territories, and organizations, and has issued a set of recommendations to combat terrorist financing.[73] Task Force highlights the fact that nonprofit organizations are particularly vulnerable, and countries should ensure that they are not misused.[74]
Principles for Good International Engagement in Fragile States & Situations	A set of principles drafted by the OECD which is a set of international voluntary standards specific to humanitarian response.[75] Of particular importance and relevance in fragile states.
Joint Standards Initiative	Created, with support from the Active Learning Network for Accountability and Performance in Humanitarian Action, to bring about greater coherence amongst standards.[76]
Emergency Capacity Building Project (ECBP)	A project initiated by the directors of several leading HAs to identify major obstacles to humanitarian assistance.[77] The project has published a report on

(*continued*)

72 FATF, www.fatf-gafi.org/

73 FATF Recommendations. In *Financial Action Task Force*, www.fatf-gafi.org/topics/fatfrecommendations/documents/internationalstandardsoncombatingmoneylaunderingandthefinancingofterrorismproliferation-thefatfrecommendations.html.

74 FATF Guidance document on the misuse of NPO. *Combating the Abuse of Non-Profit Organizations*. In *Financial Action Task Force* www.fatf-gafi.org/media/fatf/documents/recommendations/11%20FATF%20SRIX%20BPP%20SRVIII%20October%202003%20-%20COVER%202012.pdf.

75 "Principles for Good International Engagement in Fragile States and Situations," *Principles* (April 2007). www.oecd.org/dac/incaf/38368714.pdf.

76 *Joint Standards Initiative*. www.jointstandards.org/.

77 *Emergency Capacity Building Project*. www.ecbproject.org/the-project/theproject.

(*continued*)

Name of initiative	Goal and governing framework
	emergency capacity[78] and launched phase two of the project to improve the "speed, quality and effectiveness of humanitarian assistance."[79] The *Good Enough Guide: Impact Measurement and Accountability in Emergencies* also stems from the ECBP.
Task Force on Accountability to Affected People (created by the Inter-Agency Standing Committee)	To encourage operational HAs to incorporate the five Commitments to Accountability to Affected People into their guidelines and policies.[80] An increasing number of donors are expected to make commitments to the IASC guidelines as part of the terms of their donor contracts.

The HAP is another attempt by the HA community to self-regulate but particularly in regard to improving quality management by improving accountability to beneficiaries. Recently, the HA community has been particularly focused on this subset of HA accountability (i.e., "beneficiary accountability") – that is, an HA's level of accountability to those whom it attempts and claims to assist. According to the Canadian Red Cross, beneficiary accountability is "an approach to aid delivery that aims to effectively meet the needs of beneficiaries, reduce the possibility of mistakes, fraud or abuse in our delivery and promote sustainability building on and strengthening the capacities of local communities." The move to improve "downward accountability," as beneficiary accountability is sometimes referred to, is particularly important given that HAs are historically more focused on accountability toward their more powerful stakeholders – that is, their funding sources. This historical focus is apparent as most HAs have more rules dealing with donor rights – in the form of donor rights manifestos and legal contracts with funders, which often specifically include the obligations an HA has toward meeting the expectations and rules set out by the funder – than with beneficiaries' rights.

None of the standards listed previously are strictly binding, yet they have been a powerful force within the HA community while it attempts to improve the responses of all agencies involved in disaster response. In weak or fragile states, these standards are often the only guidance an HA has in order to inform and guide

78 "ECBP Report on Emergency Capacity," in *Emergency Capacity Building Project* www.ecbproject.org/downloads/resources/emergency-capacity-report-ver-july-14-final.pdf.

79 *Combating the Abuse of Non-Profit Organizations.*

80 Inter Agency Standing Committee, *IASC Task Force on Accountability to Affected Persons.* www.humanitarianinfo.org/iasc/pageloader.aspx?page=content-subsidi-common-default&sb=89.

its work. And it is the responsibility of the HA to be aware of these voluntary standards and implement them accordingly. There is no global regulatory requirement for HAs to implement any of these standards.

In summary, HAs rely on an ad hoc network of legal regulation and voluntary guidance to inform their work. It is the ad hoc nature and lack of binding guidance that leaves both beneficiaries and HAs at great risk during their humanitarian responses.

COMPARING TRANSNATIONAL CORPORATE REGULATION

There are other instances of attempts to regulate autonomous, internationally operational entities. For example, there are hundreds of private, intergovernmental, and multi-stakeholder initiatives in the world of corporate social responsibility (CSR) that focus on transnational and multinational corporations. These include legal regulations and voluntary standards adopted by corporations or governments in order to promote best practices in corporate culture. Due to the size and scale of corporate interest around the world, there have been several influential and very well written documents that form the basis of many global CSR initiatives.[81]

There is a consistent theme, though, in discussions regarding. Most CSR initiatives and guidelines are based on voluntary enrollment and compliance, and include few binding enforcement mechanisms. It has been recognized in CSR discussions that it is domestic and international law that actually binds transnational corporations; the same can obviously also be said for intergovernmental organizations.[82] If a law does not exist that can hold an entity accountable for a particular grievous action, there are few non-legal meaningful ways by which to hold the entity responsible.

As previously mentioned, international HAs similarly struggle with structurally weak accountability mechanisms. This weakness has historically also been reinforced and magnified by a lack of any type of formal mechanism imposing accountability and the ad hoc nature of the laws that apply to them. It is important to note that, this absence of structural accountability does not mean that NGOs are less accountable.[83] For example, the NGO sector, in comparison with other international corporations, leads on consulting with less powerful stakeholders like its beneficiaries before, during, and even after it has undertaken project work. In recent years, engaging with beneficiaries has become almost standard practice even

[81] These include the UN Global Compact, OECD Guidelines for Multinational Enterprises, and the ILO Tripartite Declaration of Principles Concerning Multinational Enterprises and Social Policy.

[82] OECD, "OECD Guidelines for Multinational Enterprises," (2011): 17. www.oecd.org/corporate/guidelinesformultinationalenterprises/48004323.pdf.

[83] Aswal, *Introduction to NGO Accountability*, 194.

during the most difficult humanitarian emergency responses, in spite of the difficulties that arise from doing so.[84]

Enforcement

Simply put, individual states have jurisdiction over their own territory, acts committed on their own territory, and their own citizens. And, as is the case with all law, international law usually relies on the ability of the state to enforce it. Most legal systems and international legal treaties rely on this domestic implementation, which includes the creation of domestic enforcement mechanisms. Domestic laws, in turn, also rely on domestic enforcement regimes – from regulatory bodies to policing. Regulatory bodies and their corresponding sector codes rely on the resources that regulatory bodies put toward oversight, investigations, enforcement, and sanctions. The regulatory bodies also rely on the interplay between any proximate criminal and civil enforcement mechanisms that apply to particularly grievous misconduct.

The effectiveness of both laws and their enforcement mechanisms rely on the ability of states to provide the resources and leadership to be able to create coherent and well-functioning systems. If a state is weak or failing, the systems will not be created and little protection will be afforded to those who are in most need. There is a parallel that can be drawn to HAs, and the resources they allocate toward ensuring that they themselves are abiding by both their own rules and the laws of the countries in which they operate (including international and country of origin law). If HAs do not allocate sufficient resources, time, and energy to this facet of their work, then, they, too, will fall short of meeting the protection needs of those with whom they work, as well as fail to meet other legal, ethical, and professional responsibilities. So when one considers that the majority of rules governing the delivery of services provided by HAs are guiding documents and have little binding authority, one can see the difficulty associated with ensuring compliance.

Solutions to Current Gaps

The framework regulating the conduct of humanitarian actors primarily rests (1) at an international level on non-binding self-regulation, and (2) at the domestic level, in the Canadian context, on disparate regulations of a fiscal, criminal, and, to a lesser extent, administrative nature. Few if any binding regulations exist in terms of ensuring as a matter of law that clear enforceable standards are applied, in particular when responding to an emergency in a fragile state.

In many ways, the efforts of the humanitarian community to develop standards of self-regulation have been quite remarkable. Standards like the Sphere Project have

[84] Ibid.

allowed humanitarian actors to establish a global set of professional standards, which are now a staple in the humanitarian community. Many times donors, whether private or public, will ask that such standards or other humanitarian standards – such as the Inter-Agency Standing Committee guidelines and principles on humanitarian activities or the 1994 Code of Conduct for the International Red Cross and Red Crescent Movement and Non-Governmental Organizations (NGOs) in Disaster Relief – be respected as conditions for funding to humanitarian agencies.[85] This has greatly contributed to the professionalization of the humanitarian community and has increased to an extent the accountability of humanitarian agencies toward both their donors and, ultimately, their beneficiaries. But as this is not yet a universal requirement being advocated for by all donors, there is still quite a bit of work to do to fill these accountability gaps.

Humanitarian actors are called upon to work in situations that are extremely delicate and complex, making even the Sphere standards, though voluntary, onerous and difficult for humanitarian agencies to fully comply with due to the unimaginable plethora of issues HAs face on the ground. In many circumstances, in particular large-scale responses, it has been difficult for humanitarian agencies to adhere fully to some of the guidelines that state quantitative expectations. This has notably been a problem in terms of various health responses,[86] water and sanitation responses,[87] and shelter responses.[88]

In spite of the success of initiatives such as Sphere, the HA community has recognized that it is difficult to set universal objective standards for humanitarian responses. Indeed, as Greikspoor and Collins put it: "The Sphere handbook defines minimum service standards from the perspective of beneficiaries. However, the assumptions behind these standards, such as unhindered access and adequate resources, are rarely met during large scale humanitarian emergencies. Such constraints restrict the effectiveness of all humanitarian interventions."[89] The issue of unhindered access and adequacy of resources is outside the scope of this paper but it is a critical corollary discussion that HAs must engage in and also wrestle with.

This is one reason why simply crystallizing the standards as binding, either through national or supranational law, would prove difficult. Legislating

[85] Good Humanitarian Donorship, *GHD Intiative 2003, 23 Principles and Good Practice (Principle 16)*. www.goodhumanitariandonorship.org/gns/principles-good-practice-ghd/overview.aspx .

[86] André Griekspoor and Steve Collins, "Raising Standards in Emergency Relief: How Useful are Sphere Minimum Standards for Humanitarian Assistance?" *British Medical Journal* (2001): 740–742.

[87] Jean-Francois Pinera and Lee Bosher, "Sphere and Sustainability: A Matter of Time," *The Journal of Humanitarian Assistance* (January 2011). http://sites.tufts.edu/jha/archives/874.

[88] International Federation of the Red Cross and Red Crescent Societies, *An Evaluation of the Haiti Earthquake 2012 Meeting, Shelter Needs: Issues, Achievements and Constraints*. http://wwww.ifrc.org/docs/Evaluations/Evaluations2011/Global/HTShelterClusterReview11.pdf.

[89] For the purposes of the paper we will focus particularly on HAs originating from Canada. Griekspoor and Collins' "Raising Standards in Emergency Relief," 742.

deliverable quantitative objectives would be unrealistic. One can, however, establish binding processes and principles by which humanitarians should abide. This has already been done by some HAs in some of their responses, as noted above, contractually – usually by donor states who have signed the Good Humanitarian Donorship initiative. The contractual integration of these processes and principles helps ensure that the services and their provision meet particular standards; this is an attempt to provide a "guarantee" of sorts to the beneficiaries regarding the assistance which they are receiving. Principles for the delivery of humanitarian aid can indeed be anchored in law and be made binding. In fact, at the international level the International Court of Justice, in the Nicaragua case,[90] has already issued a definition of sorts for humanitarian aid:

> In the view of the Court, if the provision of "humanitarian assistance" is to escape condemnation as an intervention in the internal affairs of Nicaragua, not only must it be limited to the purposes hallowed in the practice of the Red Cross, namely "to prevent and alleviate human suffering" and "to protect life and health and to ensure respect for the human being;" it must also, and above all be given without discrimination to all in need in Nicaragua, not merely to the *contras and* their dependents.[91]

Although not applicable to HAs in all responses, the decision sets out principles that have long been key in guiding humanitarian agencies as to the manner in which relief is to be administered and could be used as an anchor in the legal debate. Due to the specific requirements of each humanitarian response and the corresponding difficulty in creating quantitative standards (as has, in spite of this difficulty, been done by the Sphere standards), these standards are inadequate to ensure adequate disaster relief as the process by which the relief is handled, administered and dispersed is of equal importance. To address just the quantitative standards without addressing the questions related to the flow and distribution of aid is to miss a large part of the debate. Currently, HAs use the principles laid out in the Nicaragua decision to guide their decision making in an attempt to bridge these regulatory gaps. At the international level, there are a number of current initiatives focused on these overlapping issues.

THE INTERNATIONAL LAW COMMISSION

Most notably, in 2007, the International Law Commission (ILC) began work on draft articles on the topic of Protection of Persons in the of Disasters.[92] Since then,

[90] *Case Concerning* Military and Paramilitary Activities in and against Nicaragua. (*ICJ v. United States of America*), International Court of Justice, Rep. 27 June 1986.

[91] Ibid., par. 243

[92] United Nations, "Text and Titles of Draft Articles 1,2,3,4, and 5 as Provisionally Adopted by the Drafting Committee," UN Doc A/CN.4/L.758; United Nations, "Text and Titles of Draft Articles 6, 7, 8 and9 Provisionally Adopted by the Drafting Committee on 6, 7, and 8 July

the ILC has produced five reports and drafted fifteen draft articles. The primary focus of the draft articles is to outline the obligation of a state affected by a disaster by crystallizing a duty to seek assistance[93], the requirement for the affected state to consent to assistance[94], and, in general terms, the role of the affected state to direct control and coordinate assistance.[95] As a departure from other State-negotiated documents, the draft articles also include and outline a role for and obligations of humanitarian actors. Draft Article 5 sets out not only a duty for states to cooperate with each other, the United Nations, and other competent intergovernmental organizations, but also with the members of the International Red Cross and Red Crescent Movement.

Additionally, draft Article 6 reiterates the obligation to respond to disasters in accordance to the principles of humanity, neutrality, and impartiality. Although draft Article 6 does not mention its application to intergovernmental organizations or HAs (unlike draft Article 7, which specifically names and compels such actors to respect the inherent dignity of the human person) the "Report of the ILC to the UN 6th Committee" commenting on the articles[96] seems to suggest that this article also might encompass HAs by stating that such principles "were characterized as important safeguards for the relationship between relevant actors"[97] and not states alone. This language would provide more well-defined guidance within the realm of international law in terms of articulating norms governing the manner in which humanitarian agencies would have to conduct themselves. Such measures would arguably help fill, although not entirely or in detail, part of the current normative gaps.

Nonetheless, at this stage, the articles remain in draft form and the project has yet to post any comments from states. Additionally, it is unclear what final form the draft articles will be proposed in. On this point the IFRC made itself clear: "If this text will indeed result in a new set of nonbinding guidelines, then it makes sense for the ILC to avoid re-inventing existing instruments of a similar nature, like the Guidelines for the domestic facilitation and regulation of international disaster relief and initial recovery assistance (also known as the IDRL Guidelines), which were adopted by the State Parties to the Geneva Conventions in 2007. However, if

2010," UN Doc A/CN.4/L.776 ; United Nations. "Text and Titles of Draft Articles 10 and 11 Provisionally Adopted by the Drafting Committee on 19 July 2010." UN Doc A/CN.4/L.794; United Nations. "Text and Titles of Draft Articles 5 bis, 12, 13, 14 and 15, Provisionally Adopted by the Drafting Committee from 5 to 11 July 2012." UN Doc A/CN.4/L.812.

[93] Ibid., Art. 10.

[94] Ibid., Art. 11 and 13. (The duty to seek assistance refers to the duty of the state to ask for assistance from the international community. The duty to consent refers to the duty of the state to accept assistance from the international community, for example, by allowing them to enter into the country to deliver the aid.)

[95] Ibid., Art. 9.

[96] *Report of the ILC to the UN 6th Committee* A/65/10, Chapter VII.

[97] Ibid., 317.

it is intended as a basis for an eventual treaty, then it is important that key aspects of existing international instruments be reflected in this new text as well."[98]

THE IDRL GUIDELINES

The other initiative at the international level focusing on getting the right aid at the right time is the previously mentioned IDRL Guidelines.[99] As Gostelow puts it: "Size doesn't matter in effective humanitarian response: timing, access and competence do."[100] Timing, access, and competence of humanitarian response are key pillars in the IDRL Guidelines.

Though nonbinding, the IDRL Guidelines are meant to create a set of guidelines for governments on how to prepare for disasters by enabling states to adopt laws and plans for common problems in large-scale relief operations. Although the IDRL Guidelines are primarily designed to ensure that aid is able to effectively flow to those who need it the most, they also pave the way for a more legally robust approach to ensuring humanitarian accountability.

The difference between the IDRL Guidelines and other initiatives in terms of humanitarian accountability is that they are designed to engage states in developing national disaster planning/management legislation. As such, when implemented by states, they should form a binding set of domestic provisions that increase both the efficiency and accountability of HAs. One of the central elements of the guidelines is that they establish that domestic actors have a primary role and that international relief providers and HAs "be held responsible for abiding by certain minimum humanitarian standards in disaster assistance."[101]

The IDRL Guidelines further encourage the development of national standards to grant qualified humanitarian agencies legal facilities and ensure that visas are granted without delay to qualified humanitarian personnel, including foreign medical professionals, architects, engineers, and other types of licensed professionals. Having effective procedures in place to both facilitate the arrival of humanitarian personnel and ensure their qualifications increase HAs' overall accountability including toward its beneficiaries. The IDRL Guidelines in many ways are the most overarching approach toward decreasing the accountability gap and ensuring

[98] Comments on the International Law Commission draft articles on protecting persons in the event of disasters. Statement by Ms. Elyse Mosquini of the IFRC Delegation to the United Nations, at the Sixth Committee of the United Nations General Assembly, in New York. *Comments on the International Law Commission Draft Articles on Protecting Persons in the Event of Disasters*. October 29, 2010. www.ifrc.org/en/news-and-media/opinions-and-positions/speeches/2010/comments-on-the-international-law-commissions-draft-articles-on-protecting-persons-in-the-event-of-disasters/.

[99] For the purposes of the paper we will focus particularly on HAs originating from Canada.

[100] Lola Gostelow, "The Sphere Project: The Implications of Making Humanitarian Principles and Codes Work Disasters," *Disasters* (1999): 316–325.

[101] Ibid. "International Disaster Response Law Guidelines."

that broader binding standards exist to regulate humanitarian actors when working in a foreign jurisdiction.

At the global level, the IDRL Guidelines have been referenced many times in the ILC work on the Draft Articles[102] as well as by a number of international and regional organizations. Of particular note are seven UN General Assembly resolutions[103] calling on states to use the IDRL Guidelines to strengthen their domestic legal frameworks applicable in times of disaster. The guidelines have also been the focus of various commonwealth meetings and the World Customs Organization[104]. In fact, as of 2011, nine countries had adopted new disaster management laws drawing on the IDRL Guidelines.[105] In sum, the IFRC, somewhat humbly, in our opinion, has summarized its success so far as a "good start."[106] Although a huge number of states has not adopted laws in line with the IDRL Guidelines, the IFRC has managed to build momentum and bring this important topic to international attention. It would be important for the IFRC and National Red Cross and Red Crescent Societies to continue to focus their efforts on helping to create domestic legal frameworks in fragile states, as it is precisely in these states where the accountability gap is largest and most needed.

And the IFRC has proven thus far that it is not only the substance of the guidelines that can ensure their success but also the approach used by the Movement. This approach relies on National Red Cross and Red Crescent Societies which are uniquely placed in their home jurisdictions to work with their states, supported by the IFRC, to integrate these guidelines into domestic legislation. All National Red Cross and Red Crescent Societies, as a precondition of their admission to the Movement, must be recognized on the basis of national legislation as auxiliary to the public authorities in the humanitarian field.[107] States and the Movement recognized "that the public authorities and the National Societies as auxiliaries enjoy a specific and distinctive partnership, entailing mutual responsibilities and benefits, and based on international and national laws, in which the national public authorities and the National Society agree on the areas in which

[102] United Nations, "Text and Titles of Draft Articles 10 and 11 Provisionally Adopted by the Drafting Committee on 19 July 2010," UN Doc A/CN.4/L.794.; United Nations, "Text and Titles of Draft Articles 5 bis, 12, 13, 14 and 15, Provisionally Adopted by the Drafting Committee from 5 to 11 July 2012," UN Doc A/CN.4/L.812.

[103] UN General Assembly Resolutions A/RES/65/264, A/RES/65/133, A/RES/64/251, A/RES/64/76, A/RES/63/141, A/RES/63/139, A/RES/63/137.

[104] IFRC. "Progress in the Implementation of the Guidelines for the Domestic Facilitation and Regulation of International Disaster Relieft and Intitial Recovery Assistance." *31st International Conference of the Red Cross and Red Crescent*, 2011. www.ifrc.org/PageFiles/41203/IDRL%20progress%20report.pdf

[105] Ibid., 4. [106] Ibid., 4.

[107] IFRC. "Statutes of the International Red Cross and Red Crescent Movement, Article 4.3." *International Committee of the Red Cross*. www.icrc.org/eng/assets/files/other/statutes-en-a5.pdf.

the National Society supplements or substitutes for public humanitarian services."[108] Beyond this operational mandate, States and National Societies committed to maintaining and enhancing a permanent dialogue at all levels within the agreed framework for humanitarian action.

It is precisely this unique status of National Red Cross and Red Crescent Societies that will allow them to be a leading advocate for the implementation of the IDRL Guidelines and other legal solutions in order to create a more robust and binding framework to increase both the effectiveness of aid as well as beneficiary accountability even in failed or weak states.

CONCLUSION

As has been noted and remarked upon for decades, rather than being grounded in consistent and insightful legal frameworks (either domestically or internationally), HA accountability has been shaped by a moral obligation rooted in organizational missions and values. The degree of accountability and quality of the mechanisms HAs use to ensure accountability is different from organization to organization and, as is well documented, underdeveloped.[109] Moreover, the HAs have been left to undertake to negotiate and agree to common standards through codes of conduct, overarching accountability frameworks, and commitments to improve their ability to "do no harm."[110]

It is incorrect to say that HAs are unaccountable or ungoverned, as this suggests that they operate without checks or balances and are not subject to rules, regulations, and laws. Rather, they are accountable but their accountability has historically been skewed toward financial stakeholders and has not been guided by a uniform set of rules.[111] Instead, they have been left to weave their way through a maze of unclear rules and haphazard laws. HAs have attempted to correct these gaps on an individual basis and rely on the professionalism of their own staff and volunteers to ensure the quality of the services provided. But the need for accountability and standardization is present as the environments in which HAs work are open to misconduct.[112] Those who are affected by the HA community are

[108] 30th International Conference 2007: Resolution 2. Specific nature of the International Red Cross and Red Crescent Movement in action and partnerships and the role of National Societies as auxiliaries to the public authorities in the humanitarian field. (06 October 2011). www.icrc.org/eng/resources/documents/resolution/30-international-conference-resolution-2-2007.htm.

[109] Monica Blagescu, Lucy de Las Casas, and Robert Lloyd, *Pathways to Accountability: The Gap Framework* (London: One World Trust, 2005), 16–17.

[110] Ibid., at 17 [111] Ibid.

[112] As has been noted, HAs often work in fragile environments, meaning that rules and oversight are often stretched beyond capacity. This can increase the chances of those in positions of power taking advantage of this lack of regulation.

oftentimes already at a disadvantage because of a recent disaster, or a lack of clarity or understanding regarding rights. These vulnerabilities leave individuals and communities open to neglect and flagrant abuse. With the development of meaningful, clear, and enforceable laws to govern the response of HAs, governments can create conditions that will allow for the prevention of harm and alleviation suffering by allowing beneficiaries to receive meaningful help when they need it most.

9

Disasters, Despots, and Gun-Boat Diplomacy

Catherine Shanahan Renshaw

Cyclone Nargis struck Myanmar's Irrawaddy Delta in 2008, devastating the region, causing 50,000 deaths, and leaving 2.5 million people homeless. The United States and the European Union, chief architects of the harshest international sanctions against the dictatorial regime in Myanmar, offered humanitarian aid. Myanmar's government at first refused, and then delayed and impeded, the delivery of aid. This response confounded the West. There were calls for the United Nations Security Council to invoke the Responsibility to Protect (R2P) to authorize delivery of aid without the consent of Myanmar's government. Australian Prime Minister Kevin Rudd called on the international community to "bash the doors down in Burma" to get critical aid to the people as soon as possible.

This chapter considers disasters, and the Myanmar example, from three perspectives. First, how the international community responds to disasters when they occur within outlaw states. Second, how "pariah" states (or elements within them) deny international overtures of assistance in the aftermath of disasters. Third, how the failures and frustrations of the West's attempt to deliver aid to Myanmar were ultimately facilitated by a regional international organization: the Association of Southeast Asian Nations (ASEAN). Thus, the most effective response to disaster in such situations may come from organizations situated at the regional, rather than the global, level of governance.

INTRODUCTION

The tensions that exist between national laws and politics, and emerging international disaster relief law, is never more patent than when disaster occurs within an "outlaw" state. Using a case study of Cyclone Nargis, which struck military-ruled Myanmar in

2008, this chapter sets out to achieve three things.[1] The first is to chart the rise and rapid fall of R2P as a plausible international response to natural disaster. The second is to explain the reaction of one particular pariah state to international overtures of assistance in the aftermath of disaster. The third is to examine a theory that, for a variety of reasons, the most effective international response to disaster may come from organizations situated at the *regional*, rather than the global, level of governance.

Cyclone Nargis

This catastrophic storm, which made landfall on May 2, 2008, was the worst natural disaster in Myanmar's recorded history. With winds of 160 kilometers per hour and waves as high as four meters, Nargis swept through the Bay of Bengal and into Myanmar's Irrawaddy Delta. The wave surge reached 30 kilometers inland and washed away towns and villages in its path. Bridges, roads, power lines, and water treatment facilities were destroyed or damaged. The nation's telecommunications system was disrupted and the country's primary rice-producing region was laid to waste.[2] The cyclone left 140,000 people dead, 800,000 people homeless, and 2.4 million people in need of emergency aid. In the immediate aftermath of Nargis, lack of food, clean water, and sanitation threatened the lives of hundreds of thousands of people.[3]

At the time, the government of Myanmar was an unelected military junta known as the State Peace and Development Council (SPDC).[4] The military had ruled Myanmar, in one form or another, for forty-six years. In 2005, United States Secretary of State Condaleeza Rice declared that Myanmar was an "outpost of tyranny," along with other "rogue states" such as Iran, Cuba, North Korea, Zimbabwe, and Belarus.[5] The United States and the European Union had for

[1] In 1988, the State Law and Order Restoration Council (SLORC) changed the country's name from Burma to Myanmar. Most countries, as well as the United Nations, refer to the country as "Myanmar." The United States continues to call the country "Burma." It is accepted academic practice in Burma Studies to use "Burma" when referring to events before 1988, and "Myanmar" for the post-1988 period. This chapter follows this practice. See: Dominic Nardi, "Discipline-flourishing Constitutional Review: A Legal and Political Analysis of Myanmar's New Constitutional Tribunal," *Australian Journal of Asian Law* 12(1) (2010), 1n1. On the politics of using Burma / Myanmar, see: Lowell Dittmer, "Burma vs. Myanmar: What's in a Name?" *Asian Survey* 48(6) (2008) 885–888.

[2] "Burma's Rice Region Decimated – Food Shortage Feared," *Associated Press*, May 7 2008. www2.irrawaddy.org/article.php?art_id=11817.

[3] Human Rights Watch, *Burma: One Year After Cyclone Nargis*, (April 30 2009), www.hrw.org/features/burma-one-year-after-cyclone-nargis.

[4] In May 1990, elections were held in Myanmar and a democratic opposition party, the National League for Democracy, won an outright victory. Myanmar's military leaders refused to allow the newly elected representatives to convene and arrested many of them. The military embarked on a long process of drawing up a new constitution, which was not completed until 2008. During this period, the military maintained absolute authority and civil and political liberties were restricted.

[5] "Rice Names Outposts of Tyranny," *BBC News*, January 19 2005. from http://news.bbc.co.uk/2/hi/americas/4186241.stm.

twenty years ostracized Myanmar and pursued a policy of strict sanctions against the nation's military government.

Nonetheless, in the wake of Nargis, the international community immediately offered humanitarian aid, assistance, and relief. It was well understood, from the experience of the 2004 Indian Ocean tsunami, that an immediate and well-orchestrated response was critically important to preventing loss of life from secondary causes such as disease and lack of sanitation. Within days of the cyclone striking, a five-person UN Disaster Assessment Committee was assembled in Bangkok and ready for deployment. Military vessels from the United Kingdom and France, laden with emergency assistance supplies, had been dispatched to the Bay of Bengal.[6] United States planes and navy vessels, already present in the area for the annual Cobra Gold training exercises with Thailand, were diverted to the coastline of Myanmar and stood ready to transport aid and supplies into Myanmar by plane and helicopter. Donations of food and medical equipment were sent from around the world to airports in Bangkok and Dubai, waiting for approval from the government of Myanmar so that they could be flown into Yangon.

Myanmar's government at first refused, and then delayed and impeded, the delivery of aid. Humanitarian workers from the United Nations and other aid agencies were refused visas to enter the country.[7] Aid workers already within Myanmar were unable to obtain the permits required to visit disaster-stricken areas of the country. On May 9, Myanmar's Ministry of Foreign Affairs released a statement saying that "the country was not yet ready to receive search and rescue teams, as well as media teams, from foreign countries." The statement said that the government would accept donations, but that "it would distribute them by its own labours to the affected areas."[8] Ten days later, through the state-run media, Myanmar's rulers said that "the strings attached to the relief supplies carried by warships and military helicopters are not acceptable to the Myanmar people."[9] It was not until May 23 that the government concluded a general agreement to allow humanitarian aid workers to enter the affected areas.[10]

6 "UN supplies begin to arrive in Burma" *The Guardian*, May 8 2008. from www.guardian.co.uk/world/2008/may/08/cyclonenargis.burma4.

7 On May 7, the United Nations Office for the Coordination of Humanitarian Affairs announced that four of the five member of the Disaster Assessment Team had been granted visas to enter Myanmar. However, one member of the team, who was non-Asian, was not been granted a visa.

8 Emma Larkin, *Everything is Broken: The Untold Story of Disaster Under Burma's Military Regime* (London: Penguin Press, 2010), 12.

9 *New Light of Myanmar* (May 22, 2008).

10 "First visas issued to international relief teams," *United Nations Office for the Coordination of Humanitarian Affairs – Integrated Regional Information Networks (IRIN) Myanmar*, 7 May 2008. Most of these visas were issued to aid workers from China, Japan, Indonesia, Laos and Thailand. Once inside Myanmar, relief workers found that their movements were restricted and that the distribution of aid was controlled by the military authorities. Andrew Selth, "Even Paranoids Have Enemies: Cyclone Nargis and Myanmar's Fears of Invasion," *Contemporary Southeast Asia: A Journal of International and Strategic Affairs* 30(3) (2008): 388.

Disaster and the Outlaw State

Most accounts of post-colonial Burma present the nation's history as a sequence of defining historical moments: January 4, 1948 and independence from Britain; the military coups of 1958 and 1962; the August 8, 1988 mass demonstrations and their suppression; the "Saffron Revolution" of 2007; the "democratic" elections of 2010.[11] Through these events, the country's history is most commonly portrayed as a dyadic struggle between democracy (represented by Aung San Suu Kyi and the National League for Democracy) and authoritarianism (the generals of the Tatmadaw, Burma's armed forces, who have ruled Burma in one form or another since 1962). This reading, however, elides two crucial elements of the nation's political history.

The first is the intractable division that exists between Myanmar's majority Buddhist Bamar population, and the ethnic minorities (the Arakanese, Chin, Kachin, Shan, Karenni, Karen, and Mon peoples) who inhabit Myanmar's outlier regions.[12] Praetorianism and the failure of representative democracy, which are the recurrent themes of Burma's post-colonial history, both in certain senses derive from the core problem of attaining national unity in the face of ethnic diversity.[13] Democracy's other impediments – economic underdevelopment, the years of attempted socialism, the absence of the institutions of democracy (an independent judiciary, a free press) – exacerbated the destructive consequences of disunity.[14]

Nargis raised deep fears on the part of the generals about secession and the balkanization of the Union of Myanmar. The disaster-struck region included the ethnic Karen state, where insurgents were involved in a complicated and unresolved sixty-year conflict with the government. Representatives of the political wing of the insurgency movement, the Karen National Union (KNU), had on occasion appealed to the United States for funding and

11 For example, see: Chao-Tzang Yawnghe, "Burma: The Depoliticization of the Political," in *Political Legitimacy in Southeast Asia*, ed. Muthiah Alagappa (Stanford, Cal.: Stanford University Press, 1995):170–92; John B. Haseman, "Burma in 1987: Change in the Air?" *Asian Survey* 28(2) (1998): 223–228; Moksha Yitri, "The Crisis in Burma: Back from the Heart of Darkness?" *Asian Survey* 29(6) (1989): 543–558.

12 Alan Smith, "Burma/Myanmar: The Struggle for Democracy and Ethnic Rights," in *Multiculturalism in Asia*, ed. Will Kymlicka and Baogang He (Oxford University Press, 2005). Alan Saw U, "Reflections on Confidence-Building and Cooperation among ethnic groups in Myanmar: A Karen Case Study," in *Myanmar: State, Society and Ethnicity*, ed. Narayanan Ganesan and Kyaw Yin Hlaing (Singapore: Institute of Southeast Asian Studies, 2007), 219–35.

13 Ashley South, "Political Transition in Myanmar: A New Model for Democratization," *Contemporary Southeast Asia* 26(2) (2004): 233–255. South writes (at 238): "Until 1989, the Myanmar Army had been fighting two inter-connected civil wars – one against the ethnic nationalist insurgents and another against the Communist party of Burma."

14 Robert H. Taylor, "Burma's National Unity Problem and the 1974 Constitution," *Contemporary Southeast Asia* 1(3) (1979): 236.

military support.[15] For Myanmar's generals, the historical parallel to Nargis was the 1970 cyclone that hit East Pakistan – now Bangladesh – and killed between 300,000 and 500,000 people.[16] The government of Pakistan's inadequate response stimulated widespread indignation and contributed to the Bangladesh independence struggle a year later.[17] From the point of view of Myanmar's rulers, the 1970 Bangladesh cyclone presaged catastrophic domestic consequences: the secession and partition of the country.

In this context, one of the government's overriding concerns in the aftermath of Nargis was whether and how the disaster might affect the fate of the constitutional referendum, scheduled for May 10. The referendum was an essential part of the seven-step "roadmap" toward democracy, announced by Prime Minister Khin Nyunt in 2003. According to the roadmap, the referendum was to be followed by: (i) holding free and fair elections; (ii) convening the parliaments; (iii) "building a modern, developed and democratic nation."[18] While it was anticipated that this last stage might take some time, it was viewed as imperative, if peace was to be achieved, that the concrete steps of the roadmap be met according to schedule.[19]

From the regime's point of view, therefore, there was nothing incongruous about the government-run newspaper, *The New Light of Myanmar*, giving equal front page coverage to stories about the cyclone and also to stories about preparations for the referendum.[20] Indeed the first edition of the paper to be published after the storm contained on its front page a response from the government to the United Nations Security Council's statement about Myanmar's forthcoming constitutional referendum.[21] The response expressed "surprise" at the Security Council's interest in the referendum, and confirmed that the referendum would occur in a few days' time. The article stated that the people were eagerly looking forward to the referendum and that the government was continuing to implement its seven-stage road map toward

15 "Burma: A KNU Perspective On Ethnics, Peace Talks Origin Consulate Chiang Mai (Thailand)," (September 25, 2009) *Wikileaks Cable* 2009 09:06 UTC. http://wikileaks.org/cable/2009/09/09CHIANGMAI136.html.

16 "Bangladesh Profile,"' *BBC News South Asia*, (July 16, 2013). www.bbc.co.uk/news/world-south-asia-126514832013.

17 Palash Ghosh, "Hurricane Watch: 1970 Cyclone in Bangladesh Killed 500,000," *International Business Times*, August 22, 2011. www.ibtimes.com/hurricane-watch-1970-cyclone-bangladesh-killed-500000-302837.

18 H-E. U Khin Maung Win, Deputy Minister for Foreign Affairs of the Union of Myanmar "Myanmar Road to Democracy: The Way Forward," *Seminar on Understanding Myanmar, MICT Park, Yangon, 27–28 January 2004* (Yangon: Myanmar Institute of Strategic and International Studies, 2004).

19 In his address to the nation on Armed Forces Day in 2009, President Than Shwe noted that well-established mature democracy is the end result of two or three hundred centuries of development and cannot reasonably be made to appear overnight. Larkin, *Everything is Broken*, 255.

20 *The New Light of Myanmar*, May 5 2008.

21 UN Security Council, *Statement by the President of the Security Council*, S/PRST/2008/13, (May 2, 2008).

democracy. On May 11, *The New Light of Myanmar* published a cartoon depicting a group of smiling voters running toward a ballot box having the following conversation: "When we are confronted with natural disasters we all cooperate! When internal and external saboteurs disturb us we defend ourselves together! It is because the army and the people are on the march hands joined!" In one corner of the picture there are three figures labeled: Nargis, Internal, and External Saboteurs.[22]

The second critical aspect of Myanmar's political history is the nation's historical vulnerability to foreign invasion. Myanmar sits at the intersection of East Asia, South Asia, and Southeast Asia. Burma's first prime minister, U Nu, advised his countrymen to: "Take a glance at our geographical position – Thailand in the East, China in the North, India in the West, and stretching southward, Malaya, Singapore and so on. We are hemmed in like a tender gourd among the cactus. We cannot move an inch."[23] China, India, and Thailand have all, at various times, invaded Myanmar. Since World War II, Myanmar's foreign policy objective has been simple: to preserve territorial integrity and above all to maintain cordial relations with the Chinese behemoth to the North.[24] Myanmar's assiduous policy of nonalignment during the Cold War years left it with few friends but China. When the Cold War ended, Myanmar relinquished socialism but not authoritarianism. Myanmar's generals came to fear that they might become the recipients of an American-led invasion for democratic liberation, of the kind that occurred in Panama in 1989, Haiti in 1994, and Iraq in 2003.[25]

This fear formed the backdrop to the government's refusal to allow the naval ships of France, Britain, and the United States to dock in order to deliver their supplies. The USS *Essex* and its support ships left the coast of Myanmar on June 4, after fifteen unsuccessful attempts to gain permission to land relief supplies. The British frigate HMS *Westminster*, which had arrived off the coast on May 18 and been replaced on May 25 by HMS *Edinburgh*, abandoned its mission and departed on May 28. The French amphibious landing ship *Le Mistral* arrived off the coast on May 17, carrying 1,000 tons of supplies. It left in late May and offloaded its supplies at a port in Thailand to be flown into Burma. U.S. planes, which could have delivered emergency relief to difficult-to-reach inland areas, were initially also not permitted

[22] Selth, "Even Paranoids Have Enemies," 379.

[23] Quoted in Thakin Nu, *From Peace to Stability* (Rangoon: Ministry of Information, Government of the Union of Burma, 1951), 98–102.

[24] Determined to preserved neutrality during the Cold War years, Burma was a leading member of the Nonaligned Movement. see K. P. Misra, "Burma's Farewell to the Nonaligned Movement," *Asian Affairs* 12(1) (1981): 49–56. After the Cold War ended, Myanmar courted North Korea, which sold arms to Myanmar from the mid-1990's, and in late 2010 was reported to be collaborating with the Myanmar's generals in the development of nuclear facilities inside Myanmar. Bertil Lintner and Shawn W. Crispin, "Dangerous Bedfellows," *Far Eastern Economic Review*, November 20, 2003. Andrew Selth *Burma's North Korean Gambit: A Challenge to Regional Security?* (Canberra: Strategic and Defence Studies Centre, The Australian National University, 2004): 17–41.

[25] Selth, "Even Paranoids Have Enemies."

to enter Myanmar airspace. On May 11, after intervention by Thai Supreme Commander Boonsrang Niumpradit, the U.S. military was finally given permission for a C-130 cargo plane to fly in relief supplies. Niumpradit reported that: "They were very suspicious that the Americans would do more than just distribute relief supplies, but we helped convince the Burmese to allow the Americans in."[26]

The year Cyclone Nargis made landfall in Myanmar was not the first time that American warships had appeared in the Bay of Bengal. In August 1988, while hundreds of thousands of pro-democracy demonstrators marched through Rangoon carrying signs calling for "democracy" and the end of the twenty-six-year dictatorship of General Ne Win, the United States sent five U.S. naval vessels to the coast of Myanmar. Activists on the streets distributed leaflets saying that a U.S. invasion was imminent.[27] Three years later, a U.S. warship returned to the Bay of Bengal, when more than 200,000 Rohinghya refugees fled across the border from Myanmar into Bangladesh, citing grave violations of human rights by Myanmar's armed forces, the Tatmadaw.[28] On neither occasion did the United States invade Myanmar and there is no evidence that it had that intention.[29] Nonetheless, the U.S. history of pro-democracy interventions, from the Middle East to Grnada and Haiti, led Myanmar's generals to hold not implausible fears about democratic intervention in Myanmar.[30] When warships were again sighted off the coast of Myanmar in May 2008, one of the regime's senior generals, Maung Aye, called upon Russia to send three of its own warships to the Bay, to counter the "American threat."[31]

We must accept that the "American threat" – illusory though it might have been – was a reality to Myanmar's leaders.[32] For decades, Myanmar's rulers had heard the West's vituperative language about their country's government and

26 "Myanmar junta wary that foreign aid could herald reform" *The Guardian*, May 8, 2008, www.guardian.co.uk/world/2008/may/08/cyclonenargis.burma4.

27 Selth, "Even Paranoids Have Enemies," 382.

28 Operation Sea Angel was one of the largest military disaster relief efforts ever carried out, ultimately consisting of over 7,000 U.S. (and other) soldiers, who provided food, water, and medical care to nearly two million people.

29 Suspicion toward and resentment of the United States has deep historical roots. After the Second World War, the Truman administration feared that Mao Tse Tung was bent on the conquest of Southeast Asia. The United States viewed Burma, which was determinedly neutral, as a potential doorway through which the invading Chinese might pass. The Central Intelligence Agency provided military support to Kuomintang troops who fled China after 1949 and established bases in Burma. Eventually, Burmese pressure in the United Nations forced the United States to end its assistance to the Chinese Nationalists. Between 1962 and 1988, while Burma languished under the military dictatorship of General Ne Win, the United States used Burma –without the consent of the Burmese – as a base for spying on the Chinese.

30 Selth, "Even Paranoids Have Enemies."

31 Ibid. Myanmar's traditional ally, China, was preoccupied with its own natural disaster: on May 12, 2008, a 7.9-magnitude earthquake hit Sichuan Province, killing 70,000 people and leaving over 18,000 missing.

32 David I. Steinberg, "Aung San Suu Kyi and U.S. Policy toward Burma/Myanmar," *Journal of Current Southeast Asian Affairs* 29(3) (2010): 35–39.

listened to calls for regime change. The rhetoric had been accompanied by a complicated web of sanctions, put in place by the United States and its allies after the military's suppression of the 1988 uprisings and the general's abrogation of the 1990 election results. The aim of sanctions was to isolate the regime politically and to put economic pressure on its leaders, with a view to installing Aung San Suu Kyi's National League for Democracy, which had emerged as the clear victor in the 1990 elections.[33]

United States policy toward Myanmar, in the period from 1990 to 2010, included the following actions: opposing Myanmar's membership in various multilateral financial organizations; refusing to approve licenses for the export of military-related items to Myanmar; suspending Myanmar from the U.S. Generalized System of Preferences; implementing visa bans; restricting financial services; prohibiting goods imported from Myanmar; banning new investments; and constraining U.S. assistance to Myanmar.[34] At the time that the cyclone landed, key pieces of U.S. legislation targeting the Myanmar government included Executive Order 13047, issued by President Clinton on May 20, 1997, which banned most new U.S. investment in "economic development of resources in Burma" and the 2008 Tom Lantos JADE Act, which codified the existing Executive Orders and targeted the financial assets of key regime officials. The JADE Act, a response to the regime's violent suppression of the monk-led Saffron Revolution in September 2007, was signed into effect two days before Nargis struck.[35]

Until 2008, the Western policy of isolationism toward Myanmar extended even to humanitarian aid.[36] Intergovernmental organizations, under the influence of the

[33] In 1992, the US Senate Foreign Relations Committee refused to confirm the nomination of an Ambassador to Burma: an appointment was not made until 2012.

[34] The U.S. policy of isolating Burma was followed, to different degrees, by Britain, Canada, the European Union and Australia.

[35] In August 2007, protests originating with city dwellers spread to the Buddhist sangha (monks), who joined civilians in demonstrations against the regime. Selth writes that: "[s]uch was the popular mood that some activists and foreign journalists even began to predict the downfall of the military government." Andrew Selth, "Burma's 'saffron revolution' and the limits of international influence," *Australian Journal of International Affairs* 62(3) (2008): 283. See also: Ardeth Maung Thawnghmung and Maung Aung Myoe, "Myanmar in 2007: A Turning Point in the 'Road Map?'" *Asian Survey* 48(1) (2008): 13–19; Kyaw Yin Hlaing, "Challenging the Authoritarian State: Buddhist Monks and Peaceful Protests in Burma," *Fletcher Forum of World Affairs* 32(1) (2008): 125–144; Office of the High Commissioner for Human Rights, *Report of the UN Special Rapporteur on the Situation of Human Rights in Myanmar*, (2010) hwww.ohchr.org/en/countries/asiaregion/pages/mmindex.aspx; Charles Petrie, *End of Mission Report, UN Resident and Humanitarian Coordinator, UNDP Resident Representative for Myanmar 2003–2007*, (2008) www.pyinnya.com/wp-content/uploads/2008/06/end-of-mission-report-by-charles-petrie-april-2008.pdf.

[36] Ian Holliday, *Burma Redux: Global Justice and the Quest for Political Reform in Myanmar* (Columbia: Columbia University Press, 2012): 114–9. Holliday writes: "Reinforcing many bilateral sanctions, the dominant role of the U.S. and its allies in the UN and other formal intergovernmental organisations has long resulted in embargoes on humanitarian aid and engagement."

United States, limited the mandate of the United Nations Development Program and prohibited the World Bank and the International Monetary Fund (IMF) from operating in Myanmar.[37] Overseas Development Aid (ODA) to Myanmar fell from $435 million in the late 1980s to less than $50 million by the end of the 1990s. By then, ODA amounted to $2.88 per person,[38] the lowest figure among any of the world's fifty poorest countries. Several scholars have compared assistance provided to Myanmar before Nargis with the much higher amount of development aid spent in other repressive countries such as Zimbabwe, Sudan, and Laos.[39]

There is a near uniform consensus amongst analysts that policies of sanction and exclusion failed to produce any positive political change in Myanmar.[40] One reason for this is the political and financial support that Myanmar received from China. Another is Myanmar's own natural resources, which lessened the need for foreign trade and engagement. An overarching reason is that the core demand of Western powers – that the military relinquish power and deliver government to Aung San Suu Kyi – was impossible for the government to deliver.[41] The military occupies a central position in Myanmar politics and society and governance cannot occur without its cooperation.[42] From the mid 1990s, scholars and policy analysts brought these facts to the attention of the U.S. Congress and to the governments of other nations.[43] Nonetheless, sanctions continued, preventing (in the view of many) the two things that might have made a difference in Myanmar – "American soft power and global capitalism."[44] In 2009, U.S. embassy staff in Yangon reported that:

> [O]ur sanctions often have unintended consequences, directly and indirectly hurting innocent Burmese and expatriates living and working in Burma. Moreover, those who should be subject to sanctions often find ways to circumvent them, undermining the intent of U.S. policy. Many argue that our sanctions policy

[37] Ibid. [38] Ibid.

[39] See the testimonies of David Steinberg, Thant Myint-U, and David Williams, "Hearings on U.S. Policy toward Burma: its Impact and Effectiveness," before the Subcommittee on East Asian and Pacific Affairs of the Committee on Foreign Relations United States Senate, One Hundred Eleventh Congress, First Session September 30, 2009: 17. www.gpo.gov/fdsys/pkg/CHRG-111shrg55932/pdf/CHRG-111shrg55932.pdf.

[40] Ibid.

[41] Neil Englehart, "Regime Change Enough for Burma? The Problem of State Capacity," *Asian Survey* 45(4) (2005):622–644; Khin Zaw Win, "2010 and the Unfinished Task of Nation-building," in *Ruling Myanmar: From Cyclone Nargis to National Elections*, ed. Nick Cheesman, Monique Skidmore, and Trevor Wilson (Singapore: Institute of Southeast Asian Studies, 2009), 19–31.

[42] Li Chenyang and L.L. Fook, "China's Policies towards Myanmar: A Successful Model for Dealing with the Myanmar Issue?" *China: An International Journal* 255 (2009):7.

[43] See Derek Tonkin, "Take a tough look at scrapping Burma sanctions," *Financial Times*, May 13, 2008, www.ft.com/cms/s/0/7e4acff0-2086-11dd-80b4; "Norway critical of sanctions against Myanmar," *Xinhua News*, (May 16, 2008) http://news.xinhuanet.com/english/2008-05/16/content_8187370.htm; David Steinberg et al., "Hearings on U.S. Policy."

[44] Thant Myint-U, quoted in Holliday, *Burma Redux*, 143.

> frustrates any meaningful dialogue with the regime and cuts potential future leaders off from the "American experience," which may inhibit prospects for future progress. Broader sanctions that capture wide swaths of economic activity inevitably harm some innocents, crimp the ability of the general public to make a living, and raise cries of unfairness that make it difficult to gain the consensus necessary for effective implementation.[45]

There was, in summary, deep animosity toward the United States on the part of Myanmar's generals and a sense that U.S. overtures of assistance in the wake of Nargis were hypocritical, given that U.S. policies had for decades compounded poverty and hardship.

This animosity extended to the United Nations, which since the end of the Cold War, was viewed by Myanmar's rulers as a proxy for the United States.[46] In 2005, the Bush Administration succeeded in having Burma placed on the agenda of the UN Security Council. In May 2006, the United States proposed a formal Security Council resolution on Myanmar, arguing that Myanmar poses serious risks to peace and security in the region.[47] China and Russia vetoed the resolution and Indonesia, a non-permanent member of the Security Council, abstained on the vote. Given Myanmar's geopolitical location, its history and its ruler's deep insecurity about the nation's territorial integrity, it was foreseeable that U.S. warships in the Bay of Bengal would invite suspicion and apprehension.

A "RESPONSIBILITY TO PROTECT?"

Western governments and aid agencies responded with outrage to the delay in issuing visas and obstruction in the provision of aid.[48] A spokesman for the World Food Programme described the government's attitude as "unprecedented in

[45] Viewing cable 09RANGOON446, "U.S. Economic Sanctions on Burma – Some Reflections," (July 6, 2009) www.wikileaks.ch/cable/2009/07/09RANGOON446.html.

[46] From 1991, the UN General Assembly passed annual resolutions on the situation in Myanmar. In 1992, the UN Commission on Human Rights established a mandate on the situation of human rights in Myanmar. A series of Special Rapporteurs were appointed, with a mandate to "establish direct contacts with the Government and with the people of Myanmar, including political leaders deprived of their liberty, their families and lawyers, with a view to examining the situation of human rights in Myanmar and following any progress made toward the transfer of power to a civilian government and drafting of a new constitution, the lifting of restrictions on personal freedoms and the restoration of human rights in Myanmar." See Jürgen Haacke, "Myanmar and the United Nations," *The Adelphi Papers* 46(381) (2006): 83–96.

[47] UN Security Council, *Security Council Document S/PV.5526*, September 15 2006, www.securitycouncilreport.org/atf/cf/%7B65BFCF9B-6D27-4E9C-8CD3-CF6E4FF96FF9%7D/Myan%20SPV%205526.pdf.

[48] "Myanmar Cyclone Death Toll soars above 43,000," *Los Angeles Times*, May 15, 2008; "Foreign Powers Lean on Myanmar to Open up Aid," *Reuters*, May 15, 2008.

modern humanitarian relief efforts."[49] UN Secretary-General Ban Ki-Moon, whose telephone calls to Myanmar's leaders had not been returned, announced "deep concern and immense frustration at the unacceptably slow response to this grave humanitarian crisis."[50] British Prime Minister Gordon Brown criticized the "despicable regime" in Burma for obstructing aid and turning the cyclone into "a man-made catastrophe by the neglect and inhuman treatment of the Burmese people by a regime that is failing to act."[51] Laura Bush said that "the response to this cyclone is just the most recent example of the junta's failure to meet its people's basic needs."[52] The executive director of World Vision told the press that in "a race against time...we are on the cusp of a second wave of tragedy."[53]

On May 7, 2008, French Foreign Minister Bernard Kouchner invoked the principle of R2P and the possibility of Western forces delivering aid to the victims of Cyclone Nargis without the consent of the government of Myanmar.[54] Kouchner's proposal was that the Security Council should pass a resolution authorizing humanitarian intervention. As a preliminary step, on May 7, the French ambassador to the UN, Jean-Marie Ripert, proposed that the Security Council be briefed on the situation in Myanmar. China, supported by Russia, objected to the briefing proposal, arguing that a natural disaster was not a matter of international peace and security and therefore not a matter for the Security Council.[55] In a terse (and instructive) exchange between the Chinese government representative and Ripert, the Chinese representative inquired whether the Security Council had been apprised of France's recent response to the heat-wave deaths of dozens of its senior citizens. Ripert's response was that at least in France, government errors could be discussed, including in the press.[56] ASEAN members Indonesia and Vietnam, as non-permanent members of the Security Council, argued along practical lines that "there are other better forums to discuss the humanitarian dimension of the

49 "Burma Shums Foreign Aid Workers," *BBC News*, May 9 2008, http://news.bbc.co.uk/2/hi/asia-pacific/7391535.stm.

50 "Response to cyclone in Myanmar 'unacceptably slow'," *UN News Centre*, May 12, 2008, www.un.org/apps/news/story.asp?NewsID=26634&Cr=myanmar&Cr1.

51 "Burma Guilty of Inhuman Action," *BBC News*, May 17, 2008, http://news.bbc.co.uk/1/hi/uk_politics/7406023.stm.

52 "First Lady Condemns Junta's Response to Storm," *The Washington Post*, May 5, 2008, www.washingtonpost.com/wp-dyn/content/article/2008/05/05/AR2008050501814.html.

53 "Burma on the Cusp of Second Tragedy," *Metro UK*, May 9, 2008, www.metro.co.uk/news/146687-burma-on-the-cusp-of-a-second-tragedy.

54 International Coalition for the Responsibility to Protect, *The Crisis in Burma*, (2008), www.responsibilitytoprotect.org/index.php/crises/crisis-in-burma. Kouchner stated: "[w]e are seeing at the United Nations whether we can implement the Responsibility to Protect, given that food, boats and relief teams are there, and obtain a United Nations' resolution which authorizes the delivery (of aid) and imposes this on the Burmese government."

55 Rebecca Barber, "The Responsibility to Protect the Survivors of Natural Disaster: Cyclone Nargis, a Case Study," *Journal of Conflict and Security Law* 14(1) (2009): 3–34.

56 "China Blocking UN Responsibility to Protect Action For Burma," Burma Campaign UK, May 15, 2008, www.burmacampaign.org.uk/index.php/news-and-reports/newsstories.

Myanmar situation" and that "the last thing we would want is to give a political spin to the technical realities and the situation on the ground."[57]

There appeared to be, at least at first, support from some British and American politicians for a proposal to at least consider forceful intervention as a possible option. These politicians were responding to widespread public outrage about the intransigence of Myanmar's government in the face of its own people's anguish. The injustice of the situation seemed patent: Myanmar's people were being denied the aid that the rest of the world sought to provide, because of the paranoid self-interest of a despotic and predatory regime. The regime's actions following Nargis seemed to prove that the regime governed not in the people's interests, but in its own. In the public mind (of the West), it was incongruous that such a regime should be able to rely on the doctrine of state sovereignty in order to resist international aid.

There was also a high degree of uncertainty about the scope of R2P. After laborious negotiations, the principle had been adopted by unanimous agreement at the United Nations World Summit in 2005 and adopted by the General Assembly that same year.[58] But it had never yet been applied. Section 139 of the Summit's outcome document declared that:

> The international community, through the United Nations, also has the responsibility to use appropriate diplomatic, humanitarian and other peaceful means, in accordance with Chapters VI and VIII of the Charter of the United Nations, to help protect populations from war crimes, ethnic cleansing and crimes against humanity. In this context, we are prepared to take collective action, in a timely and decisive manner, through the Security Council, in accordance with the Charter, including Chapter VII, on a case-by-case basis and in cooperation with relevant regional organizations as appropriate, should peaceful means be inadequate and national authorities are manifestly failing to protect their populations from genocide, war crimes, ethnic cleansing and crimes against humanity.[59]

R2P clearly only applies to the four stated mass atrocity crimes: genocide, war crimes, ethnic cleansing, and crimes against humanity.[60] The government of Myanmar's actions in the wake of Nargis could not be construed as genocide or war crimes under the definitions set out in the Rome Statute. Neither was there evidence that a policy of ethnic cleansing was being pursued. Proponents of the application of R2P therefore focused their arguments on whether or not the government's actions might constitute a "crime against humanity." Particular attention was focused on Article 7(k) of the Rome Statute, which provides that a crime against humanity includes "inhumane acts…intentionally causing great

[57] Security Council Watch, *Update Report: Myanmar, No. 4*, (2008), www.securitycouncilreport.org/update-report/lookup-c-glKWLeMTIsG-b-4130257.php.

[58] UN General Assembly, 2005 *World Summit Outcome : resolution / adopted by the General Assembly*, October 24 2005, A/RES/60/1, www.unhcr.org/refworld/docid/44168a910.html.

[59] Ibid., section 139. [60] Ibid., paragraph 138.

suffering, or serious injury to body or to mental or physical health."[61] But under the Rome Statute, crimes against humanity require legal intent, and must be committed as part of a widespread or systematic attack directed against the civilian population.[62] There was no evidence that the government of Myanmar was deliberately attempting to create additional suffering for the people.[63]

Nonetheless, debate persisted around the question of whether or not the government's actions in the aftermath of natural disaster might yet constitute a crime against humanity. The original architects of the doctrine, whose views were sought as the Nargis debate intensified, seemed to have mixed ideas about its application. The initial Canadian-sponsored commission report that initiated the R2P concept anticipated "overwhelming natural or environmental catastrophes, where the state concerned is either unwilling or unable to cope, or call for assistance, and significant loss of life is occurring or threatened" as a situation that might call for the application of R2P.[64] In the wake of Nargis, former Canadian Foreign Affairs Minister and statesman, Lloyd Axworthy, who was responsible for initiating the original report, argued that: "there was no doubt in our minds that human security applies to natural disasters and pandemics just as much as it does to civil conflict and state-sanctioned violence."[65]

Another of the principle's original proponents, former Australian Minister for Foreign Affairs Gareth Evans, stated that R2P was not intended to relate to situations of human security generally, such as the impact of natural disasters. Nonetheless, Evans refused to close the door on the possible application of R2P to the situation in Myanmar. He argued that "[i]f what the generals are now doing, in effectively denying relief to hundreds of thousands of people at real and immediate risk of death, can itself be characterised as a crime against humanity, then the responsibility to protect principle does indeed kick in." Writing in *The Guardian*, Evans opined that:

> There is, as always, lots for the lawyers to argue about in all of this, not least on the question of intent. And there will be lots for the Security Council to quarrel about as to whether air drops and the like are justified, legally, morally and practically. But when a government default is as grave as the course on which the Burmese generals now seem to be set, there is at least a prima facie case to answer for their

[61] UN General Assembly, *Rome Statute of the International Criminal Court (last amended January2010)*, July 17 1998, A/CONF. 183/9, www.unhcr.org/refworld/docid/3ae6b3a84.html.

[62] Ibid., Article 30.

[63] In "Even Paranoids Have Enemies," Selth refers to speculation that the government was deliberately withholding aid from certain ethnic minorities.

[64] International Commission on Intervention and State Sovereignty, *The Responsibility to Protect* (Ottawa: International Development Research Centre, 2001), 33 at 4.20.

[65] "International community has a responsibility to protect Myanmar," *The Edmonton Journal*, May 13, 2008, www.canada.com/edmontonjournal/news/opinion/story.html?id=6056af9d-43eb-4575-81b3-9235750e3214&p=1.

> intransigence being a crime against humanity – of a kind which would attract the responsibility to protect principle. And that bears thinking about, fast, both by the Security Council, and the generals.[66]

Section 139 provides that even when circumstances of genocide, war crimes, ethnic cleansing, and crimes against humanity are proven, action under R2P requires Security Council endorsement, and that this should only be forthcoming if all other less extreme forms of intervention had been attempted. The doctrine also provides that there must be a reasonable likelihood of the proposed intervention halting or averting the harm in question, and that the response must be proportional to that harm, so that on balance more good than damage would be done by the intervention.

These provisions made the application of R2P in the context of Cyclone Nargis highly unlikely. First, China, Russia, and South Africa had made clear that they would not support Security Council action in relation to Nargis. Myanmar's fellow ASEAN members on the Security Council, Indonesia and Vietnam, held the same view. Second, most relief experts were publicly expressing the opinion that it would be neither practical nor effective to deliver aid against the will of Myanmar's government. The logistical problems appeared insurmountable. The form of intervention most commonly mentioned was airdrops of medicine and supplies. But how would these be distributed on the ground? What prospect was there that they would reach those who needed them most: the young, old, weak, and sick?

More coercive forms of intervention, such as putting foreign troops on the ground, seemed even less likely to succeed. There was a possibility that foreign forces would be met with resistance by the Tatmadaw and that the result would be armed conflict between foreign troops and Myanmar's military. This was hardly a result that would assist the beleaguered disaster-sufferers in the Irrawaddy. Third, although President General Than Shwe had refused to respond to the telephone calls of the UN secretary-general in the days following the cyclone, he had been in communication with leaders of ASEAN countries and with the Chinese. Diplomatic options were therefore still open, and the hope remained that Myanmar's generals might yet be convinced to cooperate with the international community.

Finally, reports were beginning to emerge from within disaster-stricken areas about the successful efforts of local civil society groups in dealing with the aftermath of the cyclone. These groups, many of them organized by Buddhist monks, were critical in the delivery of relief in the days and weeks following the cyclone. In these circumstances, it was by no means clear that international intervention would do more good than harm. On May 7, 2008, the UN under-secretary-general for humanitarian affairs and emergency relief said: "I'm not sure that invading them

[66] Gareth Evans, "Facing up to our Responsibilities," *The Guardian*, May 12, 2008, www.guardian.co.uk/commentisfree/2008/may/12/facinguptoourresponsbilities.

would be a very sensible option at this particular moment. I'm not sure it would be helpful to the people we are actually trying to help."[67]

As well as the concern that misapplication of the R2P doctrine in the context of Nargis would not help the people of Myanmar, there was also concern that a premature application of the doctrine would harm the doctrine itself. Evans himself recognized that:

> If it comes to be thought that R2P, and in particular the sharp military end of the doctrine, is capable of being invoked in anything other than a context of mass atrocity crimes, then such consensus as there is in favour of the new norm will simply evaporate in the global south. And that means that when the next case of genocide or ethnic cleansing comes along we will be back to the same old depressing arguments about the primacy of sovereignty that led us into the horrors of inaction in Rwanda and Srebrenica in the 1990s.[68]

Ramesh Thakur, another of the doctrine's original architects, was equally blunt:

> There would be no better way to damage R2P beyond repair in Asia and the developing world than to have humanitarian assistance delivered into Myanmar backed by Western soldiers fighting in the jungles of Southeast Asia again. If France has soldiers to spare for serious combat, they could relieve embattled Canadians in southern Afghanistan."[69]

By mid-May 2008, there were few serious voices still calling for the application of R2P in Myanmar. By that time, despite the early R2P rhetoric, it ought to have been clear to Myanmar's leaders that an invasion by U.S. troops under the legal justification of R2P was highly unlikely. But in the wake of the 2003 invasion of Iraq, Myanmar's leaders could not have been certain that the United States would necessarily be constrained from following its preferred course of action by concerns about legality under international law. After all, U.S. government representatives had stated on several occasions that they wished for regime change in Myanmar, in the same way that they had stated they wished for regime change in Iraq.[70] On

67 Quoted in: The International Coalition for the Responsibility to Protect, *Crisis in Myanmar*, www.responsibilitytoprotect.org/index.php/crises/crisis-in-burma.

68 Evans, "Facing Up to Our Responsibilities."

69 Ramesh Thakur, "Should the UN Invoke the 'Responsibility to Protect'?" *The Globe and Mail*, May 8, 2008, www.canada.com/edmontonjournal/news/opinion/story.html?id=6056af9d-43eb-4575-81b3-9235750e3214&p=1.

70 David Steinberg opines that the primary reason why Myanmar's generals were suspicious of the United States was because the United States had been calling for regime change in Myanmar. *The New York Times*, June 5, 2008, http://topics.blogs.nytimes.com/2008/05/06/one-question-on-myanmar/. The Asia Times Online (May 10, 2008) published an article by Shawn W. Crispin entitled "The case for invading Myanmar." Crispin wrote: "With United States warships and air force planes at the ready, and over 1 million of Myanmar's citizens left bedraggled, homeless and susceptible to water-borne diseases by Cyclone Nargis, the natural disaster presents an opportunity in crisis for the U.S. A unilateral—and potentially United Nations-approved—U.S. military intervention in the name of humanitarianism could easily

April 30, 2008, just days before Nargis struck, President Bush decreed an executive order worded to "block all property and interests in property of designated individuals and entities determined to be owned or controlled by the government of Burma (Myanmar)."[71] To the generals, U.S. pleas for access to assist Myanmar's citizens on humanitarian grounds had the ring of hypocrisy. The accuracy of these perceptions is not the point. What is relevant is that the suspicion and paranoia that marked domestic political reality in Myanmar left no room for the emerging idea that states may have a duty to assist in times of disaster.

REGIONAL INTERNATIONAL COOPERATION: ASEAN

ASEAN never endorsed the sanctions policy implemented against Myanmar by Western countries.[72] Indeed it is difficult to discern, apart from ASEAN's outburst of opprobrium after the suppression of the 2007 Saffron Revolution, any real moral condemnation by ASEAN of Myanmar's generals. Most ASEAN countries have, at different stages, struggled to resolve issues of authoritarianism, the role of the military within the state, the reconciliation of restless ethnic minorities, and the management of development and democracy.[73] The lesson learned by ASEAN nations is that the source of liberalization and democratization is indigenous, and that external pressure is at best irrelevant and at worst counterproductive.

In 1967, when ASEAN was created, Burma declined its offer of membership. From Burma's perspective, ASEAN was an anti-communist organization, comprised wholly of anti-communist regimes, the institutional heir to the explicitly anti-communist Colombo Plan (1951) and the Association of Southeast Asia (1961) as well as an integral part of Washington's strategy for containing communism, outlined in the Nixon Doctrine.[74] China had declared that ASEAN was a tool of

turn the tide against the impoverished country's unpopular military leaders, and simultaneously rehabilitate the legacy of lame-duck U.S. President George W. Bush's controversial preemptive military policies."

71 Executive Order 13464 expanded sanctions to permit asset freezes against designated Burmese entities.

72 Singapore's Minister George Yeo stated that ASEAN has recognized the ineffectiveness of sanctions in circumstances where "Myanmar's giant neighbours, China and India, have kept the back and side gates open." Transcript of replies by Minister George Yeo to Questions (on Myanmar) in Parliament on Oct 22, 2007. http://beyondsg.typepad.com/beyondsg/2007/10/replies-by-mini.html.

73 Termsak Chalermpalanupap, Assistant Director for Programme Coordination and External Relations ASEAN Secretariat, *Asia-Pacific Roundtable held in Kuala Lumpur*, (June 1, 1999) www.asean.org/2833.htm.

74 The Nixon Doctrine is examined in Virginia Brodine and Mark Selden (eds.), *Open Secret: The Kissinger-Nixon Doctrine in Asia* (New York: Harper & Row, 1972). On American perceptions of ASEAN as a component of the Nixon Doctrine, see *United States Foreign Policy for the 1970's: Building for Peace—A Report by President Nixon to the Congress*, February 25, 1971, Washington, DC, US Government Printing Office, 79–80.

American imperialists aimed at containing China and other communist powers.[75] Burma, committed to a precarious policy of non-alignment, had no wish to antagonize its powerful neighbor by joining ASEAN. The end of the Cold War, and the United States' withdrawal from Vietnam, mitigated regional tensions and led to Myanmar's entry into ASEAN in 1997, together with Laos and Vietnam.[76] For Myanmar, closer association with ASEAN provided Myanmar's ruling military with the benefits of legitimacy, a degree of protection from external criticism, and the prospect of greater economic interaction with Southeast Asian states.

Between 1990 and 1997, ASEAN employed a strategy of "constructive engagement" with Myanmar, meaning a mode of dialogue and persuasion, and the pursuit of strategic and economic interests, with concomitant encouragement of "moderate" reform along the lines of liberal democracy.[77] Underpinning ASEAN's policy was its uncertainty about China's growing power and regional ambitions and the imperative of bringing Myanmar within ASEAN's sphere of influence, rather than leaving the nation to China.[78] Constructive engagement meant access for Thai, Malaysian, Singaporean, and Indonesian businesses to Myanmar's raw materials and markets. The "constructive" aspect of engagement was intended to be two-pronged: (i) the socialization of Myanmar's elite toward good governance and gradual liberal reform; and (ii) financial investment, hopefully leading to socioeconomic development and the creation of a democratically disposed middle class.[79] "Constructive engagement" stood in marked contrast to the outright disapprobation, threats and sanctions employed by the United States and the European Union.[80] According to Myanmar's Foreign Minister Ohn Gyaw, constructive engagement meant that "ASEAN saw Myanmar as an equal."[81]

[75] Thailand and the Philippines both allowed US forces to prosecute the Second Indochina War from their military bases. Jürgen Haacke, "Myanmar and ASEAN," *The Adelphi Papers* 46 (381) (2006): 42.

[76] Myanmar was admitted despite the significant pressure placed on ASEAN by the US and the EU to refuse admission.

[77] Stephen McCarthy traces "constructive engagement" to the "comprehensive engagement" concept coined by the Australian foreign minister, Gareth Evans, in the mid 1980s. The idea was that by promoting trade, diplomatic and economic ties with an authoritarian regime, socioeconomic progress and the growth of a middle class would produce political liberalization. Stephen McCarthy, "Burma and ASEAN: Estranged Bedfellows," *Asian Survey* 48(6) (2008): 911–935. For skeptical views on the effectiveness of the policy, see: Deborah Toler, "Constructive Engagement: Reactionary Pragmatism at its Best," *Issue: A Journal of Opinion* 12(3/4) (1982): 11–18; Sam C. Nolutshungu, "Sceptical Notes on "Constructive Engagement," *Issue: A Journal of Opinion* 12(3) (1982): 3–7.

[78] "ASEAN prefers soft talk to threats in dealing with Yangon," *The Straights Times*, August 26 1992.

[79] Lee Jones, "ASEAN's Albatross: ASEAN's Burma Policy, from Constructive Engagement to Critical Disengagement," *Asian Security* 4(3) (2008): 271–293.

[80] Hermann Kraft, "ASEAN and intra-ASEAN relations: weathering the storm?" *The Pacific Review* 13(3): 453–472.

[81] *Far Eastern Economic Review* (August 15 1996): 36, quoted in: ibid., 464.

In the period from 1997 to 2012, Myanmar's engagement with ASEAN constituted its most important and sustained multilateral engagement. Bilateral relations with China, India, and Bangladesh, though of consequence economically and for Myanmar's security, did not generate the web of institutional and diplomatic linkages spawned by Myanmar's ASEAN membership. In October 1996, even before it became a member of ASEAN, Myanmar had formed a Steering Committee on ASEAN Affairs under the Department of Foreign Affairs. After assuming membership, Myanmar's participated in almost all activities at various institutional levels as required by ASEAN including all of the summits to ministerial meetings. This included "first-track" diplomacy between government officials and "second-track" engagement between "think tanks" such as the Myanmar Institute of Strategic and International Studies and its regional counterparts.

In the wake of the Saffron Revolution and its suppression, ASEAN engagement with Myanmar registered a new note. When ASEAN's foreign ministers gathered in New York in September 2007 for the UN General Assembly, they had a "full and frank discussion on the situation in Myanmar."[82] Afterward, the Foreign Ministers issued a statement expressing their "revulsion" over reports that demonstrations in Myanmar were being suppressed by violent force and that there has been a number of fatalities. The statement, delivered by Singapore's Foreign Minister George Yeo, stated that ministers were "appalled" at reports of automatic weapons being used against demonstrators, and demanded that Myanmar's government immediately desist from the use of violence. It strongly urged Myanmar to exercise utmost restraint and seek a political solution, to resume its efforts at national reconciliation with all parties concerned, to work toward a peaceful transition to democracy, and to release all political detainees including Taw Aung San Suu Kyi. The strength of ASEAN's September 2007 statement took many by surprise. ASEAN's traditional methods of "quiet persuasion" appeared to have been replaced by strong public condemnation.

It was against this backdrop that the events of Nargis occurred. As soon as the cyclone struck, the Philippines and Singapore sent their experts to the UN Disaster Assessment and Coordination Team that was assembling in Bangkok. Two days later, ASEAN Secretary-General Surin Pitsuwan called on all ASEAN member countries to provide urgent relief assistance to victims of the cyclone and to this end established an emergency relief fund, the ASEAN Cooperation Fund for Disaster Assistance. The fund's terms of reference required that donors not place conditions on the use of their contributions. However, ASEAN stressed that donations would only be used to alleviate the suffering of the cyclone's victims. Pitsuwan convened the ASEAN Committee on Disaster Management, which had been activated on only one other occasion in response to the crisis in Aceh following the Boxing Day

82 "Interview with Singapore Foreign Minister George Yeo," *Straits Time*, 2–3 October 2007, http://app.mfa.gov.sg/pr/read_content.asp?View,8389.

tsunami. He also invoked, for the first time, the 2005 ASEAN Agreement on Disaster Management and Emergency Response, which had been signed in the aftermath of the 2004 tsunami.[83] The agreement contains provisions for collective action by member states in the event of disasters. It emphasizes non-intervention and places the control and overall direction of relief efforts in the hands of the requesting or receiving party. Pitsuwan then called on Myanmar's minister for foreign affairs, Nyan Win, and minister for social welfare, relief and resettlement, Major General Maung Maung Swe, to allow ASEAN relief and rescue teams to enter Myanmar and assist in government efforts at disaster relief.[84]

General Than Shwe responded by allowing the dispatch of a small Emergency Rapid Assessment Team (ERAT) to assess the water, sanitation, health, logistical, and food needs of victims of Cyclone Nargis. The team arrived in Myanmar on May 13. Six days later, the ERAT reported that there was a major risk of a further large death toll as a result of disease and malnutrition. It proposed that ASEAN lead a humanitarian partnership involving the UN, neighboring countries, and NGOs to deliver humanitarian relief. On May 19, ASEAN foreign ministers met to decide on a course of action. According to Singapore's foreign minister, George Yeo, several Southeast Asian ministers pressured Myanmar's foreign minister, Nyan Win, to accept international overtures of assistance. Yeo explained that: "We had to suspend all our political reservations [about putting overt pressure on Myanmar] and say look, let's concentrate on the humanitarian efforts."[85] Eventually, the ministers issued a statement which focused on two key points: first, that relief efforts would be ASEAN-led and coordinated; and second, (at the behest of Myanmar) that international assistance should not be politicized. The statement also called on Myanmar to allow more international relief workers to enter the country.[86]

On May 20, Pitsuwan travelled to Myanmar and met with Prime Minister Thein Sein, in order to formally and personally deliver the statement of the ASEAN foreign ministers. Pitsuwan informed Thein Sein of the steps required to implement the statement and indicated that these steps were necessary in order to build

83 See ASEAN, *The ASEAN Agreement on Disaster Management and Emergency Response*, Vientiane, 26 July 2005, www.aseansec.org/17579.htm.

84 W. Sabandar, "Cyclone Nargis and ASEAN: A Window for More meaningful Development Cooperation in Myanmar," (2010) in *Ruling Myanmar: From Cyclone Nargis to National Elections*, ed. Monique Skidmore and Trevor Wilson (Singapore: Institute of Southeast Asian Studies, 2010), 197–207.

85 Transcript of Interview of Minister of Foreign Affairs George Yeo with the Local Media on the ASEAN Ministerial Meetings (July 16, 2008).

86 Jurgen Haake argues that the threat of R2P served as leverage to force the junta to accept international aid, and ultimately galvanized the junta into action, making it easier for ASEAN, acting as an intermediary, to convince the generals to cooperate with the United Nations. One has cause to be skeptical about this claim, given the history of Myanmar's response to threats in the past. Jurgen Haacke, "Myanmar, the Responsibility to Protect, and the Need for Practical Assistance," *Global Responsibility to Protect* 1 (2009): 156–184.

mutual confidence between Myanmar and the donors of relief goods and humanitarian aid. A UN pledging conference was scheduled to be held in Yangon on May 25. Myanmar consented to the establishment of the ASEAN Humanitarian Task Force for the Victims of Cyclone Nargis, which was complemented on the ground by a Tripartite Core Group (TCG) consisting of high-level representatives of the government of Myanmar, ASEAN, and the UN. The TCG was chaired by Myanmar's deputy foreign minister, Kyaw Thu, with Singapore's ambassador to Myanmar, Robert Chua, as ASEAN's focal person, and Dan Baker, the UNDP resident representative in Myanmar, as the UN system's main representative. The government of Myanmar agreed to the immediate deployment of medical teams from ASEAN countries and pledged that it was prepared to accept the assistance of international and regional experts.[87]

On May 23, UN Secretary-General Ban Ki-moon visited Myanmar to reaffirm the agreement and to stress the urgent need for officials to accept international assistance. After the visit, officials in Myanmar agreed to permit much greater access to humanitarian aid workers within Myanmar, to admit international aid workers "regardless of nationalities" and to allow ASEAN to oversee distribution of relief. The ASEAN secretary-general announced that ASEAN had managed to "establish a space, a humanitarian space, however small, to engage with the Myanmar authorities."[88] Pitsuwan told an audience in Washington that ASEAN was "trying to work around a very, very strict resistance and mentality and mindset that have been there for a long, long time."[89]

This was a mindset that ASEAN leaders were more attuned to than their western counterparts. Importantly, ASEAN leaders appreciated something that many in the West did not, which was that the mindset of the junta was not uniform, and that different political players, and their shifting power and roles, had direct policy consequences for the relief effort. Nargis exacerbated already existing tensions within the ruling junta between those who favored a degree of liberalization and engagement with the outside world, such as the Prime Minister Chair of the National Disaster Committee Thein Sein and General Thura Shwe Mann, and hardliners, such as Vice Senior General Maung Aye and Secretary-1 Tin Aung Myint Oo.[90]

Struggles between different factions within Myanmar resulted in continual recalibrations of policies about access for relief workers, as one or other side in

87 ASEAN, *A Humanitarian Call: the ASEAN Response to Cyclone Nargis*, (Jakarta: Association of Southeast Asian Nations, 2010).

88 Quoted in Sheldon W. Simon, "Southeast Asian International Relations: Is There Institutional Traction?" (2010) in *International Relations in Southeast Asia: Between Bilateralism and Multilateralism*, ed. N. Ganesan and Ramses Amer (Singapore: Institute of Southeast Asian Studies, 2010), 48.

89 Aung Hla Tun, "Foreign Powers Lean on Myanmar to Open up Aid," *Reuters*, May 15, 2008.

90 US Embassy viewing cable 08RANGOON471, *Burma: struggles at the top affecting relief effort, Rangoon*, www.wikileaks.ch/cable/2008/06/08RANGOON471.html.

complicated internal power struggles pushed to gain control of the relief agenda. Thein SeinTaw, emerged from these struggles with a reputation amongst the people, the military, and the international community as a moderate and a reformer, with an outward looking agenda. In contrast, Maung Aye, who on several occasions attempted to roll back agreements about access for the UN and international NGOs to affected areas and to tighten media reporting about the extent of the devastation, emerged as recalcitrant and backward looking.

By May 29, Pitsuwan reported that there were signs of progress in the relief activities. The World Food Programme was using helicopters to deliver food and supplies and the amount of international relief in the Irrawaddy region had increased significantly. The feared outbreak of disease and starvation did not occur. Over the course of June 2008, advance assessment teams were deployed to the farthest reaches of the Irrawaddy region. Partnerships between local NGOs and international organizations sprang up and community education programs were organized on practical topics such as first aid and survival strategies, but also on issues such as vulnerability analysis, community leadership, and the importance of networking and building relations.[91]

Fieldworkers writing in *The Journal of the Medical Association of America* foresaw that the cyclone's devastation would serve to reduce Myanmar's isolation from the outside world and create opportunities for regional and international collaboration in the non-governmental sector.[92] A wide range of experts now agree that "Nargis was a turning point for civil society in the country."[93] The disaster enhanced community solidarity and generated a sense of self-reliance and resourcefulness among the people, together with an awareness that people had the capacity to meet their own needs without the direct assistance of either the military or the international community.[94] Regional commanders, who had long wielded unrestrained power, had little role in the relief effort "other than to maintain civil order and follow orders given from Nay Pyi Taw."[95]

CONCLUSION

Against a history of ethnic conflict and disunity, and the West's long-standing disapprobation of Myanmar's military rulers, Nargis "succeeded in doing what no other situation has done before: uniting the West and Burma's Asian neighbors to

[91] E. Axelrad, "Disaster Entrepreneurs: Policy Entrepreneurs and Crisis Exploitation as Catalysts for Political Change in Authoritarian Countries," www.theglobaldiscussion.org/wp-content/uploads/2011/05/Axelrad-Disaster-Entrepreneurs.pdf; Jurgen Haacke, "Myanmar, the Responsibility to Protect, and the Need for Practical Assistance," *Global Responsibility to Protect* 1 (2009): 156–184.

[92] E. Stover and P. Vinck, "Cyclone Nargis and the politics of relief and reconstruction aid in Burma (Myanmar)," *The Journal of the American Medical Association* 300(6) (2008): 729–731.

[93] Axelrad, "Disaster Entrepreneurs."

[94] US Embassy viewing cable 08RANGOON471. [95] Ibid.

bring unprecedented pressure on the regime to open up and allow international relief workers unfettered access."[96] From this opening, there have followed several consequences for Myanmar's prospects for undergoing a successful transition to democracy.

First, the military regime's narrative of the Tatmadaw as the savior of the people, and the Western world as bent on the destruction of Myanmar, could not be sustained in the wake of Nargis. The cyclone made obvious the incompetence and corruption of the regime and its inability to meet the needs of the people. This had long been plain to the people of Myanmar, but crucially, after the cyclone, it also became plain to the generals themselves.[97] This awareness provoked two different responses from within the ruling military elite. The first was the hard line response of Maung Aye and his supporters, who saw the cyclone as a threat to order and advocated tighter controls over communication and even more restricted access to the country by foreigners. The second was the response of Prime Minister (now President) Thein Sein, which was to gamble on engagement with ASEAN, and later the West, and ever-increasing measures of liberalization.

By 2010, the furor over Nargis had subsided and Myanmar held its first national elections under the new constitution. Thein Sein who had prevailed on opening Myanmar to international offers of disaster relief became president of the union of Myanmar. In his inaugural address to parliament, on March 30, 2011, the president pledged legislative reform in areas such as human rights, health care, press freedom, and environmental protection. In the period from March 2011 to February 2012, the government passed a raft of legislation aimed at legalizing trade unions, releasing political prisoners, increasing the pension rate, allowing public political gatherings, easing press censorship, permitting the teaching of ethnic minority languages in schools across Myanmar, and facilitating opposition political party registration and candidacy in elections.

On August 19, 2011, President Thein Sein met with democratic opposition leader Aung San Suu Kyi who told foreign diplomats "that she is confident about the future and optimistic about the possibility of genuine change" and that "Thein Sein can be trusted, he is genuinely trying to reform the country, and needs international support."[98] In Myanmar's April 2012 by-elections, in which the National League for Democracy participated, the party's iconic leader, Aung San Suu Kyi, stood as a candidate for the national parliament and won a decisive victory.

[96] Ibid.

[97] "Cyclone Could Unleash Political Upheaval," *The Irrawaddy*, May 5, 2008, www2.irrawaddy.org/article.php?art_id=11766&Submit=Submit (June 5, 2012). The Irrawaddy quotes Win Min, a Burma expert at Thailand's Chiang Mai University, who said: "I heard people are getting angry at the authorities because of lack of prevention and disaster relief. It is not good for the junta's referendum plan."

[98] Larry Jagan, "What Thein Sein promised Suu Kyi," *Online Asia Times*, September 30, 2011, http://www.atimes.com/atimes/Southeast_Asia/MI30Ae01.html.

How much of Thein Sein's success in achieving a degree of liberalization can be traced to his actions in the weeks and months following Nargis? Thein Sein emerged from Nargis with the reputation of being someone "less corrupt than most of the former junta leaders and a good listener."[99] It is reasonable to assume that his success in managing the relief effort and engaging with ASEAN and the West increased his credibility, with the people of Myanmar, the outside world, and those in the ruling military junta who were undecided about whether to support a path of liberalization.

Of these three groups, the most critical to the future of reform was the military. According to Railway Minister Aung Min, within parliament, "some 20% of ministers are liberal and 20% are hardline, with 60% sitting on the fence waiting to see who wins."[100] It is plausible to credit a large part of President Thein Sein's courage in persisting with liberalization in the face of resistance to the demonstrated success of the Nargis relief effort. Around this success, momentum for reform gathered, and moves toward liberalization drew yet more support, increasing the sense of inevitability about the process toward democratization. This was enough to convince waverers that support for reform would place them on the winning side of history.

Second, Nargis threw into sharp relief the failures of the West's policies of isolation and exclusion, and amplified the voices of those who had long been calling for a change in policy toward Myanmar. Twenty years before the cyclone struck, UNICEF's country director in Burma had warned of a "silent emergency" affecting the country's children, and argued that help for the poorest should not have to wait for democratic change.[101] It was not until after the cyclone that these views gained traction. Initially, aid organizations operating in the Irrawaddy Delta after Nargis directed their efforts to emergency relief. But once in the country, the need for longer term development projects became apparent.

At present, a significantly higher number of international aid organizations operate within Myanmar than did before the cyclone. Crucially, the cyclone made clear to organizations and their donors that productive engagement with the regime was difficult, but not impossible. Organizations that had a presence in Myanmar predating the cyclone were already aware that the military junta was not a uniform bloc and that there were elements within the military conducive to engagement and reform.[102] Even before the cyclone, significant work on the ground had been achieved by many of these groups.

99 Aung Zaw, "The Rut and Roar Begins in Burma," *The Irrawaddy*, July 8, 2011, www.irrawaddy.org/article.php?art_id=21661.

100 Larry Jagan, "Bitter struggle puts reform process at risk," *Bangkok Post*, February 7, 2012, www.bangkokpost.com/opinion/opinion/278587/bitter-struggle-puts-reform-process-at-risk.

101 Than Myint-U, *Where China Meets India* (London: Faber and Faber Ltd, 2009), 101.

102 Interview with John Prideaux-Brown, Oxfam, Yangon, 2011, copy on file with author.

Most importantly, in terms of the development of the international law of disaster relief, Nargis had consequences for our understanding of the scope and potential of the doctrine of R2P. Prior to Nargis, with the exception of the Philippines and Thailand, Southeast Asia as a whole was equivocal about the merits of R2P.[103] Some states, such as Vietnam and Myanmar, were implacably opposed to the doctrine.[104] Vietnam indicated prior to the 2005 World Summit that it would not support the principle. It stated that:

> Article 51 of the Charter is clear and restrictive in the sense that the inherent right of individual or collective self-defense can be employed only if an armed attack occurs against a Member State. We do not believe, therefore, that Article 51 provides an expanded scope for permitting States to take military action on the basis of a perceived imminent threat. Nor are we convinced that responsibility to protect is an emerging norm of international law.[105]

Myanmar was equally hostile to R2P. Other states within the region adopted the cautious attitude of Indonesia. At the 2005 World Summit, President Susilo Bambang Yudhoyno stated that "we need a consensus on the responsibility to protect people from genocide, ethnic cleansing and crimes against humanity. To this end, force should be used only when all other means have failed."[106]

Some expected that the careless invocation of the doctrine after Nargis would create further distance between the sovereignty-conscious states of the region and acceptance of R2P.[107] But that does not seem to have been the case. In the 2009 General Assembly debate on R2P, a mere year after Nargis had brought French and U.S. warships to the Bay of Bengal and talk of invasion under the banner of R2P was rife, Indonesia, the Philippines, and Singapore all voiced strong support for the principle.[108] Even Vietnam affirmed the World Summit Outcome Document of 2005 and emphasized that the international community need no longer debate the necessity of R2P or its scope. Advocating that debate move toward rendering the Outcome Document operational, Vietnam suggested that the

[103] UN Security Council, Fifty-ninth Year, Official Records: S/PV.5015 (July 30, 2004): 10–11; UN Security Council, Fifty-ninth Year, Official Records: S/PV.5040 (September 18, 2004): 10. Singapore signaled a degree of support by joining the "Group of Friends" of R2P, which was composed of like-minded Permanent Missions to the UN in New York.

[104] UN General Assembly, Fifty-ninth Session, Official Records, *United Nations World Summit Outcomes Document*, A/59/PV.89, (April 8, 2005):22; United Nations World Summit Outcomes Document A/60/P.7, September 15, 2005, p. 5.

[105] Ibid.

[106] UN General Assembly, Fifty-ninth Session, Official Records, *United Nations World Summit Outcomes Document*, A/60/P.7, (September 15, 2005): 5. See Alex J. Bellamy and Sara E. Davies, "The Responsibility to Protect in the Asia–Pacific Region," *Security Dialogue* 40(6) (2009): 1–28.

[107] A.J. Bellamy and M. Beeson, "The Responsibility to Protect in Southeast Asia: Can ASEAN Reconcile Humanitarianism and Sovereignty?" *Asian Security* 6(3) (2010): 262–279.

[108] UN General Assembly reaffirmed R2P in 2009: A/RES/63/2009, 14 September 2009. www.un.org/apps/news/story.asp?NewsID=32047.

secretary-general's report represented an excellent ground for discussion. Myanmar voiced its support for the principal of R2P.[109]

These developments suggest that rather than undermining R2P, Nargis served to clarify its scope. Nargis made clear that in practice, the invocation of R2P by external actors will almost certainly prove counterproductive to disaster relief efforts. It is now widely understood that R2P is not generally applicable to natural disasters, that it is not part of the corpus of international disaster relief law, and that R2P is first and foremost a directive to states themselves to take the measures necessary to protect their populations.

ASEAN's role after Nargis also confirmed the important role of regional organizations in mediating between states and international bodies in disaster contexts. The Third ASEAN-UN Summit held in Hanoi in October 2010 saw the launch of the Post Nargis Recovery and Preparedness Plan as well as the adoption of the Joint Declaration on ASEAN-UN Collaboration in Disaster Management. The two organizations agreed to prepare and implement the ASEAN-UN Strategic Plan of Action on Disaster Management 2011–2015, for coordinating the relevant ASEAN bodies and established UN mechanisms.[110] In 2011, the secretary-general published a report on "The role of regional and subregional arrangements in implementing the responsibility to protect,"[111] which acknowledged that:

> States and civil society groups that are closer to the events on the ground may have access to more detailed information, may have a more nuanced understanding of the history and culture involved, may be more directly affected by the consequences of action or inaction, and may be critical to the implementation of decisions taken in New York. Timely and decisive response is most likely when intergovernmental bodies at both the global and regional levels favour similar courses of action. In such cases, decision-making at both levels is mutually reinforcing in terms of political legitimacy.[112]

The rhetoric of Western states in the wake of Nargis was that Myanmar's rulers rejected aid because they cared more about legitimizing their regime in the forthcoming constitutional referendum than they cared about the welfare of the people. This chapter has sought to complicate this explanation, pointing to the regime's rational fears about secession, disunity, and the West's long-standing interests in regime change. In the weeks following Nargis, some within the West demonstrated a profound lack of understanding about the reasons for the Myanmar government's suspicion and hostility.

[109] Asia-Pacific Center for the Responsibility to Protect, *Implementing the Responsibility to Protect: Asia-Pacific in the 2009 General Assembly Dialogue*, October 2009.

[110] "Statement by Co-Chairs of the Third Association of Southeast Asian Nations-United Nations Summit," Hanoi, 29 October 2010, www.aseansec.org/25496.htm.

[111] UN Security Council, *The role of regional and subregional arrangements in implementing the responsibility to protect: report of the Secretary-General*, A/65/877–S/2011/393, June 28, 2011, www.unhcr.org/refworld/docid/4e7713912.html.

[112] Ibid., 3.

During that period, the intervention of ASEAN was critical to promoting cooperation between Myanmar and the international community. Nargis demonstrated three things: First, the assumption that "burdened states" will be happy to receive assistance from any donor state in the wake of natural disasters is misplaced; second, the idea that "outlaw states" should be forced to receive assistance has no legal justification; and third, regional organizations who understand the political culture of the region are able to impart a critical legitimacy to operations conducted in the wake of disasters.

10

Hunger Without Frontiers: The Right to Food and State Obligations to Migrants

Katie Sykes

Disasters lead to displacement and migration. Natural disasters and armed conflict drive people from their homes, often forcing them to look for refuge in other countries, sometimes because they are at risk of political persecution, often because their means of livelihood have been destroyed. The "permanent disaster" of food insecurity is one of the most significant drivers of global migration. Migrants who are forced to flee due to hunger and material deprivation typically do not meet the requirements to be recognized as refugees under the Refugee Convention. But a complementary body of law has developed side by side with traditional refugee law, based on the principle that people cannot be expelled or deported into situations where their human rights would be violated – for example, to a state where they would be tortured or killed. The same logic that underlies this doctrine of "complementary protection" supports its extension to migrants who face starvation, notwithstanding the traditional view that so-called economic migrants do not deserve any of the protections afforded to true refugees. The right to freedom from hunger is an internationally recognized human right, and to expel or deport someone in such a way as to deprive them of that right is contrary to international law.

INTRODUCTION

This chapter explores a new, or at least unorthodox, proposition: When famine resulting from a disaster drives people to move across international borders, the receiving state has an obligation under international law not to send, expel, or deport those people if doing so would put them at risk of being deprived of the right to food.[1] Such an obligation might follow a natural or human-made disaster (such

[1] The "right to food" refers to the "right of everyone to . . . adequate food" as recognized in Article 11(1) of the International Covenant on Economic, Social and Cultural Rights, Dec. 16, 1966, S. Treaty Doc. No. 95–19, 6 I.L.M. 360 (1967), 993 UNT.S. 3 (ICESCR) and "the fundamental right of everyone to be free from hunger" recognized in Article 11(2). See

as a drought or other extreme weather event, a generalized armed conflict, or a combination of catastrophic circumstances) causing famine, where the domestic authorities either cannot or will not protect people from hunger, and where the needs of the affected population are not sufficiently met with on-site assistance from the international community.

The proposition that people fleeing such a situation might be entitled not to be deported or sent back is based on the international legal principle of non-refoulement. Non-refoulement means the duty not to expel or return international migrants if doing so would result in a deprivation of their rights under international law. This obligation is a key provision of the International Convention relating to the Status of Refugees (the Refugee Convention),[2] but it has also been recognized as extending beyond the Refugee Convention, with an independent basis in international human rights law. The category of non-refoulement obligations to migrants who do *not* fall under the protection of the Refugee Convention has become known as the doctrine of complementary protection: that is, protections for migrants derived from international legal obligations other than (complementary to) the Refugee Convention.[3] This chapter follows the usual practice in international law of referring to people who have been driven from their home countries but are not refugees within the meaning of the Refugee Convention as "migrants," although in common parlance they would typically be described as refugees.

The conceptual foundation of the argument made here is the connection between the development of the doctrine of complementary protection as an outgrowth of refugee law, and certain discernible themes that run through the law, principles, and norms that have grown up around the prevention of and responses to disasters. An overall theme of this book is to understand the latter not just as a disparate jumble of unrelated legal phenomena but as cohering, at least to a degree, around core principles of humanity and human dignity. Dug Cubie brings out the implications of that theme by proposing that transnational processes are creating (if not a fully coherent or settled framework of international disaster law) an emerging, identifiable body of law, principles, and practice that Cubie calls an *acquis humanitaire*.[4] This chapter borrows Cubie's concept of an *acquis humanitaire*, which is fully laid out in Chapter 17 of this book, as a heuristic device for identifying the fundamental legal principles at work in the context of disasters.

discussion of the content and interpretation of this right in Sections 2 and 3. The argument that human rights guarantees imply a corollary obligation of non-refoulement is not novel (see discussion in Section 2), but there has been little discussion of its implications for socio-economic rights, and particularly not in the context of disasters.

2 Convention relating to the Status of Refugees, July 28, 1951, 189 UNT.S. 150, Art 33(1).

3 Jane McAdam, *Complementary Protection in International Refugee Law* (Oxford University Press, 2007), 2–3. See the discussion of complementary protection and examples in Section 2.

4 See Dug Cubie, "Clarifying the "Acquis Humanitaire": A Transnational Legal Perspective on the Internalization of Humanitarian Norms" in Chapter 17 of this volume.

Those fundamental principles, and in particular the cornerstone principle of respect for human dignity, have implications that go further than on-site disaster response and aid donation. They also apply in situations of migration, which is an all too common consequence of disasters. In other words, the doctrine of complementary protection and the principle of non-refoulement, as they come into play with respect to migrants fleeing disaster and its aftermath, should be understood as part of the *acquis humanitaire* and extensions of its underlying rationale. Looking at the law and quasi-law of disaster prevention, relief and assistance as having a degree of coherence around organizing principles suggests that it is worthwhile to think about what the implications are for migration, because migration is one of the survival strategies people resort to in response to disasters, and to connections with developments in the law of migration that are rooted in similar principles of humanity and human rights.

DISASTERS, MIGRATION, AND RIGHTS: THE ROLE OF COMPLEMENTARY PROTECTION

There is tension between traditional state sovereignty and the individual-focused principles of humanity and respect for the needs and rights of individual people affected by disaster. From this tension, difficult and contentious questions have arisen, including when, if ever, it is legitimate for "outside" states to intervene for humanitarian purposes over the objections of the state where a disaster occurs (an issue that was brought into sharp focus by the controversy over the international response to Cyclone Nargis). The same question of constraints on traditional state sovereignty that follows from the imperative of respecting individual human dignity in a disaster situation can also be looked at from another direction – and this way round it may be a less welcome question from the perspective of the outside state. If domestic authorities fail to, or cannot, alleviate the catastrophic effects of a natural or human-made disaster, including famine, when must the sovereignty of other states yield to the needs of migrants whose coping strategies include fleeing across international borders?

What limits apply here to the traditional core discretion of states to control their own borders and populations? The standard answer to this question is straightforward: essentially none, so far as people fleeing the generalized effects of disaster are concerned. Under the Refugee Convention, certainly, states have accepted some constraints on their authority to allow or deny entry, but only when it comes to people who are defined as refugees within the meaning of that Convention. By contrast, "mere" migrants, and in particular so-called economic migrants – typically thought of as people who leave home to escape material and economic deprivation and to find better conditions – have been traditionally seen as having little or no protection under international law. Convention refugees must have a well-founded fear of persecution on specific grounds: race, religion, nationality, membership of a particular social group, or political opinion.[5]

Natural and human-made disasters are, generally, impersonal and unintended; they are not like the targeted social and political persecution that defines a Convention refugee (although it is perfectly possible for people to be victims of both disaster and persecution, or indeed for non-provision of disaster relief to be used as a way of persecuting political opponents or marginalized groups). To argue that migrants fleeing economic or material deprivation, however extreme or urgent, are entitled to international legal protection and could have a right not to be sent home runs counter to the well-established distinction between economic migrants and political refugees. It is accepted that genuine refugees are forced to leave by the loss of the protection of their home state and need protection under international law; by contrast, economic migrants are usually considered to have left home voluntarily to take advantage of better opportunities in another country.[6] The general rule that people who move for economic reasons do not merit refugee-like protection is deeply entrenched.

But the Refugee Convention does not exhaust the sources of duties that states may have to people displaced across borders by disaster. Also relevant are principles that connect migration law and the *acquis humanitaire*: the shared responsibility of the wider community of states, triggered when a state is unwilling or unable to protect its own citizens from catastrophe, including starvation;[7] the cornerstone principle of humanity; and the responsibility of states not to act in ways that deprive people of their fundamental human rights whether or not those people are their own nationals.

The doctrine of complementary protection is a bridge between the realms of refugee and migration law, on the one hand, and international human rights, on the other. Complementary protection is the general term for the protection of migrants based on their human rights, and the corollary duty of states not to deport or expel them, separate and apart from (although sharing conceptual common ground with) Refugee Convention obligations. As the International Law

5 Refugee Convention, Art 1(A)(2) (as amended by the Protocol relating to the Status of Refugees, Jan. 31 1967, 606 UNT.S. 8781).

6 A useful discussion of this traditional dichotomy is in Michelle Foster, *International Refugee Law and Socio-Economic Rights: Refuge from Deprivation* (Cambridge University Press, 2007), 2–21. As Foster observes, a sense of indignation toward 'economic migrants' is sometimes used as a pretext for narrowing the interpretation of obligations under the Refugee Convention. So for example people who are escaping both material deprivation and persecution on Refugee Convention grounds may be classified as economic migrants and denied asylum.

7 International Commission on Intervention and State Sovereignty, *The Responsibility to Protect: Report of the International Commission on Intervention and State Sovereignty* (ICISS Report), International Development Research Centre, December 2001, at viii. See discussion in Dug Cubie, "An Enchanted Tool? Humanitarian Assistance and the ILC Draft Articles on the Protection of Persons in the Event of Disasters" (2009–2010) IV–V *Irish Yearbook of International Law* 119–151, at 131–134.

Commission (ILC) has stated, the principle of non-refoulement "has passed beyond the bounds of international refugee law and . . . is also deemed to be an integral part of international human rights protection."[8]

The Convention Against Torture (CAT)[9] is a useful starting point for discussing the concept of complementary protection. This is a helpful place to begin because the CAT expressly spells out an obligation of non-refoulement – even though this particular obligation is clearly not directly applicable to famine-driven migration, it inarguably establishes the principle that non-refoulement applies in some situations not covered by the Refugee Convention. Article 3(1) of the CAT provides that states have a duty not to expel, return, or extradite a person to a state "where there are substantial grounds for believing that he would be in danger of being subjected to torture." The Committee Against Torture, which can consider communications on behalf of individuals who claim violations of their rights under the CAT by a state party that has recognized the competence of the committee to receive such communications, has become an important venue for protecting people threatened with deportation to torture and addressing some of the specific and practical questions involved in giving effect to the non-refoulement principle, including the standard of proof that a claimant has to meet.[10]

The CAT explicitly establishes an obligation not to expel or return based on international rights and duties beyond the scope of the Refugee Convention.[11] The principle underlying that obligation could be expressed as states having certain responsibilities not to send people back to a situation where they face being deprived of rights (here, the right not to be tortured) that are enshrined in international law. That same principle can also be implicit, even if not expressly stated, in other contexts and as a corollary of other rights.

One such context, where authoritative (although not universally accepted) interpretations have consistently found that there is an implicit non-refoulement obligation, is the International Covenant on Civil and Political Rights[12] (ICCPR), in particular the right to life under Article 6 and the right not to be subjected to torture or cruel, inhuman, or degrading treatment or punishment

[8] International Law Commission, "Expulsion of Aliens: Draft articles on protection of the human rights of persons who have been or are being expelled, as restructured by the Special Rapporteur, Mr. Maurice Kamto, in the light of the plenary debate during the first part of the sixty-first session," UN Doc. A/CN.4/617 (2009), 5 n. 8.

[9] Convention Against Torture (hereinafter: "CAT"), Dec. 10, 1984, 1485 UNT.S. 85.

[10] See Brian Gorlick, "The Convention and the Committee Against Torture: A Complementary Protection System for Refugees" (1999) 11 *International Journal of Refugee Law* 479–495.

[11] There is overlap between the protections of the Refugee Convention and of the non-refoulement obligations in Article 3(1) of the CAT; a situation where someone faces torture may well also be the basis of a well-founded fear of persecution. But someone might be entitled to the latter and not the former if, for example, he or she were at risk of being tortured but not on the basis of one of the five grounds specified in the Refugee Convention.

[12] Dec. 16, 1966, S. Treaty Doc. No. 95–20, 6 I.L.M. 368 (1967), 999 UNT.S. 171.

under Article 7.[13] The United Nations Human Rights Committee has stated that the obligation of states parties to respect and ensure ICCPR rights "for all persons in their territory and all persons under their control" entails "an obligation not to extradite, deport, expel or otherwise remove a person from their territory, where there are substantial grounds for believing that there is a real risk of irreparable harm" such as that contemplated by Articles 6 and 7.[14]

Similarly, the Committee on the Rights of the Child has found an obligation of non-refoulement implicit in the Convention on the Rights of the Child[15] (CRC). Noting the obligation of states to respect established non-refoulement obligations under international law, including under the Refugee Convention and the CAT, the Committee went on to state that in fulfilling their obligations under the CRC:

> States shall not return a child to a country where there are substantial grounds for believing that there is a real risk of irreparable harm to the child, such as, but by no means limited to, those contemplated under Articles 6 [the child's inherent right to life, and the obligation of States Parties to ensure to the maximum extent possible the survival and development of the child] and 37 [the rights to life and liberty, and the right not to be subjected to cruel, inhuman or degrading treatment or punishment] of the Convention.[16]

The assessment of whether there is a risk of serious violations, the Committee went on to say, "should, for example, take into account the particularly serious consequences for children of the insufficient provision of *food* or health services."[17] As Jason Pobjoy argues, the Committee's open-textured language and reference to risks including but not limited to those contemplated under Articles 6 and 7 suggest that the more general principle recognized in the comment is that the risk of a serious violation of the paramount CRC principle of the best interests of the child can give rise to a non-refoulement obligation.[18]

Although these interpretive statements are not in themselves binding law, and states may be reluctant to accept such limits on their right to expel non-nationals that are not expressly agreed to by treaty, many states do expressly recognize the principle of complementary protection in their domestic legal systems. Canada's Immigration and Refugee Protection Act,[19] for example, provides that the same protection available to Convention refugees shall also be extended to "persons in need of protection,"[20] defined as persons whose removal would subject them to a danger of torture, or a risk to life or a risk of cruel and unusual treatment or

[13] See discussion in Jason Pobjoy, "Treating Like Alike: The Principle of Non-Discrimination as a Tool to Mandate the Equal Treatment of Refugees and Beneficiaries of Complementary Protection," *Melbourne Law Review* 34(2010): 181–229 at 190–191.

[15] Convention on the Rights of the Child, Nov. 20, 1989, 1577 UNT.S. 3.

[17] Ibid. (emphasis added). [18] Pobjoy, "Treating Like Alike," 192.

[19] S.C. 2001, c.27. [20] S. 95(1)(b).

punishment.[21] There is also extensive European law on complementary protection derived in part from Article 3 of the European Convention for the Protection of Human Rights (ECHR),[22] which is similar to Article 7 of the ICCPR.[23]

What part might the principles of complementary protection and non-refoulement play as a component, or extension, of the constellation of rules and norms that make up an *acquis humanitaire*? International norms on disaster response and the principle of complementary protection share a common foundation in respect for human dignity and human rights and the duty to respond to individual human needs. These core principles are reflected in Article 2 of the ILC's Draft Articles on the Protection of Persons in the Event of Disasters, which states that the overall purpose of the draft articles is "to facilitate an adequate and effective response to disasters that meets the essential needs of the persons concerned, with full respect for their rights";[24] in Article 7, which calls on states and other entities responding to disaster to "respect and protect the inherent dignity of the human person";[25] and in Article 8, which confirms that "[p]ersons affected by disasters are entitled to respect for their human rights."[26]

Honoring the dignity, rights, and essential needs of people affected by a disaster are the overarching guiding principles for on-site response and assistance on the part of domestic authorities, non-governmental and intergovernmental organizations, and outside states. But the implications of those principles and the responsibilities that they entail do not end at the site of a disaster; the individual people affected still have their human rights, and are still entitled to respect for those rights, when they flee the disaster and its consequences, including when they move across international borders.

It is not coincidental that the terminology of *protection* is associated both with non-refoulement obligations (complementary protection) and with disaster assistance. Humanitarian assistance reflects concentric circles of responsibility for

21 S. 99.

22 Nov. 4, 1950, Europ.T.S. No. 5; 213 UNT.S. 221. Article 8 is the right not to be subjected to torture or to inhuman or degrading treatment or punishment.

23 The European Court of Human Rights has determined that there is an obligation of non-refoulement in connection with Article 3 and also Article 8, the right to respect for private and family life. See summary of the case law, and discussion of the potential for other ECHR rights to form the basis of a similar obligation, in McAdam, *Complementary Protection*, 136–72. As McAdam points out, the recognition of a principle of non-refoulement under Article 3 has a long history, going back to the early 1960s. Ibid., 137. In 2004 the EU adopted the *Qualification Directive* (Council Directive 2004/83/EC of 29 April 2004 on minimum standards for the qualification and status of third country nationals or stateless persons as refugees or as persons who otherwise need international protection and the content of the protection granted, OJ L 304, 30/09/2004 P. 0012 – 0023), which makes making certain benefits available to persons "genuinely in need of international protection" who are not covered by the Refugee Convention.

24 International Law Commission Sixty-first Session (A/CN.4/L.758, July 24, 2009).

25 International Law Commission Sixty-second Session (A/CN.4/L.776, July 14, 2010).

26 Ibid.

protecting the needs and rights of those affected. Sovereign states have the first responsibility to protect their own citizens in a catastrophe, as the ILC Draft Articles underline.[27] When disaster forces people to move within state borders, this includes, of course, their responsibilities to internally displaced persons in accordance with the UN Guiding Principles on Internal Displacement.[28]

Residual responsibility lies with the "broader community of States"[29] to protect those who are not adequately helped by the domestic authorities, whether because of generalized state failure, lack of resources, deliberate policy choices, or a combination of causes. Whatever extra-territorial obligations or standards of good practice this responsibility entails in connection with providing disaster assistance, donating emergency aid, and other help that flows into the affected state, and whatever degree of intervention in the home state's internal affairs it may or may not justify, it is centered on the individual human beings who need protection. The responsibility of the "broader community of states" towards those people does not disappear if they leave their own state, precisely because it is a responsibility to *people* and not, ultimately, to the geographical location or political entity they inhabit. This implies that other states have a responsibility to give protection, in a manner appropriate to the context, when those displaced by disaster come within their borders.

In certain disaster situations, sending migrants back would involve exposing them to conditions that could amount to cruel, inhuman, or degrading treatment, or to the risk of being deprived of the right to life. The principle of non-refoulement implied by Articles 6 and 7 of the ICCPR would then be at issue. In famine situations, the right to freedom from hunger under Article 11 of the International Covenant on Economic, Cultural, and Social Rights (ICESCR) would also be relevant. Although there has not so far been much attention to the idea of a non-refoulement principle grounded in the right to freedom from hunger, there is no reason in principle why rights enshrined in the ICESCR should not be a valid basis

[27] ILC Draft Article 9(1) states that "[t]he affected State, by virtue of its sovereignty, has the duty to ensure the protection of persons and provision of disaster relief and assistance on its territory." Draft Article 9(2) further specifies that "[t]he affected State has the primary role in the direction, control, coordination and supervision of such relief and assistance." Note 25 above..

[28] UN Economic and Social Council, *Guiding Principles on Internal Displacement*, July 22 1998, E/CN.4/1998/53/Add.2.

[29] ICISS Report, p. vii. This framework – the primary responsibility lying with the state to protect its own people, and international responsibility triggered when the state in question is unwilling or unable to remedy a situation where the population suffers serious harm – is reflected in the basic principles of the responsibility to protect doctrine as set out in the ICISS Report (see Principles 1(A) and 1(B) (p. xi)). The extent to which these principles support intervention by the international community for humanitarian reasons and their implications in the context of disasters remain highly controversial. See discussion in Cubie, "An Enchanted Tool."

for complementary protection, in a manner analogous to rights recognized in other human rights instruments.

Article 11 is of particular relevance in disaster situations in light of the interpretation of the interpretation of this right by Committee on Economic and Social Rights (CESCR), the body of experts that is responsible for monitoring implementation of the ICESCR. In its General Comment No. 12 interpreting Article 11, the CESCR notes that "immediate and urgent steps may be needed"[30] to ensure the fundamental right to freedom from hunger and malnutrition under Article 11(2), and that "States have a core obligation to take the necessary action to mitigate and alleviate hunger as provided for in paragraph 2 of article 11, even in times of natural or other disasters."[31] These interpretive statements underline the specific and pressing nature of state responsibilities to respond to famine and hunger caused by disasters. While these are in the first instance responsibilities of the affected state to its own people, there is a secondary responsibility of the "broader community of states" to protect the right to food in disaster situations. This obligation can be the foundation, in specific circumstances, of a duty not to expel migrants.

HUMAN RIGHTS AND ASSOCIATED STATE OBLIGATIONS: IMPLICATIONS FOR DISASTER RESPONSE

The expansion of non-refoulement protection for migrants has been associated more with civil and political rights than with social and economic rights, perhaps because of the strong influence of the refugee/economic migrant dichotomy, which suggests that migrants fleeing economic or material deprivation are fundamentally different from refugees who deserve protection. But the dichotomy implies a sharp distinction between civil-political and socioeconomic rights that is questionable, and can be difficult to apply to real-world circumstances. Disasters, for example, may give rise to a risk of deprivation of the right to life (an ICCPR right) and also of CESCR rights that are indispensable to sustaining life, or a life of any quality – including the right to freedom from hunger.

More than thirty years ago, Henry Shue argued compellingly against the strain in U.S. foreign policy thinking (and beyond) that economic, social, and cultural rights are "less genuine rights with less binding duties."[32] For Shue, rights to the minimum requirements for subsistence are "basic rights," in that "enjoyment of them is essential to the enjoyment of all other rights."[33] Minimal economic security, or

[30] General Comment No. 12, May 12, 1999, 20th Sess., UN Doc. E/C.12/1999/5, par. 1.

[31] Ibid., par. 6

[32] Henry Shue, *Basic Rights: Subsistence, Affluence and U.S. Foreign Policy*, 2nd ed. (Princeton University Press, 1996), 6 (originally published 1980). I am grateful to Professor Joanna Harrington for drawing my attention to Shue's work.

[33] Ibid., 19.

subsistence, including adequate food, is a basic right in the same sense that physical security is, and by the same logic.[34]

Shue argued that there are three duties that correlate to every basic right, which must be performed (by someone – not necessarily always the same someone) if the right is to be honored. They are the duty to avoid depriving people of the right, the duty to protect people from being deprived of the right, and the duty to aid those who have been deprived.[35]

The influence of Shue's three-level framework of state duties is evident in the widely accepted concept of state obligations in respect of human rights and especially of social, economic, and cultural rights: that is, the obligations to respect, protect, and fulfill rights. On the right to food in particular, Asbjørn Eide, who was the UN Special Rapporteur on the Right to Adequate Food as a Human Right, presented a detailed discussion of the tripartite respect–protect–fulfill framework in his 1989 report, Right to Adequate Food as a Human Right.[36] Eide split the third level of duty, the obligation to fulfill, into two sub-components: first, facilitation of access to adequate food, and second, direct provision of food. Eide's framework was in turn adopted by the CESCR in General Comment No. 12,[37] as well as by other human rights actors, academic commentators, and domestic courts.[38]

The CESCR interpretation of Article 11 provides further insight into the practical responsibilities that these obligations entail. The obligation to respect is a negative duty of states "not to take any measures that result in preventing" access to food.[39] The obligation to protect "requires measures by the State to ensure that enterprises or individuals do not deprive individuals of their access to adequate food."[40] The obligation to fulfill (facilitate) "means that the State must pro-actively engage in activities intended to strengthen people's access to and utilization of resources and means to ensure their livelihood, including food security."[41]

The obligation to *fulfill* by *providing* food directly arises "whenever an individual or group is unable, for reasons beyond their control, to enjoy the right to adequate food by the means at their disposal," an obligation that applies "for persons who are victims of natural or other disasters."[42] The CESCR also notes that, while some obligations of states parties are "of a long-term character, to achieve progressively the full realization of the right to food," the right also implies obligations of an immediate nature (including, it can be surmised, in times of disaster and emergency).[43]

[34] Ibid., 23. [35] Shue, *Basic Rights*, 52. [36] United Nations, 1989.

[37] General Comment No. 12, May 12, 1999, 20th Sess., UN Doc. E/C.12/1999/5, par. 1.

[38] Kerstin Mechlem, "Harmonizing Trade in Agriculture and Human Rights: Options for the Integration of the Right to Food into the Agreement on Agriculture," *Max Planck Yearbook of United Nations Law*, 2006: 127–190 at 136–137.

[39] General Comment No. 12, par. 15. [40] Ibid. [41] Ibid.

[42] Ibid. [43] Ibid., par. 16.

These state responsibilities, especially those associated with the obligation to fulfill the right to food and its two aspects of facilitation and direct provision, have special relevance for disasters. Fulfillment-facilitation (strengthening access to food) when disaster strikes can be achieved in part through strategies of disaster risk reduction and the development of resilient systems for food security and access to food, lowering the probability of famine in the wake of disaster. Fulfillment-provision, the obligation to provide food directly, becomes relevant when planning and risk reduction have failed, and disaster assistance is needed to ensure that disaster victims will not be deprived of the right to food.

The primary responsibility for disaster prevention and disaster response, and for the enjoyment of human rights including the right to food, rests with the affected state. This understanding is a common thread evident both in the ILC Draft Articles[44] in the context of disasters, and in international human rights law. Eide has written that the "primary duty-bearers" under the ICESCR are states, and that the primary responsibility "rests with each State in relation to everyone living on its territory."[45] The wider international community has a secondary responsibility to ensure protection of the right to food when the home state cannot or will not. The text of Article 11(2) refers to obligations to take measures both individually and through international cooperation to improve food production and the equitable distribution of food supplies, confirming that state commitments go further than duties owed to their own citizens and also encompass an obligation of and to the wider international community.

Eide locates this second level of responsibility under Article 11 "with the community of States, which are obligated to ensure the enabling conditions required to ensure the enjoyment of the right everywhere, to abstain from any policy which harms or prevents the enjoyment of the right to food in other countries than their own, and to take separate and joint action to assist States that have difficulties in ensuring the right to food for everyone within their territory."[46] He contends that it was "clearly envisaged from the earliest drafting of the Charter of the United Nations" that states should have obligations "to the international community and to the peoples of other States."[47] The CESCR states in General Comment No. 12 that "States parties should take steps to respect the enjoyment of the right to food in other countries, to protect that right, to facilitate access to food, and to provide the necessary aid when required."[48]

[44] A/CN.4/L.776, Article 9.

[45] *The Right to Food and the Impact of Liquid Biofuels (Agrofuels)* (Rome: Food and Agriculture Organization of the United Nations, 2008): 27. www.fuhem.es/media/ecosocial/file/Boletin%20ECOS/ECOS%20CDV/Bolet%C3%ADn%204/Right__Food_and_Biofuels.pdf, 27.

[46] Ibid., 27. [47] Eide, *Right to Adequate Food as a Human Right*, 38.

[48] General Comment No. 12, par. 36.

More recently, the Maastricht Principles on Extraterritorial Obligations of States in the Area of Economic, Social and Cultural Rights[49] (adopted by a group of academics and non-governmental organizations in 2011) have further emphasized and elaborated on the commitment of states under international law to take positive steps, including through international assistance and cooperation, for the universal realization of social, economic, and cultural rights.

In disasters, states often act in accordance with these responsibilities by providing disaster relief assistance at a state-to-state or institutional level. But the common conceptual foundation of disaster relief principles and human rights law in the dignity of the human person suggests that ultimately it is the needs of individuals affected by disaster that underpin state obligations regarding disaster risk prevention and also in the aftermath of disaster. If that is correct, then state-to-state cooperation and assistance do not necessarily amount to a complete fulfillment of state responsibilities. If institutional efforts at disaster relief fail, or are executed inadequately, and if individual people slip through holes the net of protection in their home countries, they still have rights enshrined in international law, and states still have (jointly and severally) responsibilities to ensure their protection.

It is an unfortunate reality that disaster relief efforts do fail, or fail to reach everyone in need, and, as a result, people move. These are the circumstances in which inadequate disaster relief and international protection for migrants become conceptually intertwined.

DISASTERS, FAMINE, AND MIGRATION

Poverty and hunger are strong drivers of international migration. In many cases the background cause is general conditions of poverty and underdevelopment. But often it is a sudden calamity – a catastrophic weather event, the outbreak of armed conflict, a depletion of food supplies caused by misguided government policy, or an urgent situation brought on by a combination of such factors – that brings the situation to the point of crisis and puts people in such a desperate situation that they are forced to move.

The doctrinally clear legal distinction between Convention refugees (targets of persecution who are deprived of the protection of their home state) and economic migrants (supposedly choosing to seek a better life) can be hard to reconcile to the reality of migration. As United Nations High Commissioner for Refugees António Gutteres reported to the General Assembly in 2007, economic deprivation is "the strongest driving force" of migration, with

[49] Final version (February 29, 2012), www.rtfn-watchfian.org/uploadsfileadmin/media/publications/2012.02.29_-_Maastricht_ETO_Principles__EN_on_Extraterritorial_Obligations.pdf.. Maastricht_ETO_Principles__EN_on_Extraterritorial_Obligations.pdf.

"thousands of poor migrants resorting to illegal and increasingly desperate routes to the global economy."[50]

In 1994, the then-High Commissioner Sadako Ogata observed in a 1994 Note on International Protection that there is a "lack of a complete correspondence between the categories of persons covered by the [Refugee Convention] and the broader class of persons in need of international protection," noting that those fleeing armed conflict are often driven from home by knock-on effects "including the destruction of homes, harvests, food stocks and the means of subsistence."[51] As Gutteres indicated, the challenge for the international community is: "How to answer people who have left their country to find food? If they are not refugees, can we knowingly send them back to extreme deprivation?"[52]

One example that illustrates the connections between disaster, famine, migration, and failure of protection at the domestic and international levels is the recurrent problem of famine in North Korea, which was particularly dire in the 1990s.[53] The 1990s food emergency in North Korea was the result of a combination of human and natural causes, floods and droughts in 1995 exacerbating disastrous policy decisions by the North Korean government and the shock of changes in trading relationships with China and the former Soviet Union.[54]

Tens of thousands of North Koreans driven by "desperate hunger and extreme poverty" crossed the border into China.[55] For some time, these border-crossers were quietly absorbed as undocumented immigrants, until the international media and non-governmental organizations began pressuring China to grant them asylum, following which the Chinese government cracked down on the influx and began deporting border-crossers back to North Korea.[56] At least some of these escapees from a notoriously politically oppressive regime could have had a good claim for asylum on Refugee Convention grounds; access to food rations through the government's public distribution system was determined in

[50] Statement by Mr. António Guterres, United Nations High Commissioner for Refugees, to the Third Committee of the United Nations General Assembly, 62nd Session, New York, November 8, 2007. www.unhcr.org/476132d84.html.

[51] UN Doc. A/AC.96/830, p. 15.

[52] Ibid. Note 50 above.

[53] The 1990s famine in North Korea is estimated to have killed as many as one million people, about 5 percent of the population. Steven Haggard and Marcus Noland, U.S. Committee for Human Rights in North Korea, *Hunger and Human Rights: The Politics of Famine in North Korea* www.hrnk.org/uploads/pdfs/Hunger_and_Human_Rights.pdf , 9.

[54] Ibid., 18.

[55] Human Rights Watch, *The Invisible Exodus: North Koreans in the People's Republic of China* (2002), 9. www.hrw.org/reports/2002/northkorea/norkor1102.pdf. See also Elisa Ghang, "North Korean Border-Crossers in Yanbian: The 'Protection Gap' Between the Economic Migrant and Refugee Regimes," *Georgetown Immigration Law Journal* 24 (2010): 361–376.

[56] Ghang, "North Korean Border-Crossers," 367.

part by party loyalty and geographic location,[57] so it could be argued that starving North Koreans had been persecuted based on political opinion or membership of a particular social group.

But China would also have quite credible grounds for taking the position that these people, or the majority of them, were victims of generalized, impersonal conditions – government ineptitude, geopolitical forces, and natural disaster – rather than deliberate and targeted state persecution. In that case, the Refugee Convention would not require China to grant them asylum. Yet sending those migrants back to North Korea almost inevitably meant sending them to deprivation of the right to food, and in some cases of the right to life.

Another example is the severe drought in East Africa in 2011 and 2012. Somalia was especially hard hit by a food crisis caused by four years of failed rains, and exacerbated by tensions between international aid donors and the Al Shabaab militants who controlled much of the south.[58] The UN declared a famine in the Bakool and Lower Shabelle regions of southern Somalia in July 2011.[59] The food crisis drove mass migration, mainly by land into neighboring Djibouti, Ethiopia, and Kenya.[60]

A smaller, but still significant, number of Somalis fleeing famine and armed conflict have made their way to Europe to seek asylum, many by traveling to North Africa and then undertaking the hazardous boat trip across the Mediterranean to Malta or the Italian island of Lampedusa.[61] The factors driving this stream of migration are multiple and complex, with the food crisis part of the mix and for some people the primary impetus for fleeing. If lack of food is their main reason or even a significant reason for leaving home, migrants are likely to face considerable difficulty making an asylum claim in accordance with the Refugee Convention.

Those who are not granted asylum and are deported back to Somalia face the risk of being sent into an ongoing humanitarian crisis that has rendered a large proportion of the population dependent on food aid, with thousands confined in displaced persons' camps where there is limited access to the necessities of life, including food and water.[62] In the 2011 case *Sufi and Elmi v. The United Kingdom* (where the claimants' asylum claims had been unsuccessful), the European Court of Human Rights found that being sent back to these conditions amounted to a

57 Haggard and Noland, *Hunger and Human Rights*, 9, 14.

58 Council on Foreign Relations, "Al-Shabaab and Somalia's Spreading Famine" (Interview with Rashid Abdi) (August 10, 2011), www.cfr.org/somalia/al-shabaab-somalias-spreading-famine/p25630.

59 UN News Centre, "UN declares famine in two regions of southern Somalia" (July 20, 2011), www.un.org/apps/news/story.asp?NewsID=39086.

60 UNHCR, "Crisis in Horn of Africa: A Worsening Humanitarian Situation," retrieved August 19, 2013 from www.unhcr.org/pages/4e1ff4b06.html.

61 BBC News, "Lampedusa and Malta take in African migrants," August 21, 2012, www.bbc.co.uk/news/world-europe-19334609.

62 *Sufi and Elmi v. the United Kingdom*, ECHR (2011) 1045, pars. 284–292.

deprivation of the right to be free from inhuman or degrading treatment or punishment under Article 3 of the ECHR.[63] To the extent that these conditions still prevail, refoulement to Somalia involves a real risk of being deprived of human rights, including the right to food.

In situations like these, the Refugee Convention does not always respond adequately to the loss of basic rights experienced by migrants who are driven from their homes by a disaster; nor does it exhaust the international legal responsibilities of the states to which they flee.

STATE OBLIGATIONS TO RESPECT, PROTECT, AND FULFILL AS APPLIED TO DISASTER-RELATED MIGRATION

Part of the value of examining rules, guidelines, principles, and practices that have emerged on disaster response holistically (as an "*acquis*") and analyzing the conceptual links between them lies in the insight that such examination reveals further implications of the underlying concepts, and how they might extend to situations that often ensue from a disaster. The full range of human rights, including the right to food, are, insofar as they have to do with the needs of people affected by disaster, in principle part of the *acquis humanitaire*, or at least conceptually linked to it. Disasters force people to migrate. The international rights and responsibilities applicable to such migrations should be considered as part of or connected to the architecture of the *acquis humanitaire*, and interpreted and applied in light of its animating principles.

If respect for the human rights of people affected by disaster is the cornerstone of the *acquis humanitaire*, then socioeconomic rights, including the right to food, must be part of the legal framework for responding to disasters. The requirement to respect, protect, and fulfill the right to food is still applicable when people in need migrate out of their home countries. The doctrine of complementary protection recognizes that human rights can ground an obligation of non-refoulement toward migrants who are not entitled to the protection of the Refugee Convention. If this is the case for rights enshrined in the ICCPR, then in principle it must also be potentially true for ICESCR rights. Although socioeconomic rights in general raise analytical issues that can be different from those associated with civil and political rights, and cannot necessarily be simply dropped into the same template, neither can they be categorically excluded as a possible basis for implicit non-refoulement obligations. A hard categorical distinction would be inconsistent with the principle that human rights are indivisible; interdependent; and related; to be treated in fair and equal manner; on the same footing; and with the same emphasis.[64]

63 Ibid.

64 Vienna Declaration of the 1993 World Conference on Human Rights, July 12 1993, UN Doc. A/CONF.157/23 at par. 5

In the context of disaster, the need for complementary protection might arise when the response in the affected country has failed, or has left some people's basic needs unmet. Such situations drive people to move, both within and across borders, in search of what they need to survive, including food. The logic underlying the doctrine of complementary protection implies that a receiving state could not send someone back to the disaster-affected state if it meant that person would face starvation or severely limited access to food. By implication the background circumstances that would lead to this kind of situation would have to involve a significant failure of both the home state and the wider international community to fulfill their responsibilities of disaster prevention and response and to protect the people affected. In such a context, non-refoulement might be thought of as being in the nature of a remedy for the failure; it becomes an aspect of the receiving state's obligation to "facilitate access to food" in a situation where help is needed. Indeed, an obligation of non-refoulement would actually seem to be implied by the first-order obligation to "respect" the right to food by refraining from measures that prevent access to adequate food, here by forcing someone (back) into hunger.

Recognition that the principle of non-refoulement applies to disaster-driven migration is an incremental extension of the complementary protection obligations associated with Articles 6 and 7 of the ICCPR. As the *Sufi and Elmi* case illustrates with respect to Article 3 of the ECHR, if the conditions to which a migrant would be returned involve severe deprivation of basic necessities of life, including food, then they may amount to cruel, inhuman, or degrading treatment – or, indeed, a risk of being deprived of the right to life. The right to food would also be implicated in such a situation.

Going further, the right to food is a separate and distinct right that implies cumulative state obligations. If the situation in the disaster-affected country has improved to the point that it is just good enough for life to be sustained, but there is still a severe shortage of food, migrants should still be entitled to protection and non-refoulement; people have a right to live and also a right not to be hungry or malnourished. It is in the nature of disasters, by contrast to more general conditions of poverty and underdevelopment, that they are limited in time, with a beginning and an end. It might be expected, therefore, that any non-refoulement obligation would lapse when conditions in the home country improve enough that going back would no longer entail a risk of being deprived of basic human rights – although in practice the effects of disaster can be severe and long-lasting for some of the people affected, as has proved the case in Somalia.

Complementary protection for migrants is by no means a good solution to the problem of disaster-driven migration. It is not an answer to the deprivations of disaster that people can seek shelter in another country (even if they had a recognized legal right to do so); it would be better for them not to be driven from home at all. Complementary protection should, rather, be considered as one ingredient in the mix that makes up the *acquis humanitaire*, which includes responsibilities first to take preventative measures to stop avoidable disasters from happening at all, to plan for risk reduction and mitigation so that the consequences of disasters are reduced in severity, and to respond in an effective way when disasters do occur. Non-refoulement might be thought of as a residual protection, a last resort that applies when other efforts have failed.

Another concern is that the burden of providing this protection falls disproportionately on immediately neighboring states, often straining their capacity to respond. There is undeniably a risk of unfairness if this responsibility, which neighboring states often have little choice but to assume as a practical matter, is characterized as a legal obligation. On the other hand, it is a more principled approach to consider the obligation to accept disaster-driven migrants in the framework of international law, rather than just an *ad hoc* coping mechanism.

Under the ICESCR, a state's obligations are limited in principle by its available resources.[65] Migrants fleeing disaster do also make their way to wealthier countries that arguably have more resources available to cope with an influx of displaced people, but the typical reaction of the more privileged countries of the world is interdiction or deportation, and legal cover for that reaction is provided by the distinction between refugees and economic migrants. The doctrine of complementary protection undercuts that excuse. Recognition that the basic proposition on which complementary protection is founded – sending a person back to a serious risk of deprivation of human rights is in itself a failure to protect human rights – also extends to the severe loss of social and economic rights in the wake of disaster. As we face the probability of more frequent catastrophic weather and climate events, bringing in their wake famines and mass movements of people, the importance of dealing with the consequences in a fair and principled way only increases.

CONCLUSION

How to answer people who have left their country to find food? Does international law have no answer for them but "go home?" A solid and enduring state prerogative, that of defining the state's own population, who is in and who is out, stands in the way of their claims. Peter J. Spiro has called this state prerogative – here referring in particular to the definition of citizenship – the "last bastion of sovereign discretion,"[66] and one remaining area that is relatively undiluted (with limited exceptions, including under international refugee law) by the universalizing concepts of individual rights and shared membership in a single human family. It is

65 ICESCR Art. 2(1).

66 Peter J. Spiro, "A New International Law of Citizenship," *American Journal of International Law*, (2011): 694–746 at 694.

trite international law that human rights can in principle supersede the usual inviolability of state borders. To recognize that this principle potentially extends, in situations of catastrophic suffering and urgent need, to the right to be free from hunger would be no more than extending the logic of an already widely accepted doctrine.

11

Disasters, International Law, and Persons with Disabilities

Akiko Ito

International disability rights have developed separately from international disaster relief law. The challenges for people with disabilities are compounded when disaster strikes. Yet, there has been, up to now, little discussion of what legal obligations are necessary in the area of disaster preparedness, disaster risk reduction, and disaster relief to address the unique challenges of people with disabilities. This chapter proposes a more holistic approach to mainstreaming disability rights into the emerging discipline of disaster law.

THE INTERNATIONAL NORMATIVE FRAMEWORK ON DISABILITY AND DISASTER

The work of the United Nations in the field of disability is based on its Charter – promoting economic and social progress, human rights, and fundamental freedoms for all. Based on the notion that lasting international peace and security is possible only if the economic and social well-being of people everywhere is assured, major UN conferences and summits in recent years including Rio+20, Beijing Conference, conferences on climate change and the environment, and those on financing for development, and their follow-ups have laid out a holistic approach for development, reaffirming the importance of freedom, peace and security, and respect for all human rights for all persons.

The global framework on disability is now guided by the newly adopted outcome document from the first high-level meeting of the General Assembly on disability and development, which reaffirms the need for a realization of the Millennium Development Goals (MDGs) and other internationally agreed development goals for persons with disabilities toward 2015 and beyond.[1] The document specifically emphasizes the necessity of strengthening preparedness and disaster risk reduction

[1] UN High Level Meeting on Disaster and Development (HLMDD) Article 1–2

for persons with disabilities, among other human rights. The current eight MDGs – which range from halving extreme poverty rates to halting the spread of HIV and AIDS and providing universal primary education, all by the target date of 2015 – form a blueprint agreed to by all the world's countries and all of the world's leading development institutions.

The goals and objectives of the Convention on the Rights of Persons with Disabilities (CRPD), adopted in 2006, guides the global disability architecture, which advances the development of norms and standards relating to disability, humanitarian assistance, and peace and security UN High Level Meeting on Disaster and Development (HLMDD).

The international normative framework on disability consists of the CRPD; the World Programme of Action Concerning Disabled Persons (WPA), adopted in 1982; the Standard Rules on the Equalization of Opportunities for Persons with Disabilities, adopted in 1993; as well as other relevant human rights and disability instruments, such as the Convention on the Elimination of All Forms of Discrimination against Women (CEDAW) or the Convention on the Rights of the Child (CRC), and internationally agreed development goals like the MDGs. The Sustainable Development Goals, negotiated as a result of the Rio+20 Conference as a post-2015 replacement to the current MDG framework, will also impact on international norms and standards relating to disability and disaster.

BACKGROUND TO THE ISSUE OF DISASTER AND DISABILITY

Different populations, when exposed to similar risks of environmental and man-made disasters, may experience varying degrees of impacts that are dependent on a number of factors. These include poverty, social status, geographical location, and access to mitigation and relief resources. What we do know is that individuals with disabilities are disproportionately affected in disaster, emergency, and conflict situations due to a combination of such factors.

For instance, most shelters and refugee camps are not accessible. In addition, often persons with disabilities are turned away from such facilities due to a perception that they need "complex medical" services. Disruption to physical, social, economic, and environmental networks and support systems affect persons with disabilities much more than the general population. There is also a potential for discrimination on the basis of disability when resources are scarce. Furthermore, the needs of persons with disabilities continue to be excluded over the more long-term recovery and reconstruction efforts, thus missing another opportunity to ensure that cities are accessible and inclusively resilient to future disasters.

Experience reveals that persons with disabilities are more likely to be left behind or abandoned during evacuation in disasters and conflicts due to a number of factors including lack of preparation and planning, and lack of access to early warning and information concerning disaster responses. For example, in the

2011 Tohoku earthquake and tsunami experienced by Japan, persons with disabilities suffered disproportionately. Data shows the death rate of persons with disabilities was at least twice as high as that of the general population affected.

Incidence of disability also increases during disasters and conflicts. For example, in Haiti, approximately 200,000 people are expected to live with long-term disabilities as a result of injuries sustained in the January 2010 earthquake. In the past, attention to disability in post-conflict countries has focused on land mines but unfortunately the problem runs much deeper. Conflict is a major cause of not only physical disabilities but also psychological and sensorial disabilities. Disruptions such as suspension of immunization practices during times of conflict can also be a cause of disability and highlights the multifaceted linkages between conflict and disability.

Recognizing that persons with disabilities are at greater risk in the face of natural disaster and emergency situations, it is essential to mainstream disability in the global normative framework on disaster preparedness and emergency management to mitigate human suffering. This requires a twin-track approach of targeted programs for persons with disabilities to promote their empowerment and self-advocacy hand in hand with systematic inclusion of disability in disaster-related policy frameworks and programs.

It is also about increasing the visibility of persons with disabilities in national as well as international normative frameworks for disaster preparedness and emergency management and reconstruction. In this context, this chapter provides an overview of the work of the United Nations in promoting the inclusion of persons with disabilities in disaster and emergency planning.

THE WORK OF THE UNITED NATIONS

The status of the international normative framework on disaster and its evolving norms and standards have been reviewed, analyzed, and discussed in the context of humanitarian law, human rights, and international development cooperation. While there is no comprehensive international normative framework on disaster and disaster management, international law must respond to this situation. Currently, domestic law largely prevails when dealing with disaster relief, often resulting in a lack of expertise and defective planning due to a lack of coordination between domestic legal systems.

The outcome document of the high-level meeting of the General Assembly on disability and development on September 23, 2013 urges member states to strengthen social protection and humanitarian programming and responses in the realm of disaster risk reduction for persons with disabilities.

In 2006, the General Assembly adopted the CRPD. The CRPD is intended as a human rights instrument with an explicit, social development dimension. It clarifies and qualifies how all categories of rights apply to persons with disabilities and identifies areas where adaptations have to be made for persons with disabilities to effectively exercise their rights and areas where their rights have been violated, and where protection of rights must be reinforced. As of October 1, 2013, the CRPD has 158 country signatories and 137 ratifications and accessions.

The CRPD should be read together with other human rights treaties such as the CEDAW or the CRC since they are all interrelated and mutually reinforcing. The intersection between the provisions in these instruments for example can strengthen the basis for inclusion and effective participation of all members of society (e.g., including girls and women, children, and persons with disabilities).

Turning to the disaster dimension, the CRPD specifically addresses the rights of persons with disabilities in situations of risk and humanitarian emergencies, calling upon states parties to undertake "all necessary measures to ensure the protection and safety of persons with disabilities in situations of risk, including situations of armed conflict, humanitarian emergencies and the occurrence of natural disasters" (Article 11). The CRPD also stipulates that "States Parties undertake measures to ensure and promote the full realization of all human rights and fundamental freedoms for all persons with disabilities without discrimination of any kind on the basis of disability" (Article 4.1) and recognizes the importance of international cooperation to address the limited capacities of some states to respond to situations of risk and humanitarian crises (Article 32).

Since the adoption of the CRPD, the General Assembly has, through a series of resolutions,[2] highlighted disability as a cross-cutting development concern and reiterated the urgent need to include a disability perspective and persons with disabilities in all aspects of the global development agenda, including the MDGs. It has urged states "to take, in accordance with their obligations under international law, including international humanitarian law and international human rights law, all necessary measures to ensure the protection and safety of persons with disabilities in situations of risk, including situations of armed conflict, humanitarian emergencies and the occurrence of natural disasters..."[3]

The issue of accessible emergency and disaster relief and reconstruction is gaining more attention in recent years. The choices that are made during, before, and after, such as in the reconstruction phase, are crucial to mitigate human suffering of people in vulnerable positions during emergency situations.

In April 2012, a two-and-a-half day UN Expert Group Meeting (EGM) organized by the Department of Economic and Social Affairs of the United Nations (DESA), in close collaboration with the UN Information Center in Tokyo and The Nippon Foundation, took place in Tokyo, Japan. The EGM was held under the theme "Building Inclusive Society and Development through Promoting Accessibility in Information and Communication and Technologies (ICTs)" and

[2] UNGA 63/150, 64/131 and 65/186 [3] UNGA 63/150, par. 9.

included a special plenary session on how accessible ICTs have contributed to recovery in natural disaster and emergency responses, including in the context of the Great East Japan Earthquake and subsequent tsunami, and other selected national experiences.[4]

Another example of the work being performed in the realm of disability development is the seminar on Mainstreaming Disability in Development Cooperation organized by the UN and the Norwegian Agency for Development Cooperation (NORAD) in 2012 that examined challenges faced in mainstreaming disability concerns in development cooperation policies and programs. The CRPD affirms that not only states parties have a significant role to play in international cooperation to promote the rights of persons with disabilities but so do non-state actors. States in partnership with civil society, in particular organizations of persons with disabilities, and international and regional organizations, are encouraged to take measures to support national implementation of the CRPD. These measures could include, for example, the mutual exchange and sharing of information of good practices, technical and economic assistance, cooperative research and transfer of technologies, and cooperation in professional trainings. Disaster and emergency management should be part and parcel of all of these implementation efforts.

In addition, in January 2012, the UN contributed its expertise on disability to the Fifth Shaffalah Forum under the theme "Crisis, conflict and disability: ensuring equality." Participants at the forum deliberated on how humanitarian responses and recovery and rebuilding efforts could ensure the protection and inclusion, for example, of refugees and displaced people living with disabilities. Further, many additional issues were raised in the 2012 Four Societies meeting, on which this book is based, regarding risk management, strategic planning, domestic and international law, and reconstruction.

The UN has, since its beginning, placed the social and human dimensions of development and development cooperation at the center of its policies and programs. Mainstreaming disability in development cooperation policies and programs continues to pose implementation challenges. The Convention affirms that States parties have a significant role to play in international cooperation to promote the rights of persons with disabilities. States in partnership with civil society, in particular organizations of persons with disabilities, and international and regional organizations are encouraged to take measures to support national implementation of the CRPD. These measures could include, for example, the mutual exchange and sharing of information of good practices, technical and economic assistance, cooperative research and transfer of technologies, and cooperation in professional trainings.

[4] The full report from the Meeting is available www.un.org/disabilities.

PREPARING FOR AND RESPONDING TO NATURAL DISASTERS, ARMED CONFLICT, OR HUMANITARIAN CRISES: AN OPPORTUNITY FOR INCLUSION

In "implementation of the CRPD" and operationalizing its provision on situations at risk, we can definitively say that including the needs and voices of persons with disabilities at all stages of the disaster management process, and especially during planning and preparedness, would be the most important factor to significantly reduce their vulnerability and increase the effectiveness of government response and recovery efforts.

However, despite an increasing worldwide focus on disaster risk reduction as opposed to mere disaster response, most city and related government agencies fail to adequately plan for or include persons with disabilities in their disaster management activities. This causes severe inequities in access to immediate response, as well as long-term recovery resources for people who have disabilities prior to the disaster and those who acquire a disability as a result of the disaster.

By the same token, rehabilitation and reconstruction efforts must not only be inclusive and responsive to the needs of all people, but should include the participation of persons with disabilities to ensure that their needs and rights are respected and fully taken into consideration in all stages of planning, designing, and building of, for example, permanent shelters, schools, and health clinics, to ensure equal accessibility by all persons.[5]

While most crisis-related measures have shown short-term responses, they offer opportunities to set the stage for broader and lasting reforms for international normative frameworks on disaster and disability and to develop disability-responsive socioeconomic policies and budgeting frameworks. Behind each crisis there exists an opportunity to advance the disability-inclusive development goals to benefit all members of the society in the long run.

In 2013, the Assembly convened a High-level Meeting on disability-inclusive development under the theme "The way forward: a disability inclusive development agenda toward 2015 and beyond." The meeting presented an extraordinary opportunity to ensure accessibility for and inclusion of persons with disabilities in all aspects of the international framework for development processes, including in the global disaster reduction and management agenda.

Given the General Assembly's recent expression of deep concern at the incidence and scale of natural disasters during the past several years, it follows that it would also address the synergies among disaster, disability, and development. The General Assembly has already recognized the negative social, economic, and

[5] Governments will also find such inclusion to be financially beneficial, as a number of studies have indicated that it is more cost-effective to design a barrier-free building rather than to renovate an existing building to make it fully accessible to persons with disabilities. For further information, see www.make-development-inclusive.org/toolsen/pcm1.pdf

environmental consequences of disasters particularly for vulnerable and marginalized groups and has called for programs to improve social resilience and capabilities to cope with disaster-related challenges.[6]

CONCLUSION

The topics of disaster relief and disability rights will continue to be at the forefront of the UN's efforts to be inclusive across its policy frameworks. In October 2013, the DESA and the United Nations Office for Disaster Risk Reduction (UNISDR) co-organized a panel discussion on "Disaster Resilience and Disability: Ensuring Equality and Inclusion" in observance of the 2013 International Day for Disaster Reduction.

Preparation efforts are underway for the next World Conference on natural disaster reduction in Japan in 2015. This conference will provide an opportunity to meaningfully address the invisibility of disability and persons with disabilities in the current approach to disaster and risk management and will ensure priority is given to the need to address disability rights in the context of disaster reduction.

[6] UNGA 65/157.

PART IV

Disaster Prevention/Relief: Anticipatory Responses from State Actors

FIGURE 7 **Sagamihara Police Department Conducting a Disaster Rescue Drill in Japan**. Source: Tetsuo Nakahara, U.S. Army Garrison Japan Public Affairs, September 4, 2012.

12

Help from Above: The Role of International Law in Facilitating the Use of Outer Space for Disaster Management

Brian R. Israel[*]

This chapter explores the role of international law as well as "soft law" mechanisms in enabling the use of outer space for valuable new disaster management applications. This overall challenge is addressed in three phases, ranging from the collective action problems arising from the use of space in general, to sovereignty-based objections to observing the Earth from space, to the complex coordination challenges of harnessing existing space systems for disaster applications. One mechanism in particular, the legally nonbinding International Charter for Space and Major Disasters (Disaster Charter), serves as a remarkable case study in international cooperation because of the speed with which it progressed from a proposal by a handful of space agencies to a highly effective resource for disaster management. The chapter examines what can be learned from this case study and others about the optimal role of international legal mechanisms versus non-legal alternatives in structuring international cooperation to address global challenges. In so doing it contributes to the soft law literature an explanation of the optimal role of international legal mechanisms in addressing a class of international cooperation that is largely unaccounted for in the literature yet crucial, especially in times of strapped budgets and deadlocked law-making processes. The chapter outlines a practical approach for identifying opportunities to achieve productive cooperative outcomes through creative, relatively informal international cooperation.

INTRODUCTION

Behind the tragedy of major natural disasters in the past decade lies a remarkable story of international cooperation. On hundreds of occasions since 2000, disaster-stricken states activated an international coordination mechanism to request data from Earth observation and meteorological satellites, leading participating space

[*] The views expressed are personal and do not necessarily reflect those of the U.S. Government.

agencies to re-task satellites and follow carefully coordinated procedures to get potentially life-saving information into the hands of rescue and relief operators within hours. Contemporary space systems have much to contribute to disaster management, from monitoring changing conditions for long-term forecasting; to early warning of floods, volcanoes, and hurricanes; to imagery guiding relief efforts; as well as communications, navigation, and search-and-rescue support.

This chapter examines the role of international law in enabling the use of outer space for disaster management. It conceptualizes this overall project in three phases each defined by distinct historical circumstances (both political and technological), collective action challenges corresponding to these circumstances, and varying roles for international law in facilitating international cooperation. The first phase is defined by the collective action problems inherent in utilizing outer space in general. In this phase the United Nations treaties on outer space were concluded, supplying a legal framework for the exploration and use of outer space by an ever-growing number of states and private actors.

The second phase begins with the advent of remote sensing – the space activities most relevant for disaster management – and the ensuing political divide between the few states possessing these capabilities and those concerned about being at an informational disadvantage about their territory. After a decade-long confrontation between the promise of this new capability and sovereignty, a path forward was illuminated in the General Assembly's Remote Sensing Principles, particularly in the principle of non-discriminatory access. The Principles reflect an aspiration that "[r]emote sensing shall promote the protection of mankind from natural disasters," and direct states that have identified relevant remote-sensing-derived information in their possession to promptly provide it to states threatened or stricken by a disaster. Although not itself a source of international law, the Principles play an essential role in enabling the robust regime for remote sensing relied on today for disaster management.

The third phase brings us into the present decade, and is defined by coordination challenges that, though less political, are daunting in their complexity. Disaster management applications of space systems often require: data from sensors of multiple satellites – owned by governmental and private operators in several states as well as international organizations – re-tasking of satellites; combining the raw data with data from other systems and processing it into a format usable by relief operators; and all this must happen within hours. In this present phase, the cooperative mechanisms employed by states are comparatively less formal, and non-legal. They include the Disaster Charter, global and regional coordination groups, a UN program established by General Assembly resolution, and capacity building initiatives, among others.

This third and present phase is the focus of this chapter. It examines the nature of the challenges in effectively harnessing existing remote sensing systems for disaster management applications, and the mechanisms employed by states toward this end. The legally nonbinding Disaster Charter, in particular, presents a remarkable case study in international cooperation because of the speed with which it progressed from a proposal by a handful of space agencies to a highly effective resource for disaster management. The Disaster Charter, however, was never intended to be a panacea for all disaster applications; by design, it is temporally constrained to the period immediately before, during, and immediately after a disaster, and does not encompass all humanitarian crises. For all the success of the Disaster Charter and related mechanisms, there remains work to be done in realizing the full potential of remote sensing systems for disaster management.

The chapter explores the optimal role of international legal mechanisms in capturing this potential. For example, in retrospect, would a legally binding international agreement on remote sensing for disaster management have been superior to the nonbinding Disaster Charter? Now, twelve years into the Disaster Charter's operational life, is elaboration into a legally-binding agreement – such as the International Cospas-Sarsat Programme Agreement for satellite-based search and rescue – an inevitable (or desirable) development? In so doing, it builds on, and aims to make a modest contribution to, the rich body of interdisciplinary literature on the role international legal mechanisms versus so-called soft law mechanisms in international cooperation.

The challenges in this third phase most closely resemble coordination problems, a subset of collective action problems that has received comparatively little attention in the international cooperation literature, which tends to focus on collective action problems involving costly commitments and strong interests to defect, and the design of mechanisms to realign state interests in favor of compliance. By contrast, issues of compliance are not at the fore in this third phase. The challenge is not one of aligning state interests around a costly objective – like reducing emissions or armaments – but in coordinating efforts to overcome technical and logistical hurdles to create a valuable new application of existing resources.

The optimal role of international legal mechanisms in addressing such complex coordination challenges deserves greater attention. The swift success of the Disaster Charter serves as a compelling invitation to survey what global challenges might be addressed through creative international coordination. Strapped national budgets and deadlocked formal law-making processes provide additional incentives to explore what might be achieved with existing resources through relatively informal international cooperation.

For coordination challenges for which compliance considerations are not at the fore, I submit that a focus on the role of legal form in implementation is the key to ascertaining the optimal role of international legal mechanisms versus non-legal mechanisms in a given case. The ultimate success of an international cooperative arrangement depends on each participant taking the necessary actions at home to ensure its ability to fulfill its commitments. I propose a practical approach to identifying instances in which implementation challenges call for an international legal mechanism, as well as opportunities to achieve productive cooperative outcomes through relatively informal cooperation.

THE POTENTIAL OF SPACE SYSTEMS FOR DISASTER MANAGEMENT

Existing remote sensing systems are capable of gathering data relevant to a wide range of natural and man-made crises, including floods, droughts, earthquakes, ocean storms (cyclones and typhoons), mud and rock flows, landslides, volcanoes, forest fires, deforestation, desertification, oil spills, marine and water pollution, ice hazards, and nuclear disasters, among others.[1] The data derived from these systems has much to offer to all phases of the disaster cycle: long-range forecasting, early warning, response in the crisis phase, and post-crisis recovery. Long-range forecasting involves monitoring and analyzing changing conditions on Earth (e.g., soil conditions) and the atmosphere to predict the occurrence of natural disasters. It also involves analysis of space-derived data from prior disasters to improve prediction models, and inform land-use decisions. Early warning involves detection of an event from space – such as volcanic eruption – to provide advance warning on Earth.

At present, the most developed application of land remote sensing systems is during and immediately following the crisis. Space systems are capable of providing disaster management actors on the ground with a picture of the nature, extent, and location of damage. For example, knowledge of the location and severity of floods enables prioritization of resources, while information about the condition of roads aids route and evacuation planning. High-resolution imagery of areas affected by an earthquake allows response actors – whether in the capital city or on another continent – to assess the damage to physical infrastructure and plan accordingly. And in the post-crisis rehabilitation phase, space-derived data can inform development planning, such as prioritizing the distribution of assistance to farmers based on analyses of damage to agricultural areas.[2]

Each type of disaster and phase of the disaster cycle has unique requirements for satellites equipped with certain types of sensors; breadth and depth of imagery (i.e., how "zoomed in" or "zoomed out" an image is, and its resolution); and the window of time, in relation to the onset of the crisis, in which space-derived products are most useful, including how often they must be refreshed. For example, early warning for floods requires real-time data from meteorological satellites covering

1 See Action Team on Disaster Management, "Implementation of the Third United Nations Conference on the Exploration and Peaceful Uses of Outer Space (UNISPACE III)," Final Report: UN Doc. A/AC 105/C J/L 273, December 22, 2003, 16–22 (hereinafter "Action Team Report").

2 See Peter Zeil, *Management and Prevention of Natural Disasters – What are the Requirements for the Effective Application of Remote Sensing?* Conference Paper, ISPRS Commison VI Workshop, 2002, 55. www.isprs.org/proceedings/XXXIV/6-W6/papers/zeil.pdf.

a vast area that could include several states' territories.[3] During the crisis phase, optical imagery as well as radar imagery (which is capable of "seeing through" cloud cover) is useful for assessing the location and extent of floods – which requires a broader, lower-resolution picture – and the status of infrastructure, which necessitates a higher-resolution image.[4] Such information is most useful within one to six hours of the disaster, and when refreshed within hours or days.[5]

In many instances, creating space-derived products that are useful in disaster management requires combining data from the sensors of multiple satellites, as well as air and ground-based systems. For example, detecting, analyzing, and monitoring an oil spill frequently requires data from multiple types of space-based sensors (including synthetic aperture radar and multi-spectral imagers) in combination with ground data, such as bathymetry (i.e., depths).[6] Of course, the raw data "seen" by sensors aboard satellites is not immediately useful to disaster management actors; it must be processed, analyzed, and combined with other data to produce useful products such as maps overlain with information about the nature and extent of damage. These are generally known as value-added products, and can require significant human analysis on the ground.

THE ROLE OF INTERNATIONAL LAW IN CAPTURING THIS POTENTIAL: THREE PHASES

An examination of the role of international law in facilitating disaster applications of space systems would be incomplete without accounting for its role in enabling the existence and uses of these systems that give them their great potential for disaster management. Using satellites in outer space to aid disaster management on Earth would be little more than science fiction without the existence of satellites in orbit carrying sensors and other payloads valuable for disaster management, among other applications. Such satellites, of course, are enormously expensive, and launching and operating them in an increasingly congested space presents considerable risk. Without a regime to manage this risk and provide the legal certainty to facilitate investment in these space systems by governments and private entities, it is unlikely they would exist. And it is doubtful these systems would be of much use for disaster management if outer space was treated as an extension of national territory or air space.

The overall challenge of utilizing outer space to aid disaster management divides naturally into three phases. Disaster management uses are specific applications of a broader activity – remote sensing – that is itself a subset of a larger class of activities involving the use of outer space. Although each phase coincides with distinct historical circumstances, they are also unique in the nature of the collective action

[3] Ibid. [4] Ibid.; "Action Team Report," par.17.
[5] See "Action Team Report, par." 17. [6] Ibid., par. 41.

challenges they present. They form a spectrum from the highly political – how will an entirely new activity of inherently global concern be conducted, and how will the risks and benefits be distributed – to increasingly technical, detail-intensive coordination challenges. Common to all phases is the necessity of international cooperation. The role of international legal mechanisms – compared with other mechanisms for international cooperation – in each phase is largely a function of the nature of the collective action challenges each phase presents.

Phase 1 – Using Outer Space

As space flight became a reality with the launch of the first satellite into orbit in 1957, so too did the need for an international regime for the use and exploration of outer space.[7] Demand for such a regime was not limited to the two superpowers capable of accessing space. The vast majority of states without a near-term prospect of accessing space shared an interest in managing the long-term military and economic advantage that might accrue to the first spacefaring states. Potential military applications of space loomed large in the minds of elites throughout the world. And there was concern that the first to reach space – or celestial bodies, like the Moon – would appropriate territory or resources for themselves. More immediately, the dangers inherent in launching objects into outer space – and their eventual return – were of concern to all on Earth. A degree of unpredictability attends launches and reentry, and there was a shared concern in clarifying responsibility for damage caused on Earth.[8]

The few spacefaring states at the time had their own interests in an international regime to facilitate their immediate and future plans for space exploration, including human space flight. High among them was the rescue of astronauts and the return of space objects, wherever they may land on Earth, notwithstanding the relations between the countries concerned or sovereign rights over air space and territory. And they shared an interest in rules to facilitate the seemingly endless uses of space by a growing array of actors, without collision or interference. In a remarkably prescient account of the collective action challenges inherent to the use of space, Myers McDougal and Leon Lipson, writing in 1958, observed: "Space can be used for the simultaneous flight of more than one craft, for instance, though in time some rules of the road will of course be needed."[9]

[7] See Stephan Hobe, Bernhard Schmidt-Tedd, Kai-Uwe Schrogl, and Gerardine Goh, *Cologne Commentary on Space Law*, Vol. 1 (Cologne: Carl Heymanns Verlag, 2010), 11.

[8] See Herbert Reis, "Some Reflections on the Liability Convention for Outer Space," *Journal of Space Law* (1978). (Reflecting on the concern about international liability from the very outset of space exploration).

[9] Myres S. McDougal and Leon Lipson, "Perspectives for a Law of Outer Space," *American Journal of International Law* (1952).

Consonant with the global interest in space, the negotiation of the initial "rules of the road" did not occur bilaterally, between the United States and Soviet Union, but multilaterally, under the auspices of the United Nations.[10] Discussion of outer space was taken up in the General Assembly in November 1957, within a month of the launch of the first satellite into orbit.[11] Through its Resolution 1348 (XIII) in December 1958, the General Assembly established an ad hoc Committee on the Peaceful Uses of Outer Space, comprising eighteen members, and with a mandate that included reporting on "[t]he nature of legal problems which may arise in the carrying out of programmes to explore outer space."[12] A year later, acting on the recommendation of the ad hoc committee, the General Assembly established a permanent Committee on the Peaceful Uses of Outer Space (COPUOS).[13] COPUOS established a Scientific and Technical Subcommittee and a Legal Subcommittee, which continue today as important mechanisms for facilitating international cooperation in the use and exploration of outer space.

By December 1961, four years after the orbit of the first satellite, the General Assembly "*Commend[ed]* to States for their guidance in the exploration and use of outer space the following principles:"

(a) International law, including the Charter of the United Nations, applies to outer space and celestial bodies;
(b) Outer space and celestial bodies are free for exploration and use by all States in conformity with international law and are not subject to national appropriation.[14]

At the same time, the General Assembly established a registration regime, calling upon states to provide to the secretary-general information about the launch of objects into outer space, and for the secretary-general to establish a public registry of such information.[15] A Declaration of Legal Principles and several treaties soon followed.

Declaration of Legal Principles

The 1963 Declaration of Legal Principles Governing the Activities of States in the Exploration and Use of Outer Space (1963 Declaration)[16] represented the first major attempt to elaborate an international regime for the use and exploration of space. In its Resolution 1962 (XVIII), the General Assembly unanimously, and "*solemnly declare[d]* that in the exploration and use of outer space

[10] See "*Cologne Commentary*," par. 11. [11] See ibid., par. 12.
[12] See ibid.; G.A. Res. 1348 (XIII) par. 1(d) (Dec. 13, 1958).
[13] See ibid., par. 13; G.A. Res. 1472A IXIV) par. 1 (Dec. 12, 1959).
[14] G.A. Res. 1721 A (XVI) 1 (Dec. 20, 1961). See also Goedhuis, D. "Influence of the Conquest of Outer Space on National Sovereignty: Some Observations." *Journal Space Law* (1978): 37.
[15] G.A. Res. 1721 B (XVI) 1–2 (Dec. 20, 1961). [16] G.A. Res. 1962 (XVIII) (Dec. 13, 1963).

States should be guided by [nine] principles." Whereas the first four principles largely echoed Resolution 1721 A (XVI) of 1961, reflecting the interests "of all mankind" in space, Principles 5 through 9 broke new ground in prescribing "rules of the road" for the shared use of space.[17] Among them, state responsibility for national activities in space (Principle 5); avoidance of harmful interference to other states or space activities (Principle 6); registration, ownership, and jurisdiction and control over objects launched into outer space (Principle 7); launching state liability for damage caused on earth, in air, or space (Principle 8); and assistance to astronauts as "envoys of mankind" (Principle 9).

The statements of many delegations expressing their governments' intent to abide by these principles – and in some cases suggesting the 1963 Declaration reflected or would over time precipitate customary international law – drew much attention to the legal status of the 1963 Declaration.[18] Eminent jurists such as Judge Manfred Lachs and Bin Cheng have explored whether the rapid development of space exploration condensed state practice sufficiently enough to support rules of customary international law into a short period of time[19] – or whether the 1963 Declaration and accompanying statements amounted to expressions of *opinio juris* that obviated any requirement of state practice.[20] Whatever the state of customary international law at the time or the role of the declaration in developing it, the unequivocal statements[21] by the spacefaring states that they would conduct their space activities in accordance with the principles suggest the declaration was the foundation of an international regime[22] for the exploration and use of outer space. In any event, the question of the declaration's legal status soon became primarily of academic concern, with the conclusion of the Outer Space Treaty less than four years later.

[17] See ibid.; Karen Traunmüller, "The Declararion of Legal Principles Governing the Activities of States in the Exploration of Outer Space: The Starting Point for the United Nations' Law of Outer Space" in *Soft Law in Outer Space: The Function of Non-binding Norms in International Space Law*, ed. Irmgard Marboe (Vienna: Böhlau, 2012), 145.

[18] See Manfred Lachs, *The Law of Outer Space. An Experience in Contemporary Law-Making* (Leiden: Sitjhoff, 1972); Bin Cheng, "United Nations Resolutions on Outer Space: "Instant" International Customary Law?" *Indian Journal of International Law*(1965); Goedhuis, "Influence of the Conquest," 40.

[19] See Lachs, "The Law of Outer Space," 135–8.

[20] See Cheng, "United Nations Resolutions on Outer Space."

[21] Excerpts from several statements are reproduced in Bin Cheng "United Nations Resolution on Outer Space" account, . See also Paul G. Dembling and Daniel M. Arons, "The Evolution of the Outer Space Treaty," *Journal of Air Law and Commerce*, (1967): 422.

[22] See Setsuko Aaoki, "The Function of 'Soft Law' in the Developmnet of International Space Law," in *Soft Law in Outer Space: The Function of Non-binding Norms in International Space Law*, ed. Irmgard Marboe (Vienna: Böhlau, 2012), 67. Distinguishing the elements of customary international law from those of an "international regime," defined as "a set of implicit or explicit principles, norms, rules and decision-making procedures around which actors' expectations converge in a given area of international relations."

United Nations Treaties on Outer Space

Propelled by a sense of urgency among the major space powers,[23] the principles set out in the 1963 Declaration were soon enshrined and elaborated in a series of treaties negotiated in the COPUOS Legal Subcommittee. On December 19, 1966, the General Assembly reported out the Treaty on Principles Governing the Activities of States in the Exploration and Use of Outer Space, including the Moon and Other Celestial Bodies (Outer Space Treaty),[24] which was opened for signature a month later and entered into force on October 10, 1967. The Outer Space Treaty lays down the foundational elements of an international regime for the exploration and use of outer space. For example, Article II confirms:

> Outer space, including the Moon and other celestial bodies, is not subject to national appropriation by claim of sovereignty, by means of use or occupation, or by any other means.

The Outer Space Treaty addresses certain fundamental principles – such as the international liability of the launching state for damage caused on Earth, in air, or in space (Article VII), and concepts of registration and jurisdiction and control over space objects – in general terms, anticipating their detailed elaboration in subsequent agreements.[25]

The Outer Space Treaty is forward-looking in its anticipation of a future in which use of space is not the exclusive preserve of superpowers – or even governments. Article VI establishes that "State Parties to the Treaty shall bear international responsibility for national activities in outer space . . . whether such activities are carried on by governmental agencies or non-governmental entities." It assigns responsibility for assuring all national activities conform to the terms of the Outer Space Treaty, and further requires that non-governmental activities in space be authorized and supervised "by the appropriate State Party to the Treaty." This is the mechanism through which rules of international space law are made applicable to non-governmental activities in space, generally through national legislation and licensing regimes.

A year later, the Agreement on the Rescue of Astronauts, the Return of Astronauts and the Return of Objects Launched into Outer Space (Rescue and Return Agreement),[26] was concluded and entered into force in 1968. Next came the Convention on International Liability for Damage Caused by Space Objects (The Liability Convention)[27]

[23] See Dembling and Arons, "Evolution of Outer Space Treaty," 425–426.

[24] Treaty on Principles Governing the Activities of States in the Exploration and Use of Outer Space, including the Moon and Other Celestial Bodies, opened for signature January 27, 1967, entered into force December 3, 1968.

[25] See General Assembly Resolution 2222 (XXII).

[26] Agreement on the Rescue of Astronauts, the Return of Astronauts and the Return of Objects Launched into Outer Space, opened for signature April 22, 1968, entered into force December 3, 1968.

[27] Convention on International Liability for Damage Caused by Space Objects, opened for signature March 29, 1972, entered into force September 1, 1972.

in 1972. The Liability Convention elaborated on the principles of liability announced in the 1963 Declaration and Outer Space Treaty by defining "launching state," establishing a regime of strict liability for damage caused on Earth or to an aircraft in flight (Article II), and an expedited process for the presentation of claims on behalf of injured nationals, carving out the local remedies rule (Articles VIII–XI). Last among the core space treaties was the Convention on Registration of Objects Launched into Outer Space (Registration Convention),[28] which elaborated the duties of launching states to register objects launched into outer space through the UN secretary-general.

These four agreements enjoy broad membership. At the time of this writing, the Outer Space Treaty has 102 parties and 26 signatories. The Rescue and Return Agreement has 91 parties and 24 signatories. The Liability Convention has 88 parties and 23 signatories. The Registration Convention has 54 parties and 4 signatories. A fifth treaty negotiated under the auspices of the UN – the Agreement Governing the Activities of States on the Moon and Other Celestial Bodies (Moon Agreement)[29] – never gained similar acceptance, with only 13 parties and 4 signatories, a membership that does not included the most active spacefaring states. Given their broad membership and fundamental importance to the most prevalent uses of outer space, the first four agreements are sometimes referred to as "the four core treaties" on outer space.

Phase 2 – A Global Regime For Remote Sensing

In the decade following the conclusion of the four core space treaties, global concern about the use of outer space shifted to the observation of Earth from space, and analysis and dissemination of space-derived information on Earth. By the mid-1980s, the United States' Landsat remote sensing program had been in operation for a decade, bringing into focus the tremendous potential of remote sensing for understanding and managing natural resources on Earth. France's SPOT remote sensing satellite would soon come online.[30] As the United States, France, and a handful of other states invested heavily in laying the groundwork for commercial remote sensing, they shared a keen interest in developing a global market for remote sensing products. U.S. policymakers believed that remote sensing presented significant economic opportunities for American companies.[31]

[28] Convention on Registration of Objects Launched into Outer Space, opened for signature January 14, 1975, entered into force September 15, 1976.

[29] The Agreement Governing the Activities of States on the Moon and Other Celestial Bodies, opened for signature on December 18, 1979, entered into force July 11, 1984.

[30] See Kenneth D. Hodgkins, oral history recorded by Joanne Irene Gabrynowicz. *The U.N. Principles Relating to Remote Sensing of Earth from Space: A Legislative History – Interviews of Members of the United States Delegation* Oxford: NRLSSC, (2002). (Hereinafter "Hodgkins Oral History").

[31] See ibid.

It was more broadly believed that commercializing remote sensing would expand the breadth of available products.[32]

Notwithstanding the demonstrated potential of Earth applications of remote sensing information – or perhaps because of this potential – the prospect of commercial remote sensing was greeted with suspicion, even hostility, by a broad swath of the world.[33] Many developing countries feared that the concentration of remote sensing capabilities would put "sensed" states at an informational disadvantage about their own territories – and resources within – vis-à-vis the most economically powerful states and their companies. Many feared that such an informational advantage would amount to a significant bargaining advantage for companies in possession of remote sensing data in negotiations over the exploitation of a state's natural resources

These very practical concerns of a number of countries were expressed in terms of sovereignty – over territory and natural resources.[34] The Outer Space Treaty had established that outer space cannot be an extension of a state's territory. Thus, unlike airspace, the mere presence of satellites in outer space above a state's territory does not implicate sovereignty. But a number of states espoused the position that "sensing" a state's territory from space – and disseminating this information without its consent – may nevertheless undermine the sovereign prerogatives of the sensed state.[35] At the heart of these concerns was the prospect of being at an informational disadvantage or losing control over information about conditions within a state's territory.

Such sovereignty concerns found expression in the most contentious element of a proposal by Brazil and Argentina and supported by much of the developing world: a prior consent regime in which a state's territory could not be sensed without its consent.[36] Such a regime would have been practically unworkable.[37]

[32] See Douglas A. Doetsch, "Legal and Foreign Policy Implications of the LANDSAT Commercialization Act," *Columbia Journal of Transnational Law* (1985): 114.

[33] See Foreign Relations of the United States, 1969–76, Volume E-3, Document 106: Action Memorandum from the Deputy Assistant Secretary for International Organization Affairs (Buffum), the Legal Adviser (Leigh), and the Assistant Secretary of States for Oceans and International Environmental and Scientific Affairs, (Ray) to the Deputy Secretary of State (Ingersoll), Washington, April 3, 1975 (hereinafter "April 3, 1975 Action Memo"). http://history.state.gov/historicaldocuments/frus1969-76ve03/d106.

[34] For contemporaneous perspectives on the sovereignty-related concerns and legal positions advanced by a large number of States, see generally Eleanora Ambrosetti, "Remote Sensing from Outer Space: Its Significance Problems from a Third World Perspective," *N.Y.U. Journal of International Law and Politics*, (1984); David Meyers, "Remote Sensing and National Sovereignty over Natural Resources: An Assessment of the Mexican View," *California Western International Law Journal*, (1984): 22–48; Steven Parminter, "Remote Sensing of the Earth from Outer Space: Considerations Toward Development of a Functional International Regime," *Loyola of Los Angeles International and Comparative Law Review*, (1979): 157–192.

[35] See Meyers, "Remote Sensing and National Sovereignty."

[36] See "April 3, 1975 Action Memo." [37] See "Hodgkins Oral History."

Earth observation satellites do not "see" national borders. Moreover, some of the most promising remote sensing applications envisioned at the time, such as addressing desertification, involve observation of transboundary phenomena, which could not be accomplished if sensing were made to stop at national borders, even if technologically feasible.[38]

The Soviet Union did not favor a prior consent regime for remote sensing itself, but sought to extend state sovereignty over territory and natural resources to control over the dissemination of information about them.[39] That is, it favored a prior consent regime for data dissemination. This, too, would have frustrated many disaster applications, as well as many of the other promising applications of remote sensing. Apart from limiting or delaying the observation of transboundary phenomena, constraining access to data would, in turn, limit the understanding and predictive modeling of natural disasters that can be informed by remote sensing data.

Efforts to reach a global consensus on the modalities of remote sensing spanned more than a decade. In 1974, the General Assembly recommended in Resolution 3234 that the COPUOS Legal Subcommittee consider the "legal implications of remote sensing from space, taking into account the various views of States expressed on the subject, including proposals for draft international instruments,"[40] and noted that Argentina and Brazil had submitted draft articles for "a treaty on remote sensing of natural resources by means of space technology" for the consideration of the Legal Subcommittee.[41] The remote sensing debate in the UN persisted until December 1986 when the General Assembly adopted the Principles Relating to Remote Sensing of the Earth from Outer Space (Remote Sensing Principles) by consensus.[42]

The definition of "remote sensing" in Principle I(a) reveals the relatively limited scope of the debate and the principles' applicability to remote sensing activities:

> The term "remote sensing" means the sensing of the Earth's surface from space by making use of the properties of electromagnetic waves emitted, reflected or diffracted by the sensed objects, *for the purpose of improving natural resources management, land use and the protection of the environment*.[43]

Not encompassed by this definition are remote sensing activities for national security or military purposes.[44] What the principles do address are the remote sensing activities most relevant to disaster management applications.

At the heart of the principles is the principle of non-discriminatory access in Principle XII that "the sensed State shall have access to [data] on a non-

[38] Ibid. [39] See Parminter, "Remote Sensing of the Earth,"; "April 3, 1975 Action Memo."
[40] G.A. Res. 3234 (XXIX) (1974) par. 6(c) [41] Ibid., par. 7.
[42] The Principles were adopted by the General Assembly in Resolution 41/65 (1986), to which the principles are annexed.
[43] Ibid. Emphasis added. [44] See "Hodgkins Oral History."

discriminatory basis and on reasonable cost terms." Principle XII addresses the concerns shared by many states about being at an informational disadvantage vis-à-vis the sensing state, a private company, or neighboring states about its territory and natural resources. This principle is found in the national laws of several states governing commercial and governmental remote sensing activities encompassed by the principles.[45] U.S. law, for example, mandates that licenses to operate private remote sensing systems obligate the operator to "make available to the government of *any country* (including the United States) unenhanced data collected by the system concerning the territory under the jurisdiction of such government as soon as such data are available and on reasonable terms and conditions."[46] Significantly for disaster applications, the U.S. regulations implementing this statutory section interpret "nondiscriminatory" and "reasonable" to permit the provision of data at reduced or no cost for certain types of "public benefit users."[47]

Of special relevance to disaster management applications is Principle XI:

> Remote sensing shall promote the protection of mankind from natural disasters. To this end, States participating in remote sensing activities that have identified processed data and analyzed information in their possession that may be useful to States affected by natural disasters, or likely to be affected by impending natural disasters, shall transmit such data and information to States concerned as promptly as possible.

Upon detailed examination, Principle XI is more limited than it may at first appear. As the United States explained in its Explanation of Vote, the principle "does not contemplate that states will screen all data for those purposes, but rather that states will alert other states when they have identified information that would assist those states in preventing or dealing with emergencies."[48] Given that remote sensing was in infancy at the time the General Assembly adopted the principles, and the absence of the practical arrangements necessary to effectively utilize remote sensing for disaster management, Principle XI represented more of an aspiration than a practical directive to states.

Why principles adopted by the General Assembly instead of an international agreement on remote sensing? As noted previously, Brazil and Argentina had proposed a draft early on. From the perspective of the developing countries interested in establishing a prior consent regime for remote sensing and extending

[45] For a discussion of other states' implementation of nondiscriminatory access, see Joanne Irene Gabrynowicz, "The UN Principles Relating to Remote Sensing of the Earth from Outer Space and Soft Law," in *Soft Law in Outer Space: The Function of Non-binding Norms in International Space Law*, ed. Irmgard Marboe (Vienna: Böhlau, 2012), 189–90.

[46] 51 U.S.C. § 60122(b)(2). [47] See 15 CFR Part 960.

[48] Statement by William J. Lowell, United States Representative to the Special Political Committee. "Explanation of Vote under Item 72, Priciples Relating to Remote Sensing of the Earth from Space." Press Release USUN 173-(86), November 28, 1986. (On file with author). See also "Hodgkins Oral History," 127.

sovereignty over resources to information about those resources, an international agreement would have placed these ideals on the strongest footing, provided that the states conducting or developing remote sensing activities signed up. But there was little prospect of the United States, France, or the Soviet Union agreeing to such an arrangement.[49] From the perspective of the states seeking broad, international acceptance of remote sensing – to facilitate a global market for remote sensing products – principles adopted by consensus in the General Assembly had the benefit of instant, *global* consensus, compared to the piecemeal acceptance by a small number of states over time that comes with a treaty.[50]

The Remote Sensing Principles are recommendatory; they do not, in and of themselves, possess legal force under international law.[51] Their importance lies in their statement of a consensus of the international community on the conduct of remote sensing activities, following more than a decade of intense debate.[52] Perhaps most importantly for present purposes, they helped pave the way for the robust array of remote sensing systems in operation today, many of which are operated by private commercial entities. And the principles precipitated an international regime[53] for remote sensing in which data can be gathered and disseminated rapidly, which is essential to disaster management applications.

Phase 3 – Disaster Management Applications

In the decade following the adoption of the Remote Sensing Principles by the General Assembly, a significant item on the multilateral space agenda was applications of space systems for the benefit of the environment, human security, development, and welfare.[54] Such was the focus of the Third United Nations Conference on the Exploration and Peaceful Uses of Outer Space (UNISPACE III), held in Vienna in July 1999. The resulting Vienna Declaration on Space and Human Development set forth a "strategy to address global challenges in the future," including:

> To implement an integrated, global system, especially through international cooperation, to manage natural disaster mitigation, relief and prevention efforts,

49 See" Hodgkins Oral History," 150.

50 See Traunmüller, "Declaration of Legal Principles," 153. (Quoting an April 17, 1963 statement by the UK Representative to the COPUOS Legal Subcommittee).

51 See Lowell, "Explanation of Vote." ("...under the Charter of the United Nations, these Principles can only be recommendatory in character; they cannot, in and of themselves, possess legal force... Nevertheless, these Principles constitute a significant statement of the views of the international community on a matter of importance to the future of the peaceful use and exploration of outer space.").

52 See ibid.

53 See Aaoki, "Function of 'Soft Law," 66–68. (Suggesting that "the 1986 Remote Sensing Principles may have developed or are developing into "an international regime.").

54 See ibid.; The Space Millennium: Vienna Declaration on Space and Human Development (July 30, 1999) (hereinafter "Vienna Declaration").

especially of an international nature, through Earth observation, communications and other space-based services, making maximum use of existing capabilities and filling gaps in worldwide satellite coverage.[55]

The principal challenge in realizing the disaster management aspiration expressed in the Vienna Declaration was one of coordination. Effective use of space systems in disaster situations requires a lot of moving parts to work in concert, and quickly. A given disaster application will often require data from the sensors of multiple satellites, which are operated by a diverse array of agencies, private commercial entities, and intergovernmental organizations. Satellites with the right sensors and orbital characteristics must be tasked to acquire data about the affected areas. Raw data from the sensors of these satellites must be processed, analyzed, and combined with data from ground systems to produce value-added products directly useful to disaster management actors. And all this must happen within a matter of hours.

Of course, advanced space-borne sensors and sophisticated value-added products are of little use in a disaster scenario if the local disaster response authorities are not aware of these space-derived products and do not request them in time, or if they do not have the proficiency to take full advantage of them. Thus, building capacity throughout the world, especially in less-developed, disaster-prone regions, is a significant piece of the overall challenge.[56]

In sum, disaster applications of space systems could only be achieved through international cooperation, as any given disaster requires the assets of multiple states or intergovernmental organizations. No state is immune from disasters, and even for the states with the most advanced and extensive space systems, the value of an international mechanism to coordinate these systems exceeded the relatively low cost of contributing data, whether from their own systems or under volume contracts with commercial satellite operators. The challenge lies not in creating incentives for advanced spacefaring states to contribute data for disaster applications, but in managing the complexity that comes with the diversity of systems and actors relevant to success.

Disaster Charter

The Charter on Cooperation to Achieve The Coordinated Use of Space Facilities In The Event of Natural Or Technological Disasters (Space Cooperation Disaster Charter)[57] is a cooperative mechanism through which satellite operators – government agencies, as well as intergovernmental organizations – provide satellite-derived data and information in support of disaster prevention and response. The Charter is a legally nonbinding arrangement. It was proposed by

[55] See ibid., par. (b)(ii).

[56] See "Action Team Report," par. 49(f); Zeil, *Management and Prevention*,. 57.

[57] Charter on Cooperation To Achieve The Coordinated Use of Space Facilities In The Event of Natural Or Technical Disasters Rev.3 (25/4/2000).2, *available at* www.disasterscharter.org/web/charter/charter.

the European Space Agency (ESA) and the French space agency, Centre National d'Etudes Spatiales(CNES) at the UNISPACE III conference in July 1999, and was joined shortly thereafter by the Canadian Space Agency (CSA) in October 2000.[58] With the combined resources of the three space agencies, and following some logistics planning and testing, the Space Cooperation Disaster Charter was declared operational on November 1, 2000. Within a matter of weeks, the Charter was activated for the first time in response to severe mudslides in Slovenia.[59]

In essence, the Space Cooperation Disaster Charter is a coordination mechanism that facilitates the rapid provision of space-derived information (e.g., maps showing the location or extent of damage), free of charge, to rescue and relief operators in response to a "natural or technological disaster."[60] The Charter is open to participation by national space agencies and other space system operators.[61] At the time of writing, government agencies of eleven states, as well as intergovernmental organizations such as ESA and the European Organization for the Exploitation of Meteorological Satellites (EUMETSAT), participate directly in the Charter. These participants are responsible for governance through the board.[62] Charter parties commit, on a "voluntary," "best endeavors" basis to collaborate to develop scenarios for streamlining the provision of space-derived information for disaster applications, and upon activation of the charter in the event of a "natural or technological disaster," to task satellites they control, or to which they have access, to gather data requested by an authorized user.[63]

The Space Cooperation Disaster Charter's scope is temporally limited to the most acute period of the disaster management cycle: Article I defines "crisis" to mean "the period immediately before, during or immediately after a natural or technological disaster, in the course of which warning, emergency or rescue operations take place."[64] Until recently, the Space Cooperation Disaster Charter was additionally limited in who may activate it. For its first twelve years, only charter parties (including the member states of international organization parties) were able to directly request services under the Charter.[65] Over time, access to the Charter's resources was expanded through arrangements with the UN[66]

58 A/AC.105/C.1/L.285.

59 See Charter Activations. www.disasterscharter.org/web/charter/activation_details?p_r_p_1415474252_assetId=ACT-000.

60 Ibid. The Disaster Charter defines "natural or technological disaster" as "a situation of great distress involving the loss of human life or large-scale damage to property, caused by a natural phenomenon, such as a cyclone, tornado, earthquake, volcanic eruption, flood or forest fire, or by a technological accident, such as pollution by hydrocarbons, toxic or radioactive substances."

61 Ibid., Article 1 (defining of "parties").

62 Ibid., Article 3.3.

63 Ibid., Articles 3.1; 4.

64 Ibid., Article 1.

65 See Disaster Charter. Note 57 above,: Articles 1, 3.4, 3.5, and 5.

66 Through arrangements with the UN Office of Outer Space Affairs (OOSA) and the UN Organization Satellite (UNOSAT), these entities may activate the Charter on behalf of UN users. See www.un-spider.org/mechanisms-guides/international-charter-space-and-major-disasters.

and regional organizations.[67] Parties may additionally activate the Charter on behalf of non-parties.

For example, the Argentine space agency, Comisión Nacional de Actividades Especiales(CONAE), a Charter party, has activated the Disaster Charter on behalf of Latin American users on a number of occasions. For the period between 2007 and 2009, Charter activations divided equally among each of these paths to charter services – direct activation by Charter parties, by parties on behalf of a non-party, and through the UN. In September 2012, Charter parties adopted a principle of Universal Access, expanding direct access to the disaster management authorities of all states.[68]

Space Cooperation Disaster Charter parties share responsibility for maintaining 24-7 operational capabilities, assuming responsibility for receiving requests, liaising with other participating agencies, and assigning tasks for data acquisition and value-added processing on a rotating basis. Operation of the Charter additionally relies on an array of non-governmental actors. For example, several of the high resolution imaging systems employed by Charter participants are owned and operated by private commercial entities.[69] Many of the value-added services – turning raw data into products useful for disaster management – are also performed by non-governmental entities.

The Charter leaves to each participant how to structure its participation. Thus, some participants rely on extensive in-house value added capabilities, such as the Center for Satellite Based Crisis Information of the German Aerospace Center, while others rely on relationships with private commercial entities and academic institutions. Some participants have budgets to pay for data or value-added services, others have reached arrangements to provide some – or all – free of charge for humanitarian applications, and some rely on arrangements other parts of the government have with commercial entities. This flexibility the Charter affords each participant to contribute in its own way is often cited by participating agencies as an important factor in the Charter's success.

At time of writing, the Charter has been activated more than 300 times to provide space-derived information in support of disaster management efforts. And its value

[67] The Asia Pacific Disaster Reduction Centre is authorized to request data through the Charter on behalf of thirty-one States in the Asia-Pacific region. See ibid.

[68] See Universal Access Brochure. www.disasterscharter.org/c/document_library/get_file?p_l_id=23109&folderId=127346&name=DLFE-4304.pdf.

[69] It is United States policy to promote the development and growth of a commercial remote sensing industry by "rely[ing] to the maximum practical extent on U.S. commercial remote sensing space capabilities for filling imagery and geospatial needs for military, intelligence, foreign policy, homeland security, and civil users." Fact Sheet: U.S. Commercial Remote Sensing Space Policy (May 13, 2003). http://georgewbush-whitehouse.archives.gov/news/releases/2003/05/20030513-8.html.

to the disaster management community is growing, with the steady growth of its membership – and resources –and its efforts to expand access worldwide.[70]

The Growth of a Disaster Management Ecosystem

The past five years have witnessed a proliferation of initiatives aimed at harnessing space systems to address different phases of the disaster management cycle (e.g., long-term forecasting or early warning) and aspects of the overall challenge (e.g., communications infrastructure and capacity building), as well as broader environmental and humanitarian ends. The development of these initiatives could be characterized as "outside-in." In contrast to a centralized initiative expanding outward to address broader parts of the disaster management cycle (i.e., inside-out), these initiatives have developed independently, in parallel, in diverse fora including the United Nations, as well as less-formal and regional coordinating groups, and are increasingly linking together with varying degrees of formality. Together these initiatives comprise what could be called a space-disaster management ecosystem.

Whereas the Space Cooperation Disaster Charter is the leading mechanism for coordinating the rapid tasking of satellites during a crisis, other initiatives aim to address phases of the disaster cycle or services not encompassed by the Charter. For example, proactive, long-term monitoring of Earth and climate conditions to aid disaster prediction and early warning is a significant item on the international disaster management and sustainable development agendas, highlighted as a priority in the Hyogo Framework Plan of Action[71] as well as the Rio+20 Outcome Document.[72] Among the initiatives to further prediction, risk assessment and early warning of natural disasters through coordination among space system operators is the Global Earth Observation System of Systems (GEOSS). GEOSS is an initiative of the Group on Earth Observations (GEO), a voluntary group facilitating coordination of Earth observation activities carried out by governments and intergovernmental organizations. GEOSS is proceeding according to a ten-year implementation plan (2005–15), which embraces land, air, and sea-based Earth observation activities as well as space, and aims to serve broader sustainable development objectives in addition to disaster mitigation. Longer term monitoring of the Earth environment is also among the aims of the Global Monitoring for Environment and Security (GMES) initiative of the European Commission. In support of this initiative, ESA and EUMETSAT are developing and will operate five new Earth observation satellite missions over the next decade.

[70] See International Charter Space & Major Disasters: 10th Annual Report at 10, 14 (explaining the Charter's arrangements with the UN and regional bodies to expand access, and Parties' exploration of additional means to further broaden access). http://charter-prod.netcetera.ch/c/document_library/get_file?uuid=85ca9f76-c840-46fb-b31145d379870c54&groupId=10729.

[71] Hyogo Framework Plan of Action, 17(k)

[72] Rio+20 Outcome Document, 187.

Other initiatives address different aspects of the overall challenge of getting relevant space-derived information into the hands of disaster response actors during a crisis. One challenge is the transmission of information – particularly high-resolution space-derived products – to regions lacking sufficient communications bandwidth or where communications infrastructure has been crippled by a disaster. GEONETCast, a part of the GEOSS strategy, is a system for disseminating critical information from satellites to receiving stations that can be constructed from widely available, relatively inexpensive (i.e., $2,000–$3,000) components.[73] The infrastructure for GEONETCast is provided by the U.S. National Oceanic and Atmospheric Administration (NOAA) for the Americas region, by the intergovernmental organization EUMETSAT for Europe and Africa, and by the Chinese Meteorological Administration (CMA) for the Asia-Pacific region.[74]

The United Nations Organization Satellite (UNOSAT), established in 2003, processes and analyzes remote sensing data to generate value-added products (e.g., a map showing the nature and location of damage) in support of UN emergency relief agencies. The scope of UNOSAT's services is broader than "natural or technological disaster[s]," supporting the work of the UN refugee and human rights agencies, and also includes initiatives to build capacity in the use of satellite-derived products.

Sentinel Asia is a regional, voluntary initiative led by the Asia-Pacific Regional Space Agency Forum (APRSAF) that aims to combine many of the previously described elements to provide comprehensive disaster management services in the Asia-Pacific region. For example, high-bandwidth, satellite-based communications are among the planned elements of the initiative. In addition to the satellites operated by APRSAF members, APRSAF is authorized to activate (i.e., request data from) the Disaster Charter on behalf of states in the Asia-Pacific region.

Last and certainly not least among the initiatives comprising the space-disaster management ecosystem is the United Nations Platform for Space-based Information for Disaster Management and Emergency Response (UNSPIDER). Following the UNISPACE III conference, the General Assembly tasked COPUOS to study "the possibility of creating an international entity to provide for coordination and the means of realistically optimizing the effectiveness of space-based services for use in disaster management..."[75] An ad hoc expert group surveyed potential applications of space systems for disaster management, the needs of user communities, existing mechanisms, and possible roles for the UN, reporting its findings to COPUOS in 2006.[76] Through its Resolution 61/110, the General Assembly *decided* to establish UNSPIDER to "provide universal access" to "all types of space-based information and services relevant to disaster management, and to support the full disaster management cycle" by "serving as a bridge to connect the disaster

[73] *GEONETCast Americas*. www.geonetcastamericas.noaa.gov. [74] Ibid.
[75] G.A. Res. 59/2 (2004) 9 [76] A/AC.105/C.1/L.285

management and space communities and being a facilitator of capacity-building and institutional strengthening, in particular for developing countries."[77]

EXPLORING THE OPTIMAL ROLE OF INTERNATIONAL LAW IN PHASE 3

Whereas international legal mechanisms were the tool of choice in addressing the collective action problems inherent in utilizing outer space – and play an essential role in the international cooperation to field the relevant space systems – they are not in the foreground of the third and present phase. Although this phase coincides temporally with what many have argued is a broader trend in space governance away from international legal mechanisms,[78] I submit that the relatively informal, non-legal nature of the space-disaster management ecosystem is not simply a sign of the times. Rather, it is a function of the specific characteristics of this challenge.

The central focus of this chapter is the optimal role for international legal mechanisms in this third phase.[79] I use *optimal* here to represent a balance of what I call *agility* and *efficacy*. Efficacy, unsurprisingly, concerns whether a mechanism is successful in achieving the desired cooperative outcome. Agility is the measure of time and resources necessary to get an efficacious mechanism up and running, and its flexibility to adapt once operational. Efficacy is a more holistic measure than *compliance*; it is oriented toward an end result dependent on actors beyond the parties to a given arrangement.

The first task in this inquiry is backward-looking. Should the space agencies that founded the Space Cooperation Disaster Charter have pursued an international agreement on remote sensing for disaster applications? Would an international legal mechanism have been optimal? These questions may seem disingenuous in light of the information presented in this chapter. In retrospect, the Charter is a stunning example of both agility and efficacy. Declared formally operational within a month of signature by the three founding space agencies, the Charter was activated for the first time only a few weeks later, in response to severe mudslides

[77] G.A. Res. 61/110 (January 15, 2007) 6.

[78] See generally M. Gennady Danilenko, "Outer Space and the Multilateral Treaty-Making Process," *Berkeley Technology Law Journal*, (1989).; Geraldine M. Goh, "Softly, Softly Catchee Monkel: Informalism and the Quiet Development of International Space Law," *Nebraska Law Review*, (2008).

[79] My approach is in line with Richard Bilder's suggestion that studies of international legal compared with non-legal mechanisms "not be limited to compliance alone," but also include "a comprehensive explanation of the development, interaction, and relative advantages and disadvantages of different kinds of international normative techniques, particularly legally-binding and non-binding norms, and how they can be used, alone or in combination, to achieve higher, more useful, and more stable levels of cooperation." Richard Bilder, "Beyond Compliance: Helping Nations Cooperate," in *Commitment and Compliance: The Role of Non-Binding Norms in the International Legal System*, ed. Dinah Shelton (Oxford University Press, 2000), 73.

in Slovenia. In a very short span of time, it has proven itself a valuable resource for disaster management, having been activated to provide satellite-derived data and information to the disaster response community more than 300 times.

However clear the optimality of the legally nonbinding charter appears through the lens of hindsight, what methodology could the participants at the negotiating table in 1999 have used to evaluate the merits of an international legal mechanism versus a non-legal mechanism? And by extension, how should states approach this question of institutional design in their efforts to address analogous coordination challenges?

Refocusing on Implementation

The international relations literature seeking to explain when and why states favor legally nonbinding – or soft law – mechanisms in place of international legal mechanisms offers a number of useful insights, but it is ultimately calibrated for a different class of problems. This literature has tended to focus on high-stakes collective action problems for which cooperative solutions involve costly commitments like reducing emissions or armaments. Functionalist analyses of the role of international legal mechanisms are oriented toward the international plane, with the success of cooperative mechanisms measured primarily in their capacity to alter states' behavior vis-a-vis one another[80] – to promote compliance. Explanations grounded in liberal theory account for how the demands of domestic political constituencies may influence the choice of legal form,[81] which, of course, presupposes domestic visibility and active, organized constituencies.

The challenge of harnessing existing space systems for disaster management applications through international cooperation falls through the cracks of these explanations. The low cost of participation relative to the value of direct access to the pooled resources of charter participants presents little incentive to defect.[82] This is not to suggest that the enterprise is entirely free of incentives to free ride; only that compliance considerations are not at the fore.

80 Kenneth W. Abbott and Duncan Snidal, "Pathways to International Cooperation," in *The Impact of International Law on International Cooperation*, ed. Eyal Benvenisti and Moshe Hirsch (Cambridge University Press, 2004), 50. (arguing that "an international agreement is only effective when it alters the behavior of participant states in ways that affect outcomes. This requires [*inter alia*, that] the participating states feel compelled to change some aspect of their behavior because of the agreement.")

81 See Kal Raustiala, "Form and Substance in International Agreements," *American Journal of International Law*, 2005: 595–597.

82 The charter is not simply a gift from the most technologically advanced states to the least developed. Even the states with the most advanced remote sensing capabilities stood to benefit from the rapid, streamlined access to data from multiple systems afforded the charter. Indeed, charter participants have activated the charter on a number of occasions in response to crises within their borders. See charter activation statistics. www.disasterscharter.org/web/charter/activations.

As explained, the principal challenge in this third phase is coordination. It is a complex variant of a "coordination problem," in which "incentives to violate commitments – once agreed upon – are quite low."[83] The international relations literature addresses coordination problems only in passing, as they are understandably "least interesting from a compliance perspective."[84] It does suggest that states can be expected to address coordination problems, as a category, through non-legal mechanisms, on account of the minimal incentives to defect and relative ease of conclusion,[85] but does not explain why states nevertheless address many such problems through international legal mechanisms. Acknowledging the diminished explanatory power of compliance in such cases, Andrew Guzman and Timothy Meyer perfectly frame the question I take up: "If expected levels of compliance are equal across forms of agreements, how do states make the choice between a soft or hard law agreement?"[86]

Answering this question in a given case necessitates a shift in focus away from compliance – and interstate relations – and an examination of the roles of international legal mechanisms in intrastate implementation. The efficacy of an international cooperative arrangement – whether legal or non-legal – depends on each participant taking the necessary actions at home to fulfill its commitments under the arrangement. This may require new authorities, funds, or personnel; the requirements for implementation for a given state can vary according to its constitutional arrangements, political realities, and the existing authorities and resources of the agencies responsible for fulfilling its commitments under the arrangement. Beyond the domestic actions each cooperating state must take to enable it to fulfill its commitments, the efficacy of international cooperative arrangements sometimes depends on the conduct of outside actors – states not party to the arrangement or non-governmental entities. The legal form of the arrangement is relevant to the success for both types of implementation challenges.

The case of the Cospas-Sarsat system for satellite-based search and rescue illustrates the relevance of legal form to the varied challenges of implementing an international cooperative arrangement. Consisting of satellites carrying search and rescue (SAR) payloads, ground stations around the world, and compatible distress beacons carried by ships, airplanes, and individuals, the system alerts nearby SAR authorities when a distress beacon is activated, and is now capable of pinpointing the location of the vessel or person in distress. It has thus been described as "taking the 'search' out of search and rescue."

The efficacy of the system depends on vessels carrying compatible beacons. Early in the life of the system, the cooperating states determined that widespread

[83] Raustiala, "Form and Substance," 591.

[84] Andrew T. Guzman and Timothy L. Meyer, "International Soft Law," *Journal of Legal Analysis*, (2010): 192.

[85] Ibid., 188–192; Raustiala, "Form and Substance," 593, 600.

[86] Guzman and Meyer, "International Soft Law," 190.

adoption of distress beacons operating at a frequency better suited for Earth-space communication was essential to the system's success.[87] This would entail some cost to vessel operators, and thus require coordinated regulatory action – not merely in participating states, but throughout the world, spurred by actions in international maritime and civil aviation organizations to promote the standard.[88] Worldwide adoption of the standard required global confidence in the continuity of the system, and the participating agencies believed that elevating their cooperation from a series of agency-level Memorandums of Understanding (MOUs) to an intergovernmental agreement would signal a credible commitment to the system.[89] But here the rationale for signaling was not about compliance; the audience was not parties to the agreement, but non-party states and international organizations whose coordinated regulatory action was essential to promoting the standard.[90]

It is worth noting here that the choice of legal form rarely rests on a single factor, and in many instances of international cooperation, including the Cospas-Sarsat case, an international legal mechanism is necessary for reasons ancillary to the overall objective. For example, apportionment liability or intellectual property rights arising from the cooperative activities can only be accomplished by legal agreement. The 1988 Cospas-Sarsat Programme Agreement and each of the agency-level MOUs that preceded it were necessarily legally binding as they included cross-waivers of liability. Through the 1988 agreement, the parties demonstrated their commitment to the system not by upgrading the legal form of their cooperation, but by moving from agency-level to a government-level agreement.

The implementation advantages of an international legal mechanism are greatest where successful implementation requires governmental action beyond the agency at the negotiating table. In the Cospas-Sarsat case, the National Oceanic and Atmospheric Administration (NOAA) led the U.S. involvement, but implementation required action by senior executive branch officials to incorporate funding requests for the search and rescue payloads, and the weather satellites that carry them, into the president's budget, as well as rulemaking by the Federal Communications Commission to promote the adoption of compatible distress beacons.[91] In many states, including the United States, negotiating and concluding a legally binding international agreement triggers a more comprehensive internal review process than non-binding arrangements. This internal coordination, in parallel with negotiations on the international plane, serves to ensure the

87 Richard J. H. Barnes and Jennifer Clapp, "Cospas-Sarsat: A Quiet Success Stort," *Space Policy*, 1995: 261–268.

88 See ibid., 264–266.

89 See ibid.

90 Keneth W. Abbott and Duncan Snidal, "Hard and Soft Law in International Governance," *International Organization*, (2000): 421–451 at 426–430. (Discussing choice of legal form and signaling).

91 See 47 CFR 80.1061.

government as a whole will be able to perform its commitments.[92] It facilitates *ex ante* buy-in by the governmental actors essential to the success of the cooperative arrangement, who might not have been aware of it if concluded through less-formal means.

In sum, where compliance considerations are not at the fore, and where apportionment of rights or liability is not essential, implementation is the key to understanding the choice between legal and non-legal mechanisms. In many cases, the choice will come down to whether successful implementation would be served by the internal coordination processes triggered by international legal mechanisms. To a certain extent, these processes are enemies of agility; non-binding arrangements, in general, require less coordination and authorization, and thus less time to conclude. But agility matters only if the arrangement is efficacious, if each participant implements its commitments. I propose a practical approach to assessing when an international legal mechanism would be advantageous for implementation.

An Operational Level Approach

An operational level approach begins by identifying the actors essential to success in each participating state, and then considers their level within or relationship to the state. To use the Space Cooperation Disaster Charter as an example: Whose action is necessary to task satellites and provide data free of charge to foreign entities in response to a disaster? Are the relevant satellites owned and operated by governments or private entities? Do the governments have value-added processing capabilities, or do they rely on the private sector? Within each government, which agencies must act? At what *level* of each agency would the necessary actions occur? Do the relevant agencies have existing authority to do this, or is legislative action or intra-executive delegation necessary? Are new budgetary appropriations necessary to add personnel or buy data?

Identifying the operational level brings into focus any implementation challenges that may favor an international legal mechanism. For example, in the Cospas-Sarsat case, the level of operations was relatively high and distributed across executive and legislative functions. Fielding the space segment of the system

[92] Abbott and Snidal address a potential role for international legal mechanisms in domestic implementation, arguing that "executive officials should look to hard international law to commit other domestic agencies (especially legislatures) or political groups when those officials are able to make international agreements without interference or control, and when their preferences differ significantly from those of competing power centers." Abbott and Snidal, "Hard and Soft Law," 430. While acknowledgement of the influence of implementation needs on legal form is an important insight, I disagree with the suggestion (at least in the United States) that the function of an international agreement is "commit" a legislature. Rather, the internal coordination processes are useful in securing *ex ante* buy-in by actors essential to implementation.

required new funds, and promoting the standard required regulatory action in each participating state. Where the operational level is high and distributed within cooperating governments, the implementation advantages of an international legal mechanism are greatest.

The operational level in the Space Cooperation Disaster Charter case is comparatively low and concentrated within the governments of participating states. The United States participates in the Charter through the United States Geological Survey (USGS), which operates the long-running LANDSAT remote sensing program and is charged with certain responsibilities in support of predicting and responding to natural hazards, and the National Oceanic and Atmospheric Administration (NOAA), which operates the nation's weather and environmental satellites. Observation of Earth and its atmosphere from space were already important functions of the agencies. To carry out these missions for the United States, they are equipped with expert personnel, as well as budgets and legal authorities to support these missions. Both operate remote sensing systems and have working relationships with other governmental operators – within and outside the U.S. – as well as with private commercial operators. To a significant extent, implementation of the Charter could be accomplished by the lead agencies, within existing budgets and legal authorities, and by existing personnel.

Examination of the operational level reveals the limited value an international legal mechanism might add to the Charter's implementation by each participant. Where an efficacious result can be achieved within existing resources and authorities – without action by the highest levels, or different branches of governments – the advantages of non-legal mechanisms in terms of agility will often predominate.

In cases where agility matters, keeping one eye on the operational level is a useful approach to identifying opportunities to scale an international cooperative arrangement under negotiation to lessen implementation challenges and delays. It is a methodical way for states to explore structuring their cooperation to maximize near-term results, even if only as an initial step in a longer-term, more ambitious plan for cooperation.[93]

Looking Ahead – Open Source Cooperation: Whether an international legal mechanism would be optimal now, twelve years into the operational life of the Space Cooperation Disaster Charter, is a slightly different inquiry. The fully operational Charter frees us to consider the merits of elaborating the nonbinding arrangement into a legally binding international agreement – or a broader

[93] Abbott and Snidal have demonstrated that a gradual deepening of cooperation over time by replacing non-legal mechanisms with international legal ones, or deepening the substance of international legal mechanisms, is a well-travelled pathway to international cooperation. Abbott and Snidal, "Hard and Soft Law," 59–60, 69–71. To their political uncertainty explanations of this common cooperative trajectory, I would add the implementation considerations discussed in this chapter.

agreement encompassing all phases of the disaster management cycle – with less concern for time.

In terms of value to end-users, there is little to be gained in the near term, and much to lose, by formalizing or centralizing the space-disaster management ecosystem, such upgrading the legally nonbinding and expressly "voluntary," "best endeavors" commitments of the Space Cooperation Disaster Charter Parties to legally binding, compulsory obligations. Any theoretical gains in the continuity of available resources must be weighed against the real possibility of declining participation if some current contributors are unable to undertake such a formal financial commitment.[94]

For disaster management end users, the value of the Space Cooperation Disaster Charter is growing. An expanding membership enhances the resources available through the Charter, and its recent shift to Universal Access dramatically expanded access. At the same time, elements of the space-disaster management ecosystem are linking together to create new disaster applications.[95] Take for example recent initiatives to leverage GEONETCast's global network of satellite communication systems to disseminate disaster management information – including through the Charter – when a disaster has crippled ground-based communications infrastructure.

The *outside-in* trajectory of these mechanisms – developed independently, in separate fora, to address different aspects of a global problem, and reaching in and linking together over time – invites another comparison with the Cospas-Sarsat case. The development of the latter system was comparatively centralized. Interoperability between the satellites contributed by different partners, and compatibility with distress beacons, is absolutely essential to its success. The communication between a distress beacon and satellites, and the relay of the beacons' distress signal to ground stations is automated; it does not work unless all components of the system talk to each other. The centrality of interoperability and compatibility to its operation favor what could be called a *closed system*.

By comparison, the space-disaster management ecosystem is an example of what might be called *open source cooperation*, as it is built upon varied contributions by a diverse range of actors.[96] The effective use of satellites in outer space in disaster situations on Earth depends more on human intervention than hardware compatibility. In the first instance, someone – whether in the local disaster response

94 The prospect that elaboration of a non-legal mechanism into an international agreement – or replacing an existing international agreement – will result in at least a temporary loss of some participants, gives successful international cooperative arrangements a certain inertia. This equilibrium in which the value of new international legal commitments to address relatively minor issues is exceeded by a loss in membership could be called *treaty stasis*.

95 See Disaster Charter, Section IV.2.

96 I borrow "open source" from the computer software realm, where it generally describes making the source code freely available so that the community of software developers may contribute improvements.

authority or an agency or organization acting on its behalf – must understand what space systems can contribute to addressing a given disaster, and must request this assistance through the Charter or a related mechanism. Another person must then make a judgment call about which of the available satellites to task. Still others must process, analyze, and combine the raw data to generate value added maps for other people – sometimes half a world away in the disaster zone – to interpret and incorporate into response plans.

Some of these people making the Space Cooperation Disaster Charter work day to day are employees of government space and science agencies. Some are at academic institutions, and others at private commercial entities. Some sell their data and services to Charter participants, others volunteer them. This open-ended architecture of the Charter and related mechanisms – inviting a diverse range of actors to contribute in their own way – significantly enhances the resources available to the disaster management community.

This relatively informal, open source approach seems to be headed in a productive direction, unlocking new opportunities to utilize existing space systems by accommodating the contributions of a diverse and growing range of actors. Challenges remain in realizing the full potential of space systems for disaster management, but at this early stage in the trajectory, solutions demand agility more than rigid commitments.

CONCLUSION

International law plays an indispensable role in enabling the use of outer space for disaster management. It is essential to the use of space by a diverse array of state and non-state actors, and more particularly, to enabling the observation of Earth from outer space, on terms conducive to understanding, predicting, and responding to natural disasters. Another important role of international law is in facilitating international cooperation to field the individual space systems, which can be impracticable for a state acting alone. Many such systems are the result of a vast body of bilateral and trilateral international agreements fixing the contributions of multiple states to a given satellite – sensors, launch services, ground stations, technical know-how – and allocating risks, rights, and responsibilities among them. This is not addressed as separate a phase because it spans the entire history of space exploration.

To date, international legal mechanisms have not been favored in addressing the challenges presented by disaster applications in the third phase, and this is unlikely to change in the near term. But none of this suggests that the role of existing international law is diminished in this third phase, or that it is declining over time. In fact, the opposite is true. The influence of the international law applicable to outer space continues to grow with the membership of the core instruments, and as a growing number of states enact national space laws, extending this body of international law to non-governmental activities in space.[97] The states seeking to harness space systems for disaster applications did not face a binary choice between international law and non-legal mechanisms. Rather, as in most areas of international cooperation, legal and non-legal mechanisms are layered, operating on different levels to address the aspects of an overall challenge for which they are best suited.[98]

The last phase presents a qualitatively different set of challenges than the initial phases. It does not require the ground-up establishment of a regime for an entirely new class of activities, with the attendant allocation of risks and rewards. It is an application of existing activities – and existing resources – and thus builds on the shoulders of the international regimes for space and remote sensing, and the systems they enabled. For this specific challenge, international legal mechanisms are not optimal at this time: they do not offer significant advantages in terms of efficacy, and non-legal mechanisms afford greater agility. This is not necessarily true of all humanitarian applications of space systems or challenges falling under the umbrella of "coordination problems." In general, a focus on implementation – and the operational level within each cooperating state – helps to illuminate the optimal role of international legal mechanisms in addressing global challenges of this kind.

97 For a compilation of information about national space laws, see the National Space Law Database, maintained by the United Nations Office of Outer Space Affairs. www.oosa.unvienna.org/oosa/en/SpaceLaw/national/state-index.html.

98 This description of "layered cooperation" first appeared in Harold Honju Koh, "Address: Twenty-First Century International Lawmaking," *Georgetown Law Review*, (2012): 1–19.

13

Utilizing International Climate Change Adaptation Funding to Reduce Risks of Natural Disasters in the Developing World

Paul Govind

This chapter examines the relationship between funding for climate change adaptation (CCA) and disaster risk reduction (DRR). The relationship between climate change and natural disasters is well documented. Broadly speaking, evidence has shown a strong link between rising global temperature and increased occurrences of extreme hydro-meteorological weather events affirming the relationship between climate change and natural disaster. Despite this connection, integration of natural disaster management and climate change policy has been very limited thus far. The existence of similar vulnerabilities has motivated attempts at greater integration between CCA and DRR. This chapter examines the extent to which successful integration between CCA and DRR is reflected in funding mechanisms. The Intergovernmental Panel on Climate Change (IPCC) maintains that "[o]pportunities exist to create synergies in international finance for disaster risk management and adaptation to climate change, but these have not yet been fully realized." Funding is an indispensable element in improving resilience from both climate impacts and natural disasters and a key policy instrument.

INTRODUCTION

The context of climate change and natural disaster management forms the basis of this chapter. This chapter examines the relationship between funding for CCA and DRR. It details the difficulties of managing the relationship in terms of integrated funding channels.

CCA has reached a crossroad. The decision of the 18th Conference of the Parties (COP 18) to the United Nations Framework Convention on Climate Change recognized the increasing likelihood of loss and damage occurring as a consequence of climate change.[1] This marked the first time that loss and damage has

[1] Conference of the Parties, United Nations Framework Convention on Climate Change. "Eighteenth Session, Held in Doha from 26 November to 7 December 2012, Submission on

been included in an international legal document and has inevitably prompted questions about the ongoing viability of CCA as an effective measure to ameliorate the impacts of climate change that are already occurring. Included within this decision was acknowledgment of natural disaster management and the "relevant knowledge" included in key instruments and documents.[2] The specific focus of this chapter is upon the integration between CCA and DRR.

This is a burgeoning area of scholarship as CCA and DRR are viewed as being highly complementary in both a conceptual and practical sense. CCA and DRR are both proactive and *ex ante* approaches to minimizing the effects of environmental impacts through elimination of risk. While the importance of post-disaster relief and reconstruction is not to be underestimated, CCA and DRR are aimed at avoiding the occurrence of events in the first place. In many instances the research communities of both climate change and natural disaster risk management are concerned with the same set of underlying vulnerabilities such as poverty and social circumstances that give rise to and are exacerbated by environmental impacts. These vulnerabilities mean certain nations and populations are simultaneously exposed to the impacts of climate change and natural disasters.

The existence of similar vulnerabilities has motivated attempts at greater integration between CCA and DDR. This chapter examines the extent to which successful integration is reflected in funding mechanisms. The IPCC maintains that "[o]pportunities exist to create synergies in international finance for disaster risk management and adaptation to climate change, but these have not yet been fully realized."[3] Funding is an indispensable element in improving resilience from both climate impacts and natural disasters and a key policy instrument.

Despite the similarities, CCA and DDR remain distinct pursuits with ultimately different objectives. One example is the difference in temporal conditions between climate impacts and natural disasters. Sea level rise, for instance, that is readily associated with climate impacts is regarded as a long term, slow onset form of damage that is constant compared with the sporadic occurrences of extreme

Information from Developed Country Parties on Resources Provided to Fulfill the Commitment Referred to in Decision 1/CP.16, Paragraph 95." FCCC/CP/2012/INF.1, August 31, 2012, 14. United Nations Framework Convention on Climate Change, Conference of the Parties *Report of the Conference of the Parties on Its Eighteenth Session, Held in Doha from 26 November to 8 December 2012 Draft Decision* /CP.18 (Herein referred to as the 'COP-18 Decision').

2 "Taking note of the relevant knowledge and ongoing work outside of the Convention, including the Special Report of the Intergovernmental Panel on Climate Change, Managing the Risks of Extreme Events and Disasters to Advance Climate Change Adaptation, the Global Assessment Report on Disaster Risk Reduction, the Hyogo Framework for Action, and the Global Framework for Climate Services of the World Meteorological Organization." Ibid.

3 Intergovernmental Panel on Climate Change. *Managing the Risks of Extreme Events and Disasters to Advance Climate Change Adaptation – A Special Report of the Working Groups I and II of the Intergovernmental Panel on Climate Change.* (New York: Cambridge University Press, 2012).

weather events. The consequences of damage also represent a potential point of departure between CCA and DRR. Impacts such as sea level rise can result in damage of a permanent nature – thereby explaining the currency of the term "loss and damage" in climate change discourse. These examples highlight key differences that can in certain circumstances also extend to conceptualization of risk and divergence in funding between CCA and DRR.

This chapter will begin by providing an overview of the relationship between climate change and natural disasters. Providing an environmental context for discussion of natural disasters is necessary to fully appreciate the importance of vulnerabilities. The different factors that contribute to vulnerability will be discussed in the second section. This constitutes the basis for the complementary relationship between CCA and DRR in particular with relation to funding. Both CCA and DRR are concerned with the underlying vulnerability associated with low levels of socioeconomic development. Poorer nations and communities that are at the forefront of both climate impacts and natural disasters therefore require funding to ameliorate risk. The subsequent section will examine how this synergy has developed in the context of international environmental law and policy. The main focus shall be on the climate change regime from the Bali Action Plan[4] in 2007 to the most recent COP 18 decision reached in Doha. Of particular importance are the influence of natural disaster management to the conceptualization of loss and damage and the relevance of funding. The final section will provide an illustration of the benefits and detriments of integrating CCA and DRR through the operation of Fast Start Finance commitments of Australia, New Zealand, the United States, Canada, and Japan.

CLIMATE CHANGE AND NATURAL DISASTERS

Extreme Events and Disasters

The IPCC provides the following breakdown of extreme weather events. It is important to note the temporal difference and the interaction certain conditions have on other environmental and social values thereby creating a disaster:

An extreme (weather or climate) event is generally defined as the occurrence of a value of a weather or climate variable above (or below) a threshold value near the upper (or lower) ends ("tails") of the range of observed values of the variable. Some climate extremes (e.g., droughts, floods) may be the result of an accumulation of weather or climate events that are, individually, not extreme themselves (though their accumulation is extreme). As well, weather or climate events, even if not

[4] United Nations. *Report of the Conference of the Parties on Its Thirteenth Session, Held in Bali from 3 December to 15 December 2007 – Addendum – Part Two: Action Taken by the Conference of the Parties at Its Thirteenth Session*, UN Doc FCCC/CP/2007/6/Add.1 (March 14, 2008) Decision 1/CP.13 (Herein referred to as the '*Bali Action Plan*').

extreme in a statistical sense, can still lead to extreme conditions or impacts, either by crossing a critical threshold in a social, ecological, or physical system, or by occurring simultaneously with other events.[5]

In the context of natural disaster, extreme weather is relevant in a number of ways. There is a misconception that weather contributes only to one-off extreme events. This overlooks important cumulative effects of weather patterns on a long-term basis. The IPCC categorizes the effect of climate upon weather extremes relevant to impacts and disasters into three categories:

1) Extremes of atmospheric weather and climate variables (temperature, precipitation, wind)
2) Weather and climate phenomena that influence the occurrence of extremes in weather or climate variables or are extremes themselves (monsoons, El Niño and other modes of variability, tropical and extra-tropical cyclones)
3) Impacts on the natural physical environment (droughts, floods, extreme sea level, waves, and coastal impacts, as well as other physical impacts, including cryosphere-related impacts, landslides, and sand and dust storms).[6]

In addition to the humanitarian and environmental consequences disasters can result in huge financial and humanitarian losses and set development goals back years. The relationship between disaster and development is fundamental. A key measurement of disaster is the impact on development goals and ambitions; conversely a key indicator of where a disaster might occur is a lack of development. The underlying point is that an environmental hazard does not by itself constitute a disaster – it's the impacts it causes that are significant, and impacts are in turn influenced by a number of variables. While it remains impossible to control the occurrence of environmental hazards steps can be taken to mitigate impacts and therefore minimize disaster. The ability to systematically cope with disasters depends on the "success or failure of the total adaptation of the community."[7]

Traditionally, responses to disasters have focused primarily upon ex post measures and channeling resources for immediate needs rather than future risk reduction.[8] This ignores the need to strengthen society through risk reduction. Focus on the effectiveness of ex post measures has caused a reevaluation of the impact of

[5] IPCC, *Managing the Risks*, 111. [6] Ibid., 115.

[7] Anthony Oliver-Smith, "What is a Disaster? Anthropological Perspectives on a Persistent Question," in *The Angry Earth: Disaster in Anthropological Perspective*, by Anthony Oliver-Smith and Susanna Hoffman, (London: Routledge, 1999), 25.

[8] Lisa Schipper, "Meeting at the Crossroads?: Exploring the Linkages Between Climate Change Adaptation and Disaster Risk Reduction," *Climate and Development*, (2009): 16–30 at 20. For further discussion see Lisa Schipper and Mark Pelling, "Disaster Risk, Climate Change, and International Development, Scope For, and Challenges to Integration." *Disasters*, (2006): 19–31.

disaster relief on development and has assisted the disaster community "in moving toward a more sustainable, vulnerability focused approach."[9]

Climate Change and Extreme Events

It is impossible to attribute the occurrence of any one environmental disaster to climate change. There are two main reasons for this: disasters have occurred in climatic conditions unaffected by anthropogenic influence, and there are a wide variety of factors that contribute to the occurrence of an extreme event and whether any such event becomes a disaster. Climate change projections can identify whether conditions that will exacerbate extreme events have increased over time. The issue then becomes whether human influence changes the likelihood of an event and as a consequence increase the risk of damage caused by environmental impacts. Whether this ultimately constitutes a disaster must be considered in conjunction with the vulnerabilities of a particular country or population that is the subject of the next section of the chapter.

In approximately the last twenty years, the overwhelming bulk of so-called natural disasters have been attributed to weather-related events; in particular floods (hydrological) and extreme storms (meteorological).[10] In the same period where hydro-meteorological events increased by 100 percent, the number of geophysical disasters such as earthquakes remained constant.[11] The distinct rise in global temperatures during the course of the twentieth and twenty-first century coincides with this increase. This has inevitably led to conclusions that climate change is at least in part responsible for the increase in extreme weather events. Broadly speaking, this is correct as the majority of climate change impacts will be hydro-meteorological. However, while this appears eminently sensible, the relationship between climate change and natural disasters requires more rigorous analysis.

The Fourth Assessment Report of the Intergovernmental Panel on Climate Change (AR4) stated that "various extreme events are very likely to change in magnitude and/or frequency and location with global warming."[12] Projections that are the focus of the IPCC reports relate to potential future events and outcomes. Therefore the level of confidence attached to each projection is influenced by a multitude of factors including "the type of extreme, the region and season and

[9] Ibid.

[10] Jorn Birkmann and Korinna von Teichmann, "Integrating Disaster Risk Reduction and Climate Change Adaptation: Key Challenges – Scales, Knowledge and Norms." *Sustainability Science*, (2010): 171–184 at 171.

[11] Ibid., 172. See also, Maarten K. Van Aalst, "The Impacts of Climate Change on the Risk of Natural Disasters." *Disasters*, 2006: 5–18.

[12] James J. McCarthy, Osvaldo F. Canziani, Neil A, Leary, David J, Dokken, and Kasey S. White, *Climate Change 2001: Impacts, Adaptation, and Vulnerability: Contribution of Working Group II to the Third Assessment Report of the Intergovernmental Panel on Climate Change* (New York: Cambridge University Press, 2007). IPCC, *Managing the Risks*, 9.

quality of observational data." The projections published by the IPCC in *Managing the Risks of Extreme Events and Disasters to Advance Climate Change Adaptation*, states that changes to climate extremes under different emission scenarios "generally do not strongly diverge in the coming two to three decades."[13] This affirmed with heightened confidence broad-based projections from the Fourth Assessment Report released in 2007 "that some weather events and extremes will become more frequent, more widespread or more intense during the 21st century."[14]

CCA and DRR are not exclusively concerned with impacts or outcomes but include within their scope all social, physical, and economic conditions that contribute to fostering the event or increasing its magnitude. The impacts that could accompany a rise in the incidence of weather extremes must be reviewed critically. One particular change in environmental conditions will most likely not result in a higher frequency or severity of impacts. The conditions that exacerbate damage are varied. It is necessary to acknowledge the cumulative effect of different environmental changes and the relationship between such changes.

Understanding the potential impacts of tropical cyclones is a particularly useful example of the benefits of adopting this approach. The IPCC projects that "it is *likely* that the global frequency of tropical cyclones will either decrease or remain essentially unchanged."[15] While an increase in tropical cyclones is considered unlikely, it is incorrect to necessarily conclude that there will be less damage caused by cyclonic impacts. The high percentage of damage associated with cyclones is caused by "storm-surge and freshwater flooding from extreme rainfall" – as opposed to high-speed winds.[16] This alludes to an important point about identifying and understanding relevant impacts. While the frequency of tropical cyclones is unlikely to increase, the severity in all probability will as the IPCC projects that it is "*likely* that tropical cyclone-related rainfall rates will increase with greenhouse warming."[17]

These particular consequences of cyclonic activity will be exacerbated by other environmental conditions affected by climate change such as sea level rise. The AR4 projected with *high confidence* that the rate of observed sea level rise increased from the nineteeth to the twentieth century.[18] These projections have since been affirmed and evidence demonstrates that the contribution to sea level rise from the loss of mass in Greenland and Antarctica is accelerating.[19] This has resulted in an

[13] Martin Parry, Osvaldo F. Canziani, and Jean Palutikof, *Technical Summary: Climate Change 2007: Impacts, Adaptation and Vulnerability. Contribution of the Working Group II to the Fourth Assessment Report of the Intergovernmental Panel on Climate Change* (New York: Cambridge University Press, 2007),. www.ipcc.ch/publications_and_data/ar4/wg2/en/tssts-4-4.html

[14] IPCC, *Fourth Assessment Report: Climate Change 2007*.

[15] IPCC, *Managing the Risks*, 113. Note that the projections included in SREX are based on the consistent level of cyclone patterns since the 1970s.

[16] Ibid., 158. [17] Ibid., 163. [18] McCarthy, *Climate Change 2001*.

[19] IPCC, *Managing the Risks*, 179.

increase in extreme coastal highwater across the globe. The AR4 concluded that it was "very likely" that "that humans contributed to sea level rise during the latter half of the 20th century" and "therefore that it was *more likely than not* that humans contributed to the trend in extreme high sea levels."[20] A nuanced approach to environmental interdependencies is vital when surveying the vulnerabilities of nations and populations to both climate impacts and weather extremes.

Climate Change Adaptation & Disaster Risk Reduction

It is vital to establish working definitions of the key concepts of CCA and DRR. DRR is defined as:

> The concept and practice of reducing disaster risks through analysis and management of their causal factors. It reduces exposure to hazards, lessens the vulnerability of people and assets, improves management of the land and environment and preparedness for adverse events.[21]

CCA is defined as:

> Adjustments in natural or human systems in response to actual or expected climatic stimuli or their effects that moderate harm and exploit beneficial opportunities. This can include: (a) adapting development to gradual changes in average temperature, sea level and precipitation; and (b) reducing and managing the risks associated with more frequent, severe and unpredictable extreme weather events.[22]

Climate change represents a new challenge to disaster risk management. Historical models that have served as the basis for predictions relating to hazards and impacts might no longer provide a viable outlook as the rate of global warming changes the behavior of the natural environment at an accelerated rate. This will inevitably influence development parameters and trajectories as new understandings of vulnerability and risk must be absorbed into decision making.[23]

The underlying object shared by CCA and DRR is the amelioration of vulnerability and the establishment of sustainable development pathways. The IPCC maintains that, "Adaptation to climate change and disaster risk management both seek to reduce factors and modify environmental and human contexts that contribute to climate-related risk, thus supporting and promoting sustainability in social

[20] Ibid.

[21] John Hay, *Risk Reduction and Climate Change Adaptation in the Pacific – An Institutional and Policy Analysis*. (New York: United Nations International Strategy for Disaster Risk Reduction, 2012), p. ii.

[22] Ibid., iii.

[23] Jorn Birkmann and Gerd Tetzlaff, Karl-Otto Zentel (eds.) *Addressing the Challenge: Recommendations and Quality Criteria for Linking Disaster Risk Reduction and Adaptation to Climate Change: DKKV Publication Series* 38 (Bonn: German Committee for Disaster Reduction, 2009), 10.

and economic development. The promotion of adequate preparedness for disaster is also a function of disaster risk management and adaptation to climate change."[24] In order to achieve this, the communities must act in tandem and build synergies between CCA and DRR. In both CCA and DRR "there is an implicit acknowledgement that risk is a part of everyday life, and thus social development plays a vital role."[25]

CCA and DDR can have an incremental or transformative effect on economic development through incorporating risk into decision making. While both CCA and DRR are aimed at risk reduction, it is through influencing development trajectories that both concepts find practical implementation. This association has given rise to the policy of mainstreaming CCA into existing development trajectories. The ultimate object of mainstreaming is to ensure that development does not exacerbate conditions that lead to climate change.

CCA and DRR operate in the broader context of economic development. The application greatly depends on the extent to which both CCA and DRR can be integrated into development trajectories or mainstreamed. Mainstreaming has attracted strong support as an efficient and effective way of implementing both CCA and DRR into international development and overseas development assistance. In order for mainstreaming to be effective in terms of reducing vulnerability and risk it is necessary that there is a greater level of integration between CCA and DRR. The IPCC has observed that:

> Neither disaster risk reduction nor climate change adaptation is as well integrated as they could be into current development policies and practices. Both climate change adaptation and disaster risk reduction might benefit from sharing of knowledge and experience in a mutually supportive and synergistic way. Climate change adaptation could be factored into all disaster risk management.

Funding is desperately needed to reduce vulnerability and the risk of impacts occurring in poorer countries. The occurrence of environmental hazards is not necessarily more frequent in poorer countries as opposed to the rest of the world. However, the conditions that exist in poorer countries mean that the translation of a hazard into a disaster is much more likely and the consequences more severe relative to richer countries. Statistics from the insurance company Munich Re show that "average fatalities for major disaster events have been approximately 40 times higher in low-income as compared to high-income countries." Further direct asset loss as a percentage of GNP is three times greater in poorer countries. This underlines the heightened vulnerability of lack of capacity to absorb and recover from environmental shocks that are characteristic of low levels of economic

[24] IPCC, *Managing the Risks*, 12. [25] Schipper, "Meeting at the Crossroads?", 16.

development. The UN Development Program has underlined the need for greater focus on DRR:

> Humanitarian assistance will remain essential, but emphasizing disaster response strategies at the expense of proactive integrated approaches to disaster risk reduction can have the effect of perpetuating vulnerability .[26]

The transformative quality of CCA means that while risk influences the parameters of development, ultimately a new development trajectory should internalize risk and by doing so reduce it. Underlying this is the cycle approach to DRR and disaster management more broadly that allows the experiences of past extreme events to inform risk management for the future. To reiterate, the disaster cycle moves through mitigation, preparedness, response, recovery, and reconstruction.[27] CCA can be integrated as part of the disaster cycle in a number of different phases including (1) the disaster mitigation phase when areas vulnerable to climate change or disasters can be identified; (2) preparedness phase; (3) response phase when adaptation can be incorporated into response strategies; and (4) recovery and reconstruction phrase.[28]

Despite the overt similarities, CCA and DRR have developed independently and the integration of CCA and DRR remains in its infancy. Several obstacles, both conceptual and practical, threaten to hinder further growth of synergy between CCA and DRR. Differences include separate intellectual development of the respective fields, the implementation of risk reduction measures, and competitions between institutions.

While CCA and DRR are not the same thing, there is an opportunity to share experiences and fill in knowledge when both activities are conducted in parallel. The extensive history of DRR can help guide CCA toward a future that is characterized by a high level of unpredictability. Conversely, while DRR is an interventionist strategy designed to reduce risk, it can benefit through integrating climate change information into such interventions and improving sustainability.

Discussion of CCA and DRR is informed by a number of related concepts particularly vulnerability, risks, and impact. However the definition and interpretation of these concepts often differs depending upon the relevant context. Vulnerability has been formulated in a variety of ways in an attempt to give effect to the number of different influences. Vulnerability is a ubiquitous term in the context of climate law and international environmental law more generally. The Bruntland

[26] United Nations Development Programme, A *Climate Risk Management Approach to Disaster Reduction and Adaptation to Climate Change: UNDP Expert Group Meeting. Integrating Disaster Reduction with Adaptation to Climate Change, Havana,* June 17–19, 2002 http://ccsl.iccip.net/riskadaptationintegrated.pdf

[27] See, *Federal Emergency Management Agency (FEMA)*. www.fema.gov.

[28] Birkmann and von Teichmann, "Integrating Disaster Risk Reduction," 181–182.

Report mentioned "vulnerability" forty-seven times across 300 pages;[29] the policy makers' summary of the IPCC Report on Impacts, Adaptation and Vulnerability referred to it thirty times in a sixteen-page document.[30] Similarly, within the context of disaster management in the Hyogo Framework for Action, "vulnerability" appears ten times within six pages.[31] The ubiquity and general acceptance of the importance of the concept has diverted attention from the need to establish a common meaning. The origins of the concept and its application are found in the disaster management community and over time four different classifications of vulnerability have been developed.

The first is termed the "pressure and release" model. This approach maintains that unsafe conditions that can lead to damage are more the result of underlying social conditions rather than the actual environmental hazard itself. It also links the shift in unsafe conditions to broader national and global patterns of resource distribution. The second category adopts a "social-ecology perspective" and stresses that vulnerability can be managed only if the interdependency of human and environmental systems is identified and understood. A holistic approach most aptly describes the current version of vulnerability subscribed to within the disaster risk management community. The focus of this third classification of vulnerability is upon differences in exposure, susceptibility, and capacity to respond as causes or factors in determining vulnerability. In this sense it operates on a continuum that views vulnerability as dynamic and something that needs to be regularly measured and reviewed.[32]

In the context of climate change a number of different definitions of vulnerability have been proposed. The most prolific being the definition produced by the IPCC that describes vulnerability "as a function of exposure, sensitivity, and adaptive capacity."[33] This definition exhibits characteristics peculiar to the climate change community. For example, the rate and magnitude of climate change is considered and as a consequence the definition embraces a long term view of trends and environmental stresses "rather than on current shock forecasting, something not explicitly excluded but rather rarely considered within the disaster risk management approaches."[34]

Definitions shape the concepts that form the basis of both CCA and DRR policy. If different meanings are attributed to a concept such as vulnerability then

[29] See World Commission on Environment and Development, *Our Common Future* (Oxford University Press, 1987)

[30] See Parry, *Technical Summary*.

[31] United Nations International Strategy for Disaster Reduction. "Hyogo Framework for Action 2005–2015: Building the Resilience of Nations and Communities to Disasters," Extract from Report of the World Conference on Disaster Risk Reduction, Jaunary 2005, Kobe, Japan. www.unisdr.org/files/1037_hyogoframeworkforactionenglish.pdf. (Herein referred to as Hyogo Framework).

[32] For a broader discussion see, Mark Pelling, *Adaptation to Climate Change From Resilience to Transformation* (Abingdon: Routledge, 2011).

[33] Parry, *Technical Summary*, 27.

[34] IPCC, *Managing the Risks*, 71.

"practical actions and coherent strategies might be at risk, since a different understanding of what vulnerability means implies consequently also different approaches for strategies and solutions."[35] In the context of DRR, vulnerability is equated to a condition determined by physical, social, economic, and environmental factors that increase the susceptibility of a population to the impacts associated with hazards or alternatively the capacity to recover from such impacts.[36] There is an emphasis on the underlying conditions within a society that give rise to vulnerability. In contrast, vulnerability as defined by the IPCC incorporates the rate and magnitude of climate change. This particular construction of vulnerability reflects an element of environmental risk that is external to the particular society. The intersection of CCA and DRR is where the external environmental risk exacerbates the underlying shortcomings within a society that expose it to vulnerability in the first place.

This outcome bridges the CCA and DRR communities. Birkmann has observed that the definition of vulnerability endorsed by the IPCC closely resembles the concept of risk used in DRR scholarship.[37] Under a DRR approach the outcome of the interaction between hazards and vulnerable conditions is characterized as risk. However, within the CCA discourse the final outcome following the assessment of environmental hazards and potential impacts is vulnerability rather than risk. Both outcomes are derived from very similar assessments despite the use of the term "risk" in the context of DRR and "vulnerability" in CCA to describe the outcome. In order to create more effective synergies these differences in approach need to be identified, acknowledged, and reconciled.

The relationship between risk, vulnerability, and development is evident through the plight of poorer countries. Environmental impacts either from climate change or natural disasters are invariably more severe in poorer countries due in large part to lack of economic development. Poverty is an example of underlying vulnerabilities that expose poorer countries to more severe environmental shocks. The IPCC has commented that:

> It has become clear that many less-developed regions will have limited success in reducing overall vulnerability solely by managing climate risk because vulnerability, adaptive capacity, and exposure are critically influenced by existing structural deficits (low income and high inequality, lack of access to health and education, lack of security and political access, etc.)[38]

The influence of such deficits means that the nature and extent of vulnerability is subject to change depending upon the relevant circumstances. Vulnerability is therefore dynamic and the exposure of different nations and communities can vary

[35] Birkmann, "Integrating Disaster Risk Reduction," 12.
[36] Hyogo Framework.
[37] See Birkmann, "Integrating Disaster Risk Reduction."
[38] IPCC, *Managing the Risks*, 48.

across temporal and spatial scales, and depend on economic, social, geographic, demographic, cultural, institutional, governance, and environmental factors.[39]

A clear distinction between CCA and DRR is subject matter. While CCA is only concerned with the impacts and dynamics of climate change, DRR applies to all types of events classified as natural disasters including geological hazards such as earthquakes. In terms of environmental focus CCA is not as broad ranging as DRR but the important similarities are found not in terms of subject matter but approach. Both CCA and DRR operate to modify, or in some cases transform, developmental patterns to more effectively measure and reduce risk. There is also discord surrounding the use of the terms "hazard" and "disaster." While it is not possible within the confines of this chapter to provide a deconstruction of each of these terms, it is important to note that the lack of consistency points to some the problems in building synergy between CCA and DRR.

Disasters relate to disruptive events. The event represents a shock that challenges the status quo and is generally short term in terms of impact. Attempting to characterize the effects of climate change as a disaster is problematic when considering the slow onset of damage. Examples such as sea level rise and drought take place over several years and can result in permanent loss and damage. Therefore CCA deals with an additional type of long-term risk that appears absent from traditional conceptualizations of DRR. Schipper notes that long-term risk management central to CCA closely resembles the idea of disaster risk accumulation that recognizes the cumulative effect of disasters over time. The presence of "accumulation" is useful in building synergy between CCA and DRR but it remains difficult to describe sea level rise or drought as being constituted by a series of disasters. The damage caused by these effects is incremental and is not represented by a singular event.

Institutional Governance

International law and policy present an opportunity to regulate and manage CCA and DRR at a global scale. As outlined earlier in the chapter, the CCA and DRR communities have developed independently and the current challenge for international law is to maintain and enhance a normative dialogue. This section will examine the existing legal and policy frameworks that regulate both CCA and DRR and highlight the potential for strengthening linkages and synergies. The discrepancies between the meanings of certain key terms reflect the divergent narratives that each community has adopted. Another factor is the different institutional bodies that influenced the different concepts. CCA has been driven primarily at the international level through regulatory regime provided under the United Nations Framework Convention on Climate Change (UNFCCC) while the

39 Ibid., 83.

direction of DRR has been influenced by a host of disparate and decentralized sources. The DRR community has not been afforded the benefit of an international treaty mandating global action on the issue. This has led to observations that countries are "hesitant to negotiate and accept far-reaching treaties that impose legally binding responsibilities with respect to disaster preparedness, protection, and response."[40]

Whether these differences can be reconciled is crucial to future implementation of both CCA and DRR. Given the more coherent, cohesive, and readily identifiable presence of climate law and regulation in the international arena, doing so presents the most solid basis for institutional dialogue. The UNFCCC embraces the principle of differentiation under the theme of "common but differentiated responsibility" and an overall commitment to equity. Under Article 4(8) parties must give full consideration to the actions necessary to meet the needs of developing country parties arising from the adverse effects of climate change especially on, "Countries with areas prone to natural disasters."[41] This indicated a broad recognition of the relevance between climate change and natural disasters but meaningful engagement was slow to develop further.

In the context of the UNFCCC, CCA has historically been overshadowed by the attention given to mitigation and the focus on the reduction of greenhouse gas emissions. There is however some key provisions that have served as the basis for further enhancement of adaptation policy. Article 4.1(e) commits parties to prepare for adaptation to the impacts of climate change.[42] The provision does not explicitly mention implementation of adaptation actions focusing on plans to manage and regulate potential future impacts. The relationship between adaptation and economic resilience is underlined in Article 4.1(f) that mandates climate change considerations are taken into account with "a view to minimizing adverse effects on the economy."[43] This echoes Article 3.4, which states that policies and measures to address climate change "should be integrated with national development programmes."[44] During negotiations preceding the UNFCCC, developing country governments insisted that climate change policy be viewed as a component of sustainable development that "aims at increasing national and regional capacity to deal with climate variability as well as long-term climate change."[45]

An early attempt to rationalize CCA and DDR at the international level was a conference organized by the United Nations Development Program (UNDP). This represented an important first step in terms of identifying the benefits of integration and creating the conceptual space for discussion between the two communities. The outcome of the conference was a document entitled "A Climate Risk Management Approach to Disaster Reduction and Adaptation to Climate Change" that

[40] David Fidler, "Disaster Relief and Governance After the Indian Tsunami: What Role for International Law?" *Melbourne Journal of International Law* 6, no. 2 (2005): 473.

[41] UNFCCC, article 4(8). [42] UNFCCC, article 4(1)(e). [43] UNFCCC, article 4(1)(f).

[44] UNFCCC, article 3(4). [45] Schipper, "Meeting at the Crossroads?"

proposed an "integrated climate risk management approach to both disaster risk reduction and adaptation."[46] From the perspective of natural disaster management closer alignment with CCA afforded an opportunity to shift focus away from response measures onto risk management. It allowed the climate change community to design measures for future climate impacts instead exclusively operating in the realm of hypothetical adaptation strategies.[47]

The continuation of the dialogue initiated by the UNDP has proven influential in terms of altering the normative basis of both CCA and DRR. The relevance of humanitarian costs to disaster management has arguably changed the emphasis of adaptation policy. In 2006, the Red Cross/Red Crescent Climate Centre organized an international conference on CCA and DRR.[48] The engagement by the Red Cross and Red Crescent Societies in the climate change debate is an example of the growing influence of humanitarian concerns on climate discourse. A particularly relevant aspect of the conference was discussion of the misconceptions that exist regarding the relationship between CCA and DRR.

The recognition of the relevance of extreme events in the context of the climate regime grew as CCA received a higher profile. The promotion of CCA received a major boost in the Bali Action Plan that identified adaptation as one of the four "building blocks" on which a future climate agreement was to be based.[49] Equally important are the linkages to strategies of implementation – including finance – and the recognition that adaptation needed to be enhanced through *inter alia* "risk management and risk reduction strategies, including risk sharing and transfer mechanisms such as insurance; disaster reduction strategies and means to address to address loss and damage associated with climate change impacts in developing countries that are particularly vulnerable to the adverse effects of climate change."[50] Complementing the explicit references to DRR, the Bali Action Plan also highlighted the need for a greater institutional dialogue and "ways to strengthen the catalytic role of the Convention on encouraging multilateral bodies, the public and private sectors, and civil society, building on synergies among activities and processes, as a means to support adaptation in a coherent and integrated manner."[51]

The 16th Conference of the Parties (COP16) in 2010 passed the Cancun Agreements.[52] This explicitly afforded equal priority to mitigation and adaptation as

[46] See UNDP, A *Climate Risk Management Approach to Disaster Reduction and Adaptation to Climate Change.*

[47] Schipper, "Meeting at the Crossroads?" 26.

[48] Red Cross and Red Crescent Climate Centre. *Red Cross, Red Crescent Climate Guide*, (The Hague: Red Cross, 2007).

[49] *Bali Action Plan.* [50] Ibid., 4. [51] Ibid.

[52] United Nations, "Report of the Conference of the Parties on Its Sixteenth Session, Held in Cancun from 29 November to 10 December 2010 – Addendum – Part Two: Action Taken by the Conference of the Parties at Its Sixteenth Session." UN Doc FCCC/CP/2010/7/Add.1 Decision 1/CP.16, March 15, 2011.

climate policy goals. Contained with the Cancun Agreement is the Adaptation Framework that calls for enhanced action on adaptation by undertaking relevantly:

> (e) Enhancing climate change related disaster risk reduction strategies, taking into consideration the Hyogo Framework for Action, where appropriate, early warning systems, risk assessment and management, and sharing and transfer mechanisms such as insurance, at the local, national, sub-regional and regional levels, as appropriate.[53]

The Cancun Agreements established the Adaptation Committee that has among its objects sharing relevant information, promoting synergies, and recommendations on finance.

One of the most prominent and comprehensive attempts to build synergy and integration has been spearheaded by the IPCC. The central role played by the IPCC, a body mandated to provide the scientific basis for climate policy, is a positive step toward further integration. This culminated in 2012 and the release of its report *Managing the risks of extreme events to advance climate change adaptation.*[54]

A residual challenge is integrating adaptation into the processes and operations of related international institutions. Within the UNFCCC and its different funding streams there remains an absence of clear indicators of how to measure the effectiveness of adaptation. Researchers have proposed a series of generic criteria aimed at improving the effectiveness of adaptation measures: effectiveness, efficiency, legitimacy, and equity.[55] Despite the generality of these criteria, they serve as an important basis for comparison between CCA and DRR by "matching the necessities of risk reduction and general development efforts, and vice versa."[56] Evaluative criteria must be applied at all levels of CCA and DRR governance including at the regional and local level.[57]

The key international response to DRR is the UNISDR— the secretariat of the International Strategy for Disaster Reduction and the body responsible for coordinating global disaster strategy in accordance with the "blueprint for disaster risk reduction" the Hyogo Framework for Action 2005–2015 (HFA). The HFA was adopted in 2005 at the World Conference on Disaster Reduction held in Kobe, Japan. The scope of the UNISDR mandate was expanded in 2001 to ensure synergies among other disaster reduction activities through the UN system in order to assist the mainstreaming of DRR into development goals.[58]

53 Ibid., 5.

54 Managing the Risks of Extreme Events and Disasters to Advance Climate Change Adaptation – A Special Report of the Working Groups I and II of the Intergovernmental Panel on Climate Change.

55 See Neil W. Adger, Irene Lorenzoni, and Karen L. O'Brien, *Adapting to Climate Change: Thresholds, Values, Governance*,(Cambridge University Press, 2009).

56 Birkmann, "Integrating Disaster Risk Reduction," 77.

57 Ibid.

58 The United Nations Office for Disaster Risk Reduction, *Our Mandate* (26 August 2013) www.unisdr.org/who-we-are/mandate.

The UNISDR and the international climate regime operate at different levels with different mandates and enjoy varying strengths and capacities. International climate law is underpinned by the constitutive document of the UNFCCC while by contrast the HFA is not legally binding and constitutes soft law. The voluntary nature of the HFA is viewed as a useful "flexible commitment, largely based on self-regulation and trust, while others regard this as its inherent weakness."[59] It remains a significant document as it encapsulates a political and technical agreement on the priorities for DRR. The HFA reflects broad consensus within the international community having been signed by 168 nations and endorsed by the UN General Assembly. The object of the HFA is complemented by the Global Platform on Disaster Risk Reduction that has as one of its objects assessment of progress in terms of implementing the HFA. However, the HFA lacks the institutional mechanisms of the UNFCCC, such as a secretariat, scientific and expert bodies, and the level of funding that is mobilized on behalf of climate change. The HFA lists five Priorities for Action:

1) Ensure that DRR is a national and local priority, with a strong institutional basis for implementation
2) Identify, assess, and monitor disaster risks, and enhance early warning
3) Use knowledge, innovation, and education to build a culture of safety and resilience at all levels
4) Reduce the underlying risk factors
5) Strengthen disaster preparedness for effective response at all levels.[60]

COP18 of the UNFCCC witnessed the emergence of a number of issues that pose significant conceptual and practical challenges to conceptions of CCA through the inclusion of loss and damage. It represents the culmination of a process initiated during COP16 in 2010 where the UNFCCC decided to establish a work program on approaches to loss and damage associated with climate change impacts in developing countries that are particularly vulnerable to climate change.[61] The COP requested the Subsidiary Body for Implementation (SBI) to take the issue of loss and damage under consideration with a view to delivering a set of recommendations at COP18. The SBI published the contours of a work program on loss and damage that includes three broad focus areas:

- Risk assessment of loss and damage associated with the adverse effects of climate change

[59] Pelling, *Adaptation to Climate Change*, 44. [60] "Hyogo Framework," 14–20.

[61] United Nations, *Report of the Conference of the Parties on Its Sixteenth Session, Held in Cancun from 29 November to 10 December 2010 – Addendum – Part Two: Action Taken by the Conference of the Parties at Its Sixteenth Session*, UN Doc FCCC/CP/2010/7/Add.1 (March 15, 2011) Decision 1/CP.16 , pars. 25–29. For further discussion on the history of loss and damage under the SBI, see, J. Hyvarien, "Loss and Damage Caused by Climate Change: Legal Strategies for Vulnerable Countries."

- Consideration of different approaches to understanding and addressing loss and damage – including both impacts related to extreme weather events and slow onset events
- The role of the UNFCCC in enhancing the implementation of approaches to address loss and damage[62]

COP17 included a report detailing the progress of the Work Program on Loss and Damage. The second thematic area under the Work Program proposed "A range of approaches to address loss and damage associated with the adverse effects of climate change, including impacts related to extreme weather events and slow onset events, taking into account experience at all levels."[63] Included is a set of questions to guide the formation of policy. Most relevantly the questions address resource requirements, links and synergies between risk reduction and how to tailor risk management to national contexts.

The UNISDR appears inherently limited in terms of absorbing climate change impacts and risks into DRR. Hazards that are considered the focus of DRR are classified as being natural in origin. Climate change and related sea level rise are however classified as being human-induced environmental hazards and therefore not the focus of DRR. Echoing the earlier concerns of inconsistent terminology the UNISDR applies the term adaptation to non-climactic effects such as soil erosion.

The phrase "slow onset change" does not appear in the terminology used by the UNISDR. The HFA does recognize the relevance of climate change to DRR by promoting "the integration of risk reduction associated with existing climate variability and future climate change into strategies for the reduction of disaster risk and adaptation to climate change." Crucially, in terms of risks, the HFA does not discern between short-term, sudden impacts and slow onset impacts that pertain to sea level rise and other climate change consequences. The reference to "future climate change" suggests that the HFA is receptive to long-term effects but the timeframe is restricted to ten years and is therefore inherently limited.[64]

The 2011 Global Assessment Report compiled by the UNISDR discerns between two elements of climate change adaptation: adapting to incremental/slow onset changes such as sea level rise and managing risks related to irregular extreme

62 See, Subsidiary Body for Implementation, United Nations Framework for the Convention of Climate Change, *Approaches to Address Loss and Damage Associated with Climate Change Impacts in Developing Countries that are Particularly Vulnerable to the Adverse Effects of Climate Change to Enhance Adaptive Capacity at its Thirty Seventh Session*, UN Doc FCCC/SBI/2012/L.44

63 United Nations, *Report of the Conference of the Parties on Its Seventeenth Session, Held in Durban from 18 November to 9 December 2011 – Draft Report of the Conference of the Parties on its Seventeenth Session*, UN Doc FCCC/CP/2011/L.1 (December 7, 2011) ('*Durban Platform*') See Decision 7/CP.17 'Work Programme on Loss and Damage.'

64 For a further discussion of the issue of slow onset damage see, Linda Siegele, *Loss and Damage: The theme of slow onset impact* (Bonn: Climate Development and Knowledge Network, 2012)

weather events – including those that display an unprecedented level of severity.[65] In this context, adaptation includes responding to rise in temperature as a result of climate change on a long-term basis.

The similarities with DRR were most apparent at the level of implementation due to the existence of relevant shared experience and the absence of a natural disaster treaty equivalent to the UNFCCC. The issue becomes to what extent DRR can be absorbed into the architecture of the UNFCCC.

Financing CCA and DRR

A challenge for the international community is how to meet the adaptation finance needs that have been identified. Despite the growing dialogue at the international level there are few examples of integration and synergy between CCA and DRR. The disparate nature of CCA and DRR funding reflects the differences in institutional structure.[66]

Funding schemes applicable to DRR are predominantly aimed at providing resources for short-term recovery measures. These responses are generally very specific focusing upon one particular sector or outcome – for example reconstruction of physical structures – rather than allowing a comprehensive analysis of vulnerability across an affected nation or community. This can also restrict perception and understanding of damage as consequences of sea level rise are identified on a longer temporal scale.

A study conducted into the relationship between CCA and DRR in the Pacific noted that one obstacle to closer integration is making more effective and efficient use of external assistance. It is recommended that recipient countries ensure "policy coherence in national planning through budget processes and aid coordination."[67] A host of financial instruments are employed to funnel climate change finance including: grants, loans, equity, loan guarantees, insurance, and debt-for-nature swaps. Currently under the UNFCCC, there is no objective of financial instruments that count as climate finance.

In terms of implementation, mainstreaming adaptation into development appears logical as opportunities to create synergies can also be an efficient use of limited funds. However, mainstreaming CCA and DRR funding has attracted some skepticism. Developing countries have expressed concerns that finance earmarked for adaptation will be absorbed into existing Overseas Development Assistance budgets.[68] Another

65 See, United Nations International Strategy for Disaster Risk Reduction, *Revealing Risk, Redefining Development, Global Assessment Report on Disaster Risk Reduction* (Geneva, United Nations Press, 2011).

66 Hay, *Risk Reduction and Climate Change Adaptation*, 14.

67 Ibid., 18.

68 Axel Michaelowa and Katharina Michaelowa, "Climate or development: is ODA diverted from its original purpose?" *Climatic Change* 84 (2007): 7. See Axel Michaelowa and Katharina Michaelowa "Does climate policy promote development," *Climatic Change* 84 (2007): 1–4.

potential problem in that merging adaptation with general development activities could make it infinitely difficult to analyze the effectiveness of finding with respect to climate change specifically.[69]

An obstacle to closer integration between CCA and DRR is the architecture and operation of funding channels. This reflects some of the deeper conceptual differences discussed earlier in the chapter that affect the implementation of CCA and DRR. Broadly speaking, funding with regard to natural disasters is short term and responsive to the occurrence of a particular event. The majority of funding that is contributed to natural disasters relates to *ex post* or recovery operations rather than risk reduction. Overall funding for natural disasters both *ex ante* and *ex post* is sourced from voluntary contributions from donor countries. The lack of predictable and reliable funding channels makes it difficult to operate DRR programs as the bulk of available funds are reserved for inevitable disaster recovery operations.

The short-term perspective prevalent in natural disaster management is problematic when trying to build synergies with CCA. CCA requires a long-term funding response that can respond to the inherent dynamics of climate change and climate impacts. The challenge therefore is to build clear linkages between short- and long-term funding objects that reflect the time frame required to implement CCA measures.[70] It is equally important that funding for CCA and DRR embraces "comprehensive vulnerability reduction and adaptation" that covers multiple sectors and different environment hazards that can either individually or cumulatively result in impacts and disaster.[71]

Funding channels should utilize the relevant experience within the recipient countries. This would require in certain contexts channeling CCA funding through existing institutions that have been established principally to manage DRR and post-disaster recovery. Such a situation could pose some problems.

Fast Start Finance

The year 2012 marked the conclusion of the two-year Fast Start Finance initiative. Fast Start Finance was initially proposed under at COP15 in 2009. The commitment reiterated in 2010 and Fast Start Finance commenced operation that year. Its immediate purpose was to raise "approaching USD 30 billion for the period 2010–2012" among developed countries for the purpose of assisting climate change policy in developing countries. A collateral benefit of the Fast Start Finance period was to build trust among the developed and developing country groups, identify potential obstacles, and design solutions that could guide the operation of the Green Climate Fund.

69 Farhana Yamin, "The European Union and Future Climate Policy: Is Mainstreaming Adaptation a Distraction or Part of the Solution?" Tearfund Climate Change Briefing Paper, (2005): 349–361.

70 Birkmann, "Integrating Disaster Risk Reduction," 42.

71 Ibid.

Following the establishment of Fast Start Finance, the COP in the following year "took note" of the collective commitment and reaffirmed that funding for adaptation will be prioritized for the most vulnerable developing countries, such as the least developed countries (LDCs), the Small Island Developing States (SIDS), and Africa. A number of other conditions were attached to the mobilization and allocation of finance, chief among these was ensuring funding was new and additional. These conditions must be viewed in the context of the considerable challenge of raising the required amount of $30 billion.

Evaluating the effectiveness of Fast Start Finance is hindered by both procedural and substantive elements. Issues associated with ascertaining the quantum of the actual amount pledged are exacerbated by a lack of guidelines in terms of reporting.

Immediately prior to COP18 in November 2012, the amount pledged under Fast Start Finance amounted to $33.92 billion. This appears at first glance to be very positive, however it must be underlined that this figure only represents the amount pledged – ensuring that the funds are allocated and subsequently disbursed could remain challenging. The World Resources Institute has noted that according to the reports submitted by the donor nations to the UNFCCC, $28.06 billion has been "requested and/or budgeted by the executive bodies of the countries during the fast-start period."[72] Most country reports do not provide plans regarding how the process of allocation and disbursement will be implemented. In 2011, it was reported that only $11.3 billion had been delivered to different funding agencies. In relation to climate finance broadly only 10 percent of funds that are initially pledged are ultimately delivered.

The information contained in the most recent UNFCCC documentation compiled for COP18 gives a mixed impression of Fast Start Finance. It must be noted that only twenty-three donor nations in total provided reports. Given the lack of guidelines detailing what information a report must include, each donor country adopted its own reporting methodology – for example, certain conditions relevant to the operation of Fast Start Finance are simply not discussed by certain countries. This is problematic, as it does not give an objective and universal basis on which to evaluate the effectiveness of funding under Fast Start Finance.

Some of the general observations provided by the UNFCCC underline the unstructured and complex nature of fund mobilization and distribution. The UNFCCC Fast Start Finance documentation notes that donor countries used bilateral and multilateral agencies as channels for delivery of finance in addition to other international organizations. Among those listed were, *inter alia*, the Global Environment Facility Trust Fund, the Special Climate Change Fund, the Least

[72] Athena Ballestros, et al., "Summary of Developed "Fast-Start" Climate Finance Pledges," (Report, World Resources Institute, November 2012) www.wri.org/publication/summary-of-developed-country-fast-start-climate-finance-pledges.

Developed Countries Fund, the Adaptation Fund, the Forest Carbon Partnership Facility, and the Climate Investment Funds. While some of these operate under the UNFCCC (for example the Least Developed Countries Fund and Adaptation Fund), other funds are still in formation and present enormous administrative burden in ensuring that money is spent effectively on projects. The following case studies will illustrate some of these issues. The predominant donor case study is Australia, but brief examples of other nations are also included.

CASE STUDIES

Australia

Australia's contributions under Fast Start Finance provide an excellent illustration of the potential benefits and obstacles of linking climate change adaptation and disaster risk reduction. As of August 2012, Australia had committed $563 million to Fast Start Finance. The Australian government confirms that it is "new and additional" and on track to deliver real outcomes. At this point, $380 million has been disbursed and it is completely grants based (as opposed to loans). The Australian government maintains that the nation's contribution to Fast Start Finance is balanced between adaptation (52 percent) and mitigation (48 percent); and is new and additional, meaning that the funds are not being diverted from existing aid programs. The 52 percent allocated to adaptation constitutes a total of $309 million.

Fast Start Finance contributions are not part of an exclusive financial mechanism or fund. Rather, Fast Start Finance flows through a variety of both bilateral and multilateral channels primarily the International Climate Change Adaptation Initiative and the International Forest Carbon Initiative. While Australia's commitments extend globally, the distribution of funds is focused mainly on Australia's geographic region – the Pacific and South East Asia.

In terms of adaptation, project funding is performed through the following multilateral channels:

- Support for the secretariat functions of the Alliance of Small Island States (AOSIS)
- Adaptation Fund
- Least Developed Countries Fund
- Community Level Climate Change fund for countries in the Mekong subregion (adaptation)
- UNFCCC workshops for the Adaptation Fund
- Green Climate Fund administrative contribution (adaptation)[73]

[73] .Australian Government, *Australia's Fast-Start Finance Update Report August 2012* (Department of Climate Change and Energy Efficiency: Canberra, 2012),10.

The policy documents detailing Australia's commitment adopt a discourse couched in the language of resilience – for example, the Fast Start Finance Update Report submitted in August 2012 describes adaptation funding as allowing Australia "to develop and support a range of effective tools which increase the resilience of communities across the region as they adapt to the unavoidable impacts of climate change."[74]

The Update Report submitted in November 2011 provides some indication of what resilience might include. The Pacific Adaptation Strategy Assistance Program (PASAP) is a $12 million initiative, in Australian dollars (AUD), with AUD $10.2 million attributed to Fast Start Finance. The outline of PASAP provided by the government is an example of mainstreaming as it is described as integrating climate change adaptation information into the planning and development strategies of Pacific nations.

The PASAP also provides an example of the linkage between climate change adaptation and disaster risk reduction under the Fast Start Finance commitments. The 2011 Update Report lists as an example of PASAP the AUD $4 million investment aimed at upgrading infrastructure in the Solomon Islands to reduce the risk from extreme weather events. Another example is funding provided through the International Climate Change Adaptation Initiative (ICCAI). In South East Asia, AUD $48 million is being invested to improve climate resilience. As part of the same program AUD $15 million is being invested in Bangladesh focusing on the reduction of vulnerability to disaster risk and "to support institutions and organizations to implement development projects with climate change adaptation co-benefits."

Moving away from the Asia-Pacific, Australia's commitments in the Caribbean provide further evidence of the possible synergies between climate change adaptation and disaster risk management. The AUD $16.5 million commitment is designated under Fast Start Finance and the adaptation projects it funds are to support regional partners and communities in the Caribbean to address vulnerabilities associated with climate change and to assist in natural disaster preparation and response planning.

More specifically, AUD $2.3 million is allocated to Caribsave, the Caribbean Community Climate Change Centre, and the Caribbean Disaster Emergency Management Agency to fund activities that will "improve the capacity of Caribbean countries to reduce their vulnerability and enhance their resilience to natural disasters."[75] In order to establish this objective, activities that will benefit from funding include establishing climate risk profiles, economic impact assessments, leadership training for national disaster offices, and restocking regional disaster response warehouses.

[74] See, for example, ibid., 3. [75] Ibid., 7.

While the dual objects of adaptation and disaster risk management are apparent, it is not clear to what extent this is the result of synergies between the two disciplines. The commitments allocated for adaptation versus disaster risk management appear to be separate and the policy documents do not suggest how the funding for disaster risk management will also serve adaption goals. Though it is outside the parameters of this presentation it is worth noting that all developed country commitments under Fast Start Finance make reference to natural disasters but with little reference to synergies it shares with climate change adaptation. While both climate change adaptation and disaster risk management often deal with the same set of vulnerabilities the outcome appears a good one. However the lack of clear linkages could potentially make it more difficult to measure the effectiveness of adaptation funding which principally Fast Start Finance is designed to accomplish.

This alludes to some broader issues regarding adaptation funding. There are no universal indicators that measure the success of adaptation projects. Different funds such as the Adaptation Fund under the Kyoto Protocol use certain criteria but this has been found to lack comparability to other funding channels such as the Global Environmental Facility. In contrast to disaster risk management the benefits attached to a successful adaptation program will not be evident for several years at the earliest. The effectiveness of adaptation must be considered in the context of mitigation and the rate of increasing or decreasing emissions. In terms of Fast Start Finance, there is a lack of agreement among the international community concerning how the outcomes should be assessed.[76]

The Update Report submitted in August 2013 describes a set of more recent adaptation projects. Interestingly disaster risk management is absent from descriptions of adaptation projects funded under Fast Start Finance.

Canada

In 2010–11, Canada provided CAD $4.5 million under Fast Start Finance to Haiti. The object was to assist communities to respond to the adverse impacts of climate change and reduce vulnerability to natural disasters. Relevant projects included the rehabilitation and protection of water sheds, promotion of sustainable management of agriculture and forestry resources, and the promotion of water conservation.[77]

Japan

Japan has devoted a share of grant aid under Fast Start Finance to the prevention of disaster and rehabilitation across twenty-five countries. This includes recovery

[76] Brown et al., *Fast-Start Finance*, 133.

[77] Conference of the Parties, United Nations Framework Convention on Climate Change, *Eighteenth Session, Held in Doha from 26 November to 7 December 2012, Submission on Information from Developed Country Parties on the Resources Provided to Fulfil the Commitment Referred to in Decision 1/CP.16, paragraph* 95, FCCC/CP/2012/INF.1 (August 31, 2012), 14.

measures against natural disasters and water supply, and adaptation measures to climate change including large-scale cyclones and typhoons.[78] Part of the Fast Start Finance provided to the Solomon Islands has improved the use of shortwave radio to broadcast emergency news relating to natural disasters through the country. In Fiji, Japan has helped to strengthen disaster prevention capacity by enhancing systems that help populations escape floods more readily.[79]

New Zealand

A particular relevant aspect of New Zealand's Fast Start Finance commitment is the New Zealand Aid Program that has provided more than NZ $1 million to improve the Tukao and Tauhunu cyclone shelters in the northern Cook Islands.[80] The relevant locations are a few levels above sea level and therefore vulnerable to sea level rise as well as extreme events. The measures were designed to provide protection in the likely increase in cyclone intensity and as a result of rising sea levels.

United States

An example of bilateral Fast Start Finance has seen the United States government invest USD $7.6 million in Guatemala into a "broad adaptation and sustainable land use program" that is designed to simultaneously build resilience to climate change impacts and reduce greenhouse gas emissions from projects involving land use change. The intersection between adaptation and development is emphasized in a number of projects funded by the United States; for example, USD $3 million was invested in Mali to reduce vulnerability to climate impacts and reduce poverty in the process. This project focused on the redistribution of seasonal forecasts together with aggregated climate data for the benefit of the agricultural sector.

As part of multilateral commitments, the United States contributed USD $10 million to the Pilot Program for Climate Resilience (PPCR). This operated in partnership with the World Bank and Africa Development Bank and has channelled $50 million in grant funding to projects including weather forecasting systems and socioeconomic infrastructure in zones highly vulnerable to climate impacts.

CONCLUSION

CCA funding has the potential to assist DRR but there is an inherent risk that it could divert attention from adaptation aims. The key is that the synergies between adaptation, disaster risk management, and economic development that lie at the

[78] Ibid., 29. [79] Ibid., 30. [80] Ibid., 39.

basis of funded projects must be made clear. The funds are provided under pre-existing funding channels, including some dedicated to disaster relief, testing the strength and durability of linkages between adaptation, disaster risk reduction, and development. Ultimately, it must be remembered that these funds are supposed to be designed for climate change purposes and success will, at least principally, be measured on that basis.

The imminent expiry of HFA means that the impetus for greater integration and synergy, particularly in terms of funding, must come from the CCA community. This chapter has provided an overview of the latest funding channels utilized under the climate regime. Problems persist particularly in terms of discerning between adaptation objects and broader development finance. There is little in terms of universality regarding definitions and systems that can effectively measure climate-specific finance while developing countries lack the capacity to register, monitor, and report on the finance received. The choice of financial instruments has also caused consternation among developing country recipients who insist that funding must take the form of grants and not loans.

The advent of loss and damage casts doubts over the long-term viability of adaptation. The meaning of "dangerous" in the context of climate change has long been debated. Adaptation can be viewed as a policy designed to remove the danger from unavoidable climate impacts but this does not necessarily mean that adaptation will prevent all incidents of loss and damage. The focus upon loss and damage though has a sense of immediacy that contrasts with the traditional projections of future impacts. The integration with DRR can assist CCA greatly in this regard but it must be recalled that each concept has different objectives. The long-term, slow-onset damage of climate change must be recognized and reflected in any measurement used to evaluate the effectiveness of CCA and DRR policy into the future. If not, then despite the obvious crossovers in terms of CCA and DRR, benefits will continue to be considered incidental rather than deliberate and neither CCA and DRR will be implemented effectively.

14

Challenges to State Sovereignty in the Provision of International Natural Disaster Relief

Amelia Telec

International and non-governmental organizations (NGOs) often play a leading role in disaster relief and recovery. However, there is also a need to rely on the provision of state-to-state assistance in the aftermath of a large-scale disaster, given the expertise and resources states possess to facilitate and provide international disaster relief. One of the most important contributions a state can make in the immediate aftermath of a disaster is the deployment of personnel to the territory of the disaster-affected state to assist in providing relief to the affected population. The deployment of personnel from an assisting state to the territory of a disaster-affected state will generally require that international legal frameworks are in place to regulate the deployment, namely a legal basis for the deployment and privileges and immunities for deployed personnel. The purpose of this chapter is to highlight how these legal arrangements can raise challenging issues in relation to state sovereignty, particularly where the disaster-affected state refuses external assistance, or where there is a power differential between the state offering assistance and the state in need of that assistance. It recommends that by institutionalizing the key legal elements of a deployment in a legal framework that generally applies to the provision of disaster assistance, international law may be able to assist, rather than hinder, rapid assistance to victims in the immediate aftermath of a natural disaster.

INTRODUCTION

Any country, regardless of resources, can be susceptible to disasters and require outside assistance in dealing with those disasters. Recent examples of large-scale disasters, including the earthquake and tsunami in Japan in 2011, the earthquake in Haiti in 2010, the cyclone in Myanmar in 2008, and the Indian Ocean tsunami of 2004, illustrate that the ability to respond effectively to a major disaster is beyond the capacity of many states. However, developing states often suffer disproportionate effects from disasters and are more likely to require outside assistance.

Apart from contributions of food and financial aid, the deployment of relief workers, technical experts, and possibly even military personnel from foreign states may be needed to help the disaster-affected state to recover, particularly in the immediate aftermath of a disaster.

While NGOs stand ready to assist in these circumstances, the sheer scale of a disaster may mean that a state's resources are required to address its humanitarian consequences. The deployment of personnel from an assisting state to the territory of a disaster-affected state will generally require that international legal frameworks are in place to regulate the deployment. This chapter focuses on how these types of deployments interact with the sovereign right of states to be protected from intervention in their domestic affairs.

The international legal frameworks that must be in place to facilitate deployments to undertake disaster assistance raise challenging issues in relation to state sovereignty, particularly where the disaster-affected state refuses external assistance, or where there is a power differential between the state or organization offering assistance and the state in need of that assistance. There are two main legal mechanisms that an assisting state will generally want in place before deploying personnel to undertake disaster relief: a basis under international law for the deployment, and privileges and immunities for its personnel relating to the law of the disaster-affected state. The legal basis for deployment will often be straightforward – the disaster-affected state will consent to assistance being provided by the assisting state and this consent will form the legal basis for the deployment. However, where the disaster-affected state is unwilling to accept help from the international community the provision of disaster assistance may be at odds with the sovereign rights of that state.

Even when disaster-affected states are willing to receive assistance on their territory, states may require deployed personnel to be accorded privileges and immunities to exempt them from the local law. While there is an international legal framework in place to provide privileges and immunities to UN personnel, for individual states sending national contingents to disaster-affected countries, privileges and immunities generally have to be negotiated on a case-by-case basis. Not only does this hinder the ability of the assisting state to provide rapid assistance, it also raises issues for the disaster-affected state with regard to accountability to the local population for actions undertaken by deployed personnel.

This chapter proceeds in three parts. It begins by discussing the concept of sovereignty and its relationship to the deployment of disaster relief personnel. It then goes on to explore possible alternative legal bases for deployment of disaster assistance personnel in circumstances where a state will not consent to external assistance in the aftermath of a disaster, including Security Council authorization and other proposals being discussed by the International Law Commission (ILC) in its work on the Protection of Persons in the Event of Disasters. Finally, it considers the provision of privileges and immunities for personnel deployed to a disaster-affected state, including the appropriate levels of privileges and immunities to be accorded to those personnel and the advantages of elaborating such privileges and immunities in an international legal instrument with general application to disaster assistance. The purpose of this chapter is to analyze critically the interaction between state sovereignty and the existing legal framework for the provision of international disaster relief to determine how international law may be able to assist, rather than hinder, rapid assistance to victims in the wake of a natural disaster.

SOVEREIGNTY IN THE MODERN SYSTEM OF INTERNATIONAL LAW

The principles of sovereignty and equality of states are the foundation on which the post–UN Charter system of international law is based.[1] Despite continued predictions about the decline of sovereignty, its enduring relevance to states in their relations with each other should not be underestimated. The concept of sovereignty is central to questions of the deployment of personnel to the territory of a disaster-affected state for the purpose of assisting that state and its population. While "sovereignty" denotes one of the most important concepts of the theory of national and international law, it has a variety of meanings.[2] Bardo Fassbender and Albert Bleckmann point out that in the modern system of international law, "sovereignty of States stands out as a legal concept which exposes one of its flanks to politics and power."[3] Sovereignty has a variety of meanings, not all of which are legal, and states have historically interpreted this concept in a variety of ways. However, sovereignty is a relationship with other states that continues in large part to be defined by the law.[4] The purpose of the following discussion is to map the core characteristics of sovereignty in the modern system of international law, in order to provide a context for the discussion of how this principle interacts with the provision of international disaster relief.

While the concept of sovereignty predates the establishment of the UN, pre-1945 notions of sovereignty should not be conflated with the post–UN Charter meaning of the principle. The principle of sovereignty prior to 1945 provided states with many opportunities for action that was not based on the authority of the law.[5] The introduction of the principle of "sovereign equality" in the UN Charter was intended to prioritize the equality of states within international relations rather

[1] Ian Brownlie, *Principles of Public International Law*, (Oxford University Press, 2008); *Corfu Channel (United Kingdom v. Albania) (Merits)*. (ICJ, 1949): "Between independent States, respect for territorial sovereignty is an essential foundation of international relations."

[2] Hans Kelsen, "Sovereignty and International Law," *Georgetown Law Journal* (1960): 627.

[3] Bardo Fassbender and Albert Bleckman, "Article 2(1),")." in *The Charter of the United Nations: A Commentary*, ed. Bruno Simma, Hermann Mosler, and Albrecht Randelzhofer (Oxford University Press, 2002): 72.

[4] Brownlie, *Principles of Public International Law*, 289.

[5] Fassbender and Bleckmann, "Article 2(1)," 91.

than reinforce the claims of legal superiority embodied in traditional notions of sovereignty.[6] Sovereign equality was to be achieved through the structure of international relations established by the UN Charter, in which a number of interdependent principles would ensure respect for sovereignty. In contrast to previous understandings of sovereignty, the system established by the UN Charter was one that sought to regulate state relations comprehensively,[7] making any state action contrary to this framework prohibited.

The principle of sovereign equality is contained in Article 2(1) of the UN Charter, which makes clear that "the Organization is *based on* the principle of the sovereign equality of all its Members."[8] While the concept was not elaborated in the Charter, the General Assembly has subsequently identified some of the core characteristics of the principle of sovereign equality in the Declaration on Principles of International Law Concerning Friendly Relations and Co-operation Among States in Accordance with the Charter of the United Nations. Those principles are: the juridical equality of states; the rights inherent in full sovereignty; the duty to respect the personality of other states; the inviolability of the territorial integrity and political independence of the state; the right to freely choose and develop the political, social, economic, and cultural systems of the state; and the duty of the state to comply fully and in good faith with its international obligations and to live in peace with other states.[9]

The inclusion of the prohibition on the use of force in Article 2(4) of the UN Charter[10] was intended to ensure the continued existence of sovereign equality in the UN system. The prohibition "provides ultimate protection for State autonomy [as] no State may be threatened by another State's decision to use force."[11] The principles of territorial integrity and political independence of the State, which are explicitly protected by Article 2(4), were considered by the drafting committee of the UN Charter as concepts that were integral to the principle of sovereign equality[12] – a connection that has been emphasized by the International Court of Justice (ICJ) in the *Corfu Channel Case* and *Military and Paramilitary Activities in and against Nicaragua*.[13]

[6] Ibid., p. 83 . [7] Ibid., p. 91 .

[8] "Charter of the United Nations," October 24, 1945, 1 UNTS XVI: Art 2(1),. (Emphasis added)

[9] "Declaration on Principles of International Law Concerning Friendly Relations and Co-Operation Among States in Accordance with the Charter of the United Nations," GA Res 2625, UN GAOR, 25th Sess., UN Doc A/8028 (1970).

[10] "Charter of the United Nations," Art 2(4): "All Members shall refrain in their international relations from the threat or use of force against the territorial integrity or political independence of any state, or in any other manner inconsistent with the Purposes of the United Nations."

[11] Mary Ellen O'Connell, "Regulating the Use of Force in the 21st Century: The Continuing Importance of State Autonomy," *Columbia Journal of Transnational Law*, 1998: 475.

[12] "Report of Rapporteur of Sub-Committee I/1/A to Committee I/1: Chapter II," UNCIO VI, (1945):717.

[13] *Corfu Channel Case*, 35; *Military and Paramilitary Activities in and against Nicaragua (Nicaragua v United States of America) (Merits)* (ICJ, 1986).

An important corollary to the principle of sovereign equality in Article 2(1) of the UN Charter is the principle of nonintervention. Article 2(7) of the Charter protects states from interference by the UN in matters that are "essentially within the domestic jurisdiction" of the state.[14] This provision is generally understood to preclude interference by the UN in matters that are in the reserved domain of domestic jurisdiction (i.e., "the domain of state activities where the jurisdiction of the state is not bound by international law.")[15] Only the sovereign state, through consent to international obligations, can narrow this reserved domain. This ability to consent to actions that would otherwise be a violation of the principle of territorial integrity and nonintervention has long been a recognized principle of international law.[16] As noted by the Permanent Court of International Justice (PCIJ) in the *Case of the S.S. Wimbledon*, the ability to enter into a treaty that places restrictions on the exercise of sovereignty is itself an attribute of that sovereignty.[17] The primary exception to Article 2(7) is the Security Council's power to determine a threat to international peace and security under Chapter VII of the Charter.

While Article 2(7) does not explicitly protect states against acts of interference by other states, it has been subsequently interpreted as embodying the general principle of nonintervention. This principle provides that "[n]o State or group of States has the right to intervene, directly or indirectly, for any reason whatever, in the internal or external affairs of any other State."[18] The Friendly Relations Declaration, adopted by the UN General Assembly in 1970, extended for the first time the principle of nonintervention in Article 2(7) to relations between states, and is thought to have given "expression to a new consensus about an enlarged concept of intervention under general international law."[19] By the time of the ICJ's judgment in *Military and Paramilitary Activities in and against Nicaragua* in 1986, the principle of nonintervention as applied to relations between states was recognized by the Court as a principle of customary international law.[20]

14 "Charter of the United Nations," Art 2(7): "Nothing contained in the present Charter shall authorize the United Nations to intervene in matters which are essentially within the domestic jurisdiction of any state or shall require the Members to submit such matters to settlement under the present Charter; but this principle shall not prejudice the application of enforcement measures under Chapter VII.."

15 Brownlie, *Principles of Public International Law*, 293; *Nationality Decrees Issued in Tunisia and Morroco (France v. Britain) (Advisory Opinion)*. (PCIJ, 1923).

16 Kelsen, "Sovereignty and International Law," 637.

17 *Case of the S.S. "Wimbledon" (U.K. v. Japan) (Judgment)*. (PCIJ), 1923. "No doubt any convention creating an obligation to this kind places a restriction upon the exercise of the sovereign right of the State, in the sense that it requires them to be exercised in a certain way. But the right of entering into international engagements is an attribute of State sovereignty."

18 Friendly Relations Declaration. See also, *Nicaragua Paramilitary Case*.

19 Georg Nolte, "'Article 2(7)," in *The Charter of the United Nations: A Commentary*, by Bruno Simma et al. (ed.)(Oxford University Press, 2002), 154.

20 *Nicaragua Paramilitary Case*.

There have been frequent predictions since the Charter's establishment about the demise of the sovereign state as the principal subject of international law. These predictions are often based on the observation that the reserved domain of states is shrinking. A prominent example of this is the development of protection of the rights of the individual against the state under international human rights law,[21] which has resulted in the continued establishment of direct links between the individual and international law.[22] Other examples include the extension of the application of international law to cover relationships between the sovereign and private commercial entities, the regulation of international trade and investments under globalized capital markets, and the establishment of regional frameworks, with the European Union being the most far-reaching example of this.

Georg Nolte has observed that the developments that have taken place in the almost seventy years since the establishment of the Charter have "led to the coverage of so many fields by (consensual or customary) rules of international law that the definition of 'matters under the domestic jurisdiction' by the PCIJ does not leave very much room for this concept any longer."[23] These are all examples of state autonomy being limited by the law. However, if international law is viewed as the "collective expression of sovereign wills,"[24] consistent with the PCIJ's reasoning in the *Case of the* S.S. *Wimbledon*, it is the concept of sovereignty that allows states to consent to such limits being placed on them. Viewed through this lens, the increasing regulation by international law of state behavior formerly in the reserved domain of state jurisdiction affirms, rather than undermines, the essential character of sovereignty, and its ongoing relevance in the modern system of international relations.

The principles of sovereign equality and noninterference in the internal affairs of states are integral to achieving the aims of the Charter to maintain international peace and security. State sovereignty in this system is composed of two seemingly incompatible concepts: supreme power and regulation of that power. However, a state's power will only be regulated to the extent that it consents to such regulation. The right to autonomy that states are accorded under the system of international law embodied in the UN Charter also entails obligations vis-à-vis other states (e.g., the obligation not to engage in threat or use of force, the obligation not to interfere in the domestic jurisdiction of the state).[25] It is the interdependence of rights and obligations in this system that ensures that all states enjoy at least formal sovereign equality in their mutual relations. It is in this context that discussion of the effects of sovereignty on the provision of assistance to victims of disasters must be situated.

[21] Ronald Brand, "External Sovereignty and International Law," *Fordham International Law Journal*, (1994–1995): 1694.

[22] Ibid., 1686. [23] Nolte, "Article 2(7)," 157.

[24] Oscar Schachter, "The Decline of the Nation-State and its Implications for International Law," *Columbia Journal of Transnational Law*, (1998): 7.

[25] Fassbender and Bleckmann, "Article 2(1)," 89.

RELATIONSHIP BETWEEN SOVEREIGNTY AND THE DEPLOYMENT OF FOREIGN DISASTER RELIEF TEAMS

International and non-governmental organizations often play a leading role in disaster relief and recovery. These organizations generally possess great expertise in responding to and addressing the humanitarian consequences of disasters. However, there is also a need to rely on the capacity of states to provide personnel with expertise to facilitate and provide international disaster relief. This need arises from the particular resources and capabilities possessed by states, including specialized equipment and personnel that possess particular skills. It is the relationship between the disaster-assisting state and the disaster-affected state that will be the focus of the following analysis.

It has become common practice for states to come to the aid of disaster-affected states by deploying civilian personnel to deliver aid and provide expertise. This can include expertise in everything from health care to engineering. In addition, the deployment of military personnel in these situations has been increasing in recent years.[26] Militaries possess expertise in logistics, rapid deployment, and search and rescue, which can prove an invaluable skill set in the aftermath of a disaster. Military assets, such as naval vessels and aircraft, are also likely to be the most appropriate assets to facilitate disaster relief and are usually able to deploy rapidly. Regardless of the civilian or military character of a deployment, assisting states will generally require certain elements to be put in place before deploying their nationals to a foreign state, the most important of which are the legal basis for deployment, and the privileges and immunities that will be granted to their personnel by the disaster-affected state. While a range of instruments exist to facilitate relief provided in the wake of disasters,[27] there is no overarching legal framework in place to regulate the provision of relief to disaster-affected states.

The provision of relief following disaster has been considered by the ILC in the context of its work on the Protection of Persons in the Event of Disasters. The objective of adding this topic to the ILC's work was:

> [t]he elaboration of a set of provisions which would serve as a legal framework for the conduct of international disaster relief activities; clarifying the core legal principles and thereby creating a legal 'space' in which such disaster relief work could take place on secure footing.[28]

[26] Secretariat of the International Law Commission. "Memorandum by the Secretariat, Protection of Persons in the Event of Disasters." ILC 60th Sess., UN Doc A/CN.4/590, 2008:114.

[27] "Tampere Convention on the Provision of Telecommunication Resources for Disaster Mitigation and Relief Operations; Framework Convention on Civil Defence Assistance; Guidelines for the Domestic Facilitation and Regulation of International Disaster Relief and Initial Recovery Assistance." June 18, 1998, 2296 UNTS 5.

[28] "Report of the International Law Commission." GAOR 58th Sess., UN Doc A/61/10, 2006: 475.

The ILC has adopted a number of draft articles on this issue, but the project is ongoing and the final outcome remains unclear at this stage.[29] The issue of legality of intervention when a disaster-affected state is unwilling to act, and the provision of privileges and immunities, will be discussed in further detail in the sections that follow.

LEGAL BASIS FOR DEPLOYMENT

The Need for a Legal Basis

In order for a foreign state to deploy military or civilian personnel to provide assistance to victims of disaster, there must be a legal basis for the deployment of those personnel. This need for a legal basis is derived primarily from three sources: (1) the right to territorial integrity inherent in the concept of sovereignty; (2) the customary law duty of nonintervention in the domestic affairs of any state; and (3) the absolute prohibition on the use of force.

Territorial integrity

Regardless of the character of the deployment, personnel crossing from the territory of one state into another without a basis to do so under international law would breach the territorial integrity of that state. While the ICJ has held that "the provision of strictly humanitarian aid to persons or forces in another country, whatever their political affiliations or objectives, cannot be regarded as unlawful intervention, or as in any other way contrary to international law,"[30] the Court was referring specifically to the provision of aid, not the deployment of personnel.

Nonintervention

The deployment of personnel to a disaster-affected state without a basis under international law to do so would not be automatically contrary to the principle of nonintervention. A prohibited intervention is one that involves matters in which each state is permitted "by the principle of State sovereignty to decide freely."[31] To constitute intervention, interference must be "forcible or dictatorial, or otherwise coercive, in effect depriving the state intervened against of control over the matter in question."[32] Therefore, not every type of deployment without consent will constitute prohibited intervention – this would have to be assessed on a case-by-case basis taking into account all the circumstances.

29 Eduardo Valencia-Ospina, "Preliminary Report on the protection of persons in the event of disasters," *International Law Commission*, UN GAOR, 60th Sess., UN Doc A/CN.4/598, 2008: 21–22.

30 *Nicaragua Paramilitary Case*, 124–5.

31 Ibid., 107–8 [205].

32 Robert Jennings and Arthur Watts (eds., *Oppenheim's International Law* (9th ed), 2 vols. (Essex: Longman, 1992): 432.

An argument may be made that the protection of victims of disasters is no longer "essentially within the domestic jurisdiction" of a disaster-affected state by virtue of its acceptance of other international law obligations, for example, human rights obligations. However, even if that was the case, it does not provide another state a right to intervene, particularly in a manner that is prevented on other grounds (i.e., territorial integrity or prohibition on the use of force).

Prohibition on the use of force

Whether or not a deployment for the purposes of disaster relief without a legal basis could constitute a prohibited use of force will depend on the character of the deployment as military or civilian. Deployment of civilian personnel to provide disaster relief assistance without a legal basis is unlikely to amount to a prohibited use of force, given that the prohibition is generally understood as a prohibition the use of "armed"[33] or "military" force.[34] However, it is clear that the movement of the regular armed forces of a state into the territory of another in the absence of a legal basis is generally considered to be a flagrant example[35] of the most grave form of a prohibited use of force (i.e., an armed attack).[36] Accordingly, the consequences of deploying military personnel without a proper legal basis are especially grave, and could give rise to a right to self-defense by the disaster-affected state under Article 51 of the UN Charter.

POSSIBLE LEGAL BASES FOR DEPLOYMENT

Consent of the Disaster-Affected State

The least controversial and most likely legal basis for the deployment of personnel for the purpose of providing disaster relief assistance is the consent of the affected state. The need to obtain a territorial state's consent to respect the sovereignty of that state and ensure compliance with international law in connection with the provision of disaster relief has been recognized on numerous occasions, including by the General Assembly,[37] and in the International Red Cross and Red Crescent (ICRC) Movement's Guidelines for the Domestic Facilitation and Regulation of International Disaster Relief and Initial Recovery Assistance (IDRL Guidelines).[38]

33 Albrecht Randelzhofer "Article 2(4)," in *The Charter of the United Nations: A Commentary*, by Bruno Simma et al., (Oxford University Press, 2002), 117.

34 Ibid., p. 119 . 35 Ibid., p. 103 . 36 Ibid., p. 103.

37 "Strengthening of the coordination of humanitarian emergency assistance of the United Nations." General Assembly Resolution 46/182, UN GAOR 78th plenary meeting, , UN Doc A/RES/46/182. 1991: "The sovereignty, territorial integrity and national unity of States must be fully respected in accordance with the Charter of the United Nations. In this context, humanitarian assistance should be provided with the consent of the affected country and in principle on the basis of an appeal by the affected country."

38 "IDRL Guidelines," Principle 10(1): "Disaster relief or initial recovery assistance should be initiated only with the consent of the affected State and in principle, on the basis of an appeal. The affected State should decide in a timely manner whether or not to request disaster relief or

The ILC Secretariat, in its 2008 memorandum on the Protection of Persons in the Event of Disasters, identified consent of the disaster-affected state as a function of the principle of respect for sovereignty, territorial integrity, national unity, non-interference in the domestic affairs of the state and cooperation.[39] The Secretariat also identified consent as a corollary of a requirement that it is the primary responsibility of the state to respond to a disaster on its own territory.[40] In addition, many of the existing instruments in the disaster relief assistance field contain specific provisions on the consent of the disaster-affected state.[41] The requirement for consent of the disaster-affected state has also been reflected in the draft articles adopted by the ILC.[42]

Consent as a legal basis for deployment has application outside the field of disaster relief. As previously discussed, it is a well-established principle of international law that states may consent to restrictions being placed on the rights inherent in their sovereignty. In its Articles on State Responsibility for Internationally Wrongful Acts, the ILC recognized consent as a circumstance precluding the wrongfulness of conduct that would otherwise not be in conformity with the international obligations of the state concerned.[43] The ILC commentary to the Articles on State Responsibility notes that "it is a daily occurrence that States consent to conduct of other States, which, without such consent, would constitute a breach of an international obligation."[44] The commentary also makes clear that the existence of circumstances precluding the wrongfulness of conduct "do[es] not annul or terminate the obligation; rather they provide a justification or excuse for non-performance while the circumstance in question subsists."[45] In order to preclude the wrongfulness of the conduct, consent must be "valid in international law, clearly established, really expressed (which precludes merely presumed consent),

initial recovery assistance and communicate its decision promptly. In order to make this decision, the affected State should promptly assess needs. Consideration should be given to undertaking joint needs assessments with the United Nations and other assisting humanitarian organisations."

39 Secretariat of the ILC, "Memorandum," 41.

40 Ibid.

41 "Tampere Convention," Art 4(5): "No telecommunication assistance shall be provided pursuant to this Convention without the consent of the requesting State Party. The requesting State Party shall retain the authority to reject all or part of any telecommunication assistance offered pursuant to this Convention in accordance with the requesting State Party's existing national law and policy"; and "Framework Convention," Art 3(a): "Only assistance requested by the Beneficiary State or proposed by the Supporting State and accepted by the Beneficiary State may take place."

42 United Nations, "Report of the International Law Commission," GAOR 63rd Sess., UN Doc A/66/10. 2011, Art 11.

43 "Articles on State Responsibility for Internationally Wrongful Acts," Art 20: "Valid consent by a State to the commission of a given act by another State precludes the wrongfulness of that act in relation to the former State to the extent that the act remains within the limits of that consent."

44 International Law Commission, "Draft Articles on Responsibility of States for Internationally Wrongful Acts," *Yearbook of the International Law Commission* II, no. 2 (2001): 72.

45 Ibid., 71.

internationally attributable to the state and anterior to the commission of the act to which it refers."[46]

For consent to a valid legal basis for deployment, the deployment must remain within the bounds of the consent given.[47] This is particularly important for the deployment of military personnel. The consequence of taking military action outside the scope of consent was recognized by the General Assembly in its Definition of Aggression, adopted in 1974, which provided that a state whose armed forces are on the territory of another state with the latter state's consent, but acting outside the conditions provided for their presence, would be committing an act of aggression.[48] Accordingly, the parameters of the consent provided will be an important factor in deciding whether it is sufficient to carry out the activities intended to be undertaken by the assisting State.

Other Possible Legal Bases

Incidents like the initial failure of Myanmar's military junta to consent to international disaster relief in the wake of 2008's Cyclone Nargis raised the question of what can be done in the event that the disaster-affected state was both unwilling and unable to provide effective relief to its population. Cyclone Nargis struck Myanmar on May 2, 2008. Winds from the cyclone reached speeds of 200 kilometers per hour at landfall and caused extensive damage in the Irrawady Delta and the nation's former capital of Yangon.[49] The cyclone killed approximately 140,000 people, and severely affected an estimated 2.4 million others.[50] More than 1 million people lost their livelihoods as a result of the damage.[51] In the days following the disaster, offers of support from other states poured in, and foreign aid agencies stood ready to provide relief, including military support for potential relief operations.[52] However, it was not until May 22, 2008,[53] twenty days after the cyclone had first hit, that Myanmar

46 Ibid., 112.

47 ILC, "Draft articles on Responsibility of States for Internationally Wrongful Acts, with commentaries," 73.

48 United Nations, "Definition of Aggression," General Assembly Resolution 3314 (XXIX), 2319th Plenary Meeting. 1974, Art 3(e).

49 "Cyclone Nargis: Two Months Later," *World Food Programs*, July 3, 2008. www.aseanpost-nargiskm.org/about/cyclone-nargis-overview.

50 "Cyclone Nargis Overview." *Post Nargis Knowledge Management Portal*. www.aseanpostnar-giskm.org/about/cyclone-nargis-overview.

51 "Cyclone Nargis: Two months later."

52 USAID. "Burma-Cyclone." September 8, 2008. www.archive.usun.state.gov/fact_sheet/ecosco_20080908_09.pdf.

53 An ASEAN Emergency Response Assessment Team, consisting of disaster response from ASEAN countries, was deployed to the Ayeyarwady delta on May 9 2008 under the ASEAN Agreement on Disaster Management and Emergency Response. The team reported on May 19, which led to the recommendation to establish a mechanism where ASEAN would facilitate aid from international agencies: ASEAN's Response in *Post Nargis knowledge management portal*, www.aseanpostnargiskm.org/response-to-nargis/aseans-response.

consented to the provision of extensive international aid, with humanitarian personnel arriving in early June.[54] The initial failure of the government of Myanmar to accept international disaster relief assistance received virulent criticism, including from then UK Prime Minister Gordon Brown, who stated that the natural disaster:

> [i]s being made into a man-made catastrophe by the negligence, the neglect and the inhuman treatment of the Burmese people by a regime that is failing to act and to allow the international community to do what it wants to do.[55]

Other instances of states refusing to accept necessary disaster assistance when their own capacities are overwhelmed include India's refusal to accept international aid after both the 2004 tsunami and 2005 earthquake in Kashmir,[56] and Sri Lanka's reluctance to provide relief to Tamil-controlled areas of Sri Lanka also after the tsunami in 2004.[57]

While the application of the principle of the Responsibility to Protect(R2P) has been considered by the ILC in its work on Protection of Persons in the Event of Disasters,[58] it is clear from the ILC's discussions that there is not widespread support for its application to international disaster relief at present.[59] R2P, as set out in the 2005 World Summit Outcome Document, is limited to genocide, war crimes, ethnic cleansing, and crimes against humanity.[60] In his 2009 report on implementing the responsibility to protect, the UN secretary-general confirmed his opinion that R2P could not be extended to cover situations of natural disasters, stating that:

> [t]o try to extend it to cover other calamities, such as HIV/AIDS, climate change or the response to natural disasters, would undermine the 2005 consensus and stretch the concept beyond recognition or operational utility. . .[61]

54 United Nations, "Update Report No. 1: Myanmar," *Security Council Report*. September 10, 2008.

55 BBC News, "Burma 'Guilty of Inhuman Action'," May 17, 2008 http://news.bbc.co.uk/2/hi/uk_news/politics/7406023.stm.

56 Somini Sengupta, "Pride and Politics: India Rejects Quake Aid," *New York Times International*, October 19, 2005.

57 "Sri Lankan Government Denies Hold Up in Aid to Tamil Areas," *Australian Broadcasting Corporation*, January 13, 2005 www.abc.net.au/pm/content/2005/s1281502.htm.

58 The Special Rapporteur noted in his Second Report on the protection of persons in the event of disasters that "... some delegations considered that the Commission should not find itself prevented from considering that notion, should the logic of its undertaking propel in that direction" (Valencia-Ospina, Eduardo. "Second Report on the Protection of Persons in the Event of Disasters," ILC 61st Sessions, UN Doc A/CN.4/615. 2009:4.

59 By its 64th Session, the Special Rapporteur and the ILC expressed agreement on the non-applicability of the concept of responsibility to protect, although some reserved the possible relevance of the concept in the future (United Nations. "Report of the International Law Commission." UN GAOR, 64th Session, UN Doc A/64/10 SUPP.).

60 United Nations, "2005 World Summit Outcome," General Assembly Resolution 60/1, GAOR 60th S, [138], UN Doc A/RES/60/1 (2005).

61 *Implementing the Responsibility to Protect*, Report of the Secretary-General, UN GAOR 63rd Sessions, UN Doc A/63/677. 2009: [10(b)].

Even if this concept were to apply to the provision of disaster relief, it does not provide a legal basis outside the framework of the Charter (consent or Chapter VII action) on which individual states could intervene.[62] Accordingly, it has not been considered further in this context. Two possible alternative legal bases for intervention will be discussed in the sections that follow.

First is Chapter VII of the UN Charter. Deployments authorized by the Security Council under Chapter VII enjoy a clear legal basis under international law. The Chapter VII powers of the Security Council are an explicit exception to the principle of non-intervention embodied in Article 2(7) and customary international law.[63] However, in order for the Security Council to be able to act under Chapter VII, the Council must determine that there is a threat to international peace and security.

It is conceivable that the Security Council could determine that a large-scale disaster that the affected state is unwilling or unable to address, and that may have unintended cross-border effects such as mass migration, is a threat to international peace and security. This was recognized by former UN Secretary-General Boutros Boutros-Ghali, who considered that "[h]umanitarian emergencies, by causing the mass exodus of people, may constitute threats to international peace and security."[64] While the ILC Secretariat has characterized this course of action as "conceivable," it also considered that the possibility of such circumstances being determined to be a threat to international peace and security as remote.[65] The Security Council has expanded the notion of what constitutes a threat to international peace and security in the years since the end of the Cold War, recognizing humanitarian emergencies as reaching the threshold of "threat to international peace and security" in a number of instances.[66]

Despite this, coercive action has not been authorized outside the context of armed conflict. There was resistance from some of the permanent five members of the Security Council when the issue of taking Chapter VII action was discussed in the context of the response to cyclone Nargis,[67] and it is unlikely that the Council will authorize the deployment of disaster-relief personnel on this basis in the near

62 The Outcome Document identified that the responsibility to protect its population lies with the State, and that the international community has the responsibility to use means in accordance with the UN Charter to help protect populations (2005 World Summit Outcome, 138–39).

63 Nolte, "Article 2(7)," 166 .

64 The report of the Secretary-General on the work of the Organization, UN GAOR 48th Session, [481], UN Doc A/48/1 (SUPP) (1994).

65 Secretariat of the ILC, "Memorandum," 148.

66 In Somalia: UN Security Council Resolution 794, 3145th Meeting. UN Doc S/RES/794 (1992); and Rwanda: UN Security Council Resolution 929, 3392nd Meeting. UN Doc S/RES/929 (1994).

67 "China, Indonesia Reject France's Myanmar Push," *Reuters*, May 8, 2008 www.reuters.com/article/2008/05/08idUSN08518240.

future. Accordingly, Chapter VII authorization cannot be considered a likely legal basis for disaster relief assistance at present.

The second alternative legal basis stems from the duty on disaster-affected states to protect their populations. The ILC in its Draft Articles on the Protection of Persons in the Event of Disaster has sought to address the failure of states to accept assistance in the wake of a disaster by including in its draft articles a duty on states to ensure the protection of persons on its territory through the provision of disaster relief and assistance.[68] This duty has previously been included in a number of instruments, including General Assembly resolutions,[69] resolutions adopted by the International Red Cross and Red Crescent Movement,[70] and the IDRL Guidelines,[71] and is derived from the obligations a state owes to its population by virtue of international human rights law. The recognition of this duty has been underpinned, both in the text of the ILC Draft Articles and its previous incarnations,[72] by the sovereign rights of states. To address the issue of disaster-affected states declining consent to outside assistance, the ILC's Special Rapporteur on the Protection of Persons in the Event of Disasters has sought to elaborate on the idea of a duty to protect in the form of two additional but related duties: the duty to seek assistance, and a duty not to arbitrarily withhold consent to external assistance.[73]

These duties have been incorporated into Articles 10 and 11 of the draft articles.[74] The draft commentary to Article 10 states that the duty to seek assistance arises from "an affected State's primary responsibilities under international human rights

68 Report of the International Law Commission, UN GAOR 63rd Session, UN Doc A/66/10 (2011): Article 9.

69 General Assembly Resolution 46/182,: "Each State has the responsibility first and foremost to take care of the victims of natural disasters and other emergencies occurring on its territory. Hence, the affected State has the primary role in the initiation, organization, coordination, and implementation of humanitarian assistance within its territory."

70 Red Cross and Red Crescent Movement, *Resolution 7: Strengthening normative frameworks and addressing regulator barriers concerning disaster mitigation, response and recovery*, 31st International Conference, (2011): "*reaffirming* that states have the primary duty to take effective action to protect their citizens from the effects of natural disasters, to provide them with any necessary humanitarian assistance in their aftermath as well as to promote their recovery..."

71 IDRL guidelines, principle 3(2): "If an affected State determines that a disaster situation exceeds national coping capacities, it should seek international and/or regional assistance to address the needs of affected persons."

72 General Assembly Resolution 43/131, UN GAOR, 75th plenary meeting, UN Doc A/RES/43/131 (1988): "Reaffirming the sovereignty, territorial integrity and national unity of States, and recognizing that it is up to each State first and foremost to take care of the victims of natural disasters and similar emergency situations occurring on its territory." This text was reaffirmed in General Assembly Resolution 45/100, UN GAOR 68th plenary mtg., PP2, UN Doc A/RES/45/100 (1990).

73 Eduardo Valencia-Ospina, "Fourth Report on the Protection of Persons in the Event of Disasters," International Law Commission, UN GAOR 63rd Sess., UN Doc A/CN.4/643, 2011, 13–24.

74 Report of the International Law Commission, GAOR 63rd Sess., UN Doc A/66/10 2011: Article 10 and 11.

instruments and customary international law."[75] Article 11 creates "a qualified consent regime in the field of disaster relief operations."[76]

However, there is clearly some resistance among states to the idea that these duties currently form part of customary international law. For example, the commentary to the ILC Draft Articles makes clear that some members of the Commission hold the view that these duties represent progressive development of the law, and not codification of it.[77] The ILC Secretariat has also noted that a review of the drafting history of pertinent General Assembly Resolutions that contain the duty to protect reveals that states have expressed their support for this duty on the basis that such resolutions were not to be interpreted as creating a right or duty to interfere in the domestic affairs of another state.[78]

If accepted by states, these duties would have the effect of increasing diplomatic pressure on the disaster-affected state to seek assistance and require it not to "arbitrarily" withhold its consent, and not allow states to use the principle of sovereignty as a basis to deny assistance to its population. However, while a breach of these obligations would result in the state committing an internationally wrongful act, it would not give rise to a corresponding right of intervention by other states.[79] In this way, such a duty would have similar limitations as the principle of R2P has in relation to instances of genocide, war crimes, ethnic cleansing, and crimes against humanity (i.e., it does not authorize action beyond the acceptable legal bases for deployment set out earlier in this chapter).

The preceding analysis demonstrates that, at present, there are no viable alternative legal bases to consent for the deployment of personnel to assist a disaster-affected state. Discussions in the ILC demonstrate that international law has not developed in a way that allows humanitarian considerations to override the right to territorial integrity inherent in the concept of sovereignty. While the ILC Draft Articles have the potential to increase pressure on states to seek assistance in the event that its domestic response capacity is overwhelmed, and not arbitrarily refuse to grant consent if such assistance is offered, this will depend on widespread acceptance of these obligations.

PRIVILEGES AND IMMUNITIES

There are a number of practical legal issues that must be settled before a deployment of personnel, either civil or military, can take place in another country. These

[75] Ibid., 262. [76] Ibid., 65. [77] Ibid., 262, 268.

[78] Secretariat of the ILC, "Memorandum," 21.

[79] The legal consequences of a disaster affected State committing an internationally wrongful would be limited to those outlined in Chapter I, Part Two of the International Law Commission's articles on Responsibility for Internationally Wrongful Acts, for example, making reparations: International Law Commission, *Responsibility of States for Internationally Wrongful Acts*, annexed to General Assembly Resolution 5683, UNGAOR 56th Session, Agenda item 162, UN Doc A/RES/56/86 (2002).

include the entry into and departure from the receiving state of personnel and goods, the recognition of professional qualifications, and the application of local import duties, taxes, and restrictions. One of the most important of these legal issues from an assisting state's perspective is the acquisition of privileges and immunities for deployed personnel. The negotiation of privileges and immunities requires a balance to be struck between the sovereign interests of the disaster-affected state in retaining jurisdiction over persons in its territory, and of the assisting state in protecting its nationals from exposure to liability and regulatory barriers when undertaking disaster assistance. A failure to satisfactorily resolve this issue has the potential to jeopardize the viability of a deployment.

Privileges and immunities are generally contained in status of visiting personnel agreements or arrangements. Such instruments can be bilateral or multilateral in nature, and take the form of treaty-level agreements or less-than-treaty-status arrangements, such as memoranda of understanding, *notes verbale*, and exchanges of letters. Treaty-level agreements may be preferred in circumstances where a binding commitment is considered necessary, for example, to obtain a higher level of assurance that privileges and immunities will be honored by the host state. Conversely, nonbinding arrangements may be preferred in certain circumstances, such as when speedy conclusion is necessary, as they may not be subject to domestic approval processes, which can be lengthy. Accordingly, non-binding instruments are likely to be more common for the purposes of disaster relief, where the timeliness of the provision of assistance to victims can be critical.

These instruments are necessary to define the legal position of deployed personnel, both civil and military, who are in the territory of another state with the latter state's consent.[80] Despite their widespread use, there has been no universally applicable regime to regulate the status of personnel in military and civilian deployments. Auriel Sari has identified three reasons why this may be the case: (1) states send their personnel abroad for a variety of purposes; (2) the operating environments differ from one deployment to another; and (3) powerful states tend to leverage more favorable conditions in the receiving state, which they would not reciprocate.[81] The instruments will also generally vary depending on the military or civilian character of a deployment.[82] The result is that these instruments, and the privileges and immunities they provide, can vary widely.

The Legal Content of Privileges & Immunities

The term "privileges and immunities" primarily relates to the exemption of the exercise of jurisdiction, including civil and criminal jurisdiction, by the host state

[80] Auriel Sari, "Status of Forces and Status of Mission Agreements Under the ESDP: The EU's Evolving Practice," *The European Journal of International Law* (2008): 68.

[81] Ibid., 69. [82] Ibid., 85.

over the visiting personnel. However, privileges and immunities clauses can cover a wide range of issues such as exemption from visa requirements and taxation provisions, and the inviolability of premises, archives, and communication. The proceeding discussion focuses on immunity from criminal and civil jurisdiction, but a broader range of privileges and immunities are undoubtedly relevant to the deployment of international disaster relief personnel, particularly waiver of normal entry and exit requirements and registration of professional qualifications in the disaster affected state.

There are three main levels of immunities from criminal and civil jurisdiction that may be accorded:

- Complete (or almost complete) immunity from civil and criminal jurisdiction of the host state
- Immunity to the extent that the sending state has jurisdiction, and
- Concurrent jurisdiction by the sending state and the host state, with a mechanism to resolve which state has jurisdiction in the particular circumstance.

Complete Immunity from Civil and Criminal Jurisdiction

At one end of the spectrum, deployed personnel may be given complete immunity from criminal and civil jurisdiction in the host state. This may be achieved simply through agreement between the sending state and the host state. Deployed personnel may also be accorded the full range of privileges and immunities set out under the Vienna Convention on Diplomatic Relations (VCDR), or status equivalent to the administrative and technical staff under that Convention.[83] Such personnel enjoy immunity from the criminal jurisdiction of the receiving state, and immunity from its civil and administrative jurisdiction, except in actions relating to private immovable property, succession, or professional or commercial activity outside official functions.[84] They are not liable to any form of arrest or detention,[85] and are not subject to measures of "execution" or enforcement of civil or criminal law (except in relation to the limited categories of civil and administrative liability outlined previously).[86]

(b) Sending State Has Exclusive Jurisdiction Where it is Able to Exercise its Own Jurisdiction

The host state and the sending state may agree that the sending state may exercise jurisdiction over its personnel to the extent permitted by its own law to the exclusion of the host state. Under these types of arrangements, the host state is still able to exercise jurisdiction over foreign personnel in its territory in circumstances where the sending state cannot. Such provisions encourage sending states to ensure

[83] Ibid., 76. [84] *Vienna Convention on Diplomatic Relations*, Art 31 (1).
[85] Ibid., Art 29. [86] Ibid., Art 31 (3).

that their personnel remain subject to its law, and that there is no gap in jurisdiction. This formulation is particularly appropriate for military personnel, who remain subject to the criminal and disciplinary jurisdiction of the sending state when deployed overseas by virtue of domestic criminal laws, and over whom a sending state has a particular interest in retaining jurisdiction.[87] However, such provisions will also be appropriate for a sending state that has in place legislation that extends its criminal jurisdiction over civilian personnel extraterritorially in specific circumstances.[88]

Concurrent Jurisdiction by the Sending State and the Host State

Concurrent jurisdiction usually arises where both the sending state and the host state are given exclusive jurisdiction over offenses that violate their own laws, but not the law of the other state, and there is a residual category where the law of both jurisdictions can apply.[89] In military contexts, the issue of concurrent jurisdiction is generally resolved by reference to whether the act in question occurred in the course of "official duties."[90] Commonly, the types of immunities described in categories (ii) and (iii) are combined in one instrument.

Privileges and Immunities Appropriate to Disaster Assistance

The level of privileges and immunities granted for each deployment of disaster relief personnel will depend on a number of factors that will inform the preferences of the assisting state and the disaster-affected state, which essentially relate to the level of risk each party faces in relation to the deployment of personnel. For example, the character of the disaster-affected state's judicial system will in large part influence what immunities the assisting state may try to secure. If there are concerns about the independence of the judicial system, or about the nature of the penalties that may be applied (for example, the death penalty), the assisting state may seek complete immunity for its personnel. The functions the personnel will be performing will also inform what immunities are sought. For example, personnel performing certain roles that may be exposed to higher risk of legal liability may negotiate a higher level of exemption from the local justice system.

In the context of disaster assistance, the ILC's Special Rapporteur on the Protection of Persons in the Event of Disasters stated that waiver of national law

87 See, for example in Australia the Australian Defence Force remains subject to the criminal jurisdiction of Australia by virtue of the *Defence Force Discipline Act 1982* (Cth), which extends criminal and disciplinary jurisdiction over ADF members extraterritorially.

88 See, for example, the *Crimes (Overseas) Act 1964* (Cth), which extends Australian criminal liability over Australian personnel or Australian contractors who have been deployed under an agreement or to a country declared under regulations to the Act.

89 Paul Conderman, "Vehicles," in *The Handbook of the Law of Visiting Forces*, ed. Dieter Fleck (Oxford University Press, 2001), 103.

90 *NATO Model Status of Forces Agreement*, Art 3(a)(ii).

"should be reasonable when balancing the affected State's duty to provide assistance and its obligation to protect its population from harm in the light of the particular circumstances."[91] Sari has noted that there is no norm of international law that would compel a state to request only such privileges and immunities that are necessary for the purposes of a deployment.[92] However, Sari also acknowledges that the practice of negotiating extensive privileges and immunities does not "sit well" with the accountability of those deployed.[93] It should be noted, for example, that diplomatic privileges and immunities serve a specific function – to facilitate the diplomatic presence of one state on the territory of another – and the level of protection they provide may not be appropriate in all circumstances. Ultimately, any privileges and immunities to be granted to visiting personnel will be the result of negotiation between the host and sending states.

Interests of the Assisting State

Privileges and immunities are granted to allow deployed personnel to carry out their role effectively and are intended to reduce the regulatory barriers to the provision of disaster assistance.[94] However, they are often also requested by sending states in order to protect their nationals, particularly where they are being deployed in fragile states. There may be domestic political considerations for the sending state in acquiring the necessary privileges and immunities. For example, it is likely that if a civilian or military expert who has been deployed to assist a disaster-affected state is subjected to an unfair trial, or severe punishment, this would engender negative reactions from the sending state.

These types of reactions may be particularly acute where the personnel in question are civilian in character, and have not assumed as high a level of professional risk as military personnel. Such a reaction may even have the effect of an assisting state refraining from providing personnel in future, which would not be a positive outcome for the assistance of victims of disaster. In this regard, a parallel can be made between the domestic backlash within sending states after the death of nationals during peace operations in the 1990s. For example, the deaths of U.S. peacekeepers in Somalia in 1993 and of Belgian peacekeepers in Rwanda in 1994 led to a general resistance of many states to contribute personnel to future peacekeeping operations.

Interests of Disaster-Affected State

A grant of privileges and immunities by the sending state circumscribes the host state's right to exercise legal jurisdiction over those in its territory and therefore

[91] Eduardo Valencia-Ospina, "Fifth Report on the Protection of Persons in the Event of Disasters," *International Law Commission*, 64th Session, UN Doc A/CN.4/652. 2012, 36 [135].
[92] Sari, "Status of Forces," 67. [93] Ibid. [94] Valencia-Ospina. "Fifth report," 35–8 .

operates as an exception to the general rules of international law that apply to territorial jurisdiction.[95] While the grant of privileges and immunities may be a voluntary limit on the sovereign rights of the host state, it is not unusual for tensions to arise as a result of "each State's inherent interest in criminal jurisdiction."[96]

Immunities clauses generally contain a provision about respect for local law.[97] As such, even extensive immunities are not incompatible per se with the respect for law in the host state.[98] As discussed, both military and civilian personnel often remain subject to the criminal law of the sending state when deployed to another country, to avoid the creation of a legal lacuna. However, justice delivered in this form is less visible to the population of the host state, and for serious criminal acts, such as those resulting in death, this can give rise to grievances among the population of the host state and be viewed as an unreasonable limitation on the host state's sovereignty.[99]

This may be compounded where a host state was in a weak negotiating position vis-à-vis the sending state. It is not difficult to imagine that the power differential may be particularly acute in a crisis situation such as the provision of disaster relief, where timely assistance may be critical, and less developed states are more likely to require assistance in responding to a disaster. While the grant of privileges and immunities has been characterized as sharing the sovereign prerogative between the receiving and sending state,[100] such a balance may clearly be tipped in favor of the sending state in certain circumstances.

Some preexisting international disaster relief instruments contain provisions on privileges and immunities.[101] For example, the Tampere Convention, which facilitates telecommunications assistance in the event of disaster, provides that persons acting pursuant to the Convention will enjoy immunity from arrest, detention, and legal process, including criminal, civil, and administrative jurisdiction of the affected state for acts or omissions "specifically and directly related" to the provision of assistance under the Convention.[102] However, the Tampere Convention makes the grant of privileges and immunities subject to the qualification that these privileges and immunities should be granted "to the extent permitted by the [affected state's] national law."[103]

95 Dieter Fleck, "Are Foreign Military Personnel Exempt from International Criminal Jurisdiction under Status of Forces Agreements." *Journal of International Criminal Justice* (2003): 656.

96 John W. Egan, "The Future of Criminal Jurisdiction over the Deployed American Solider: Four Major Ends in Bilateral US Status of Forces Agreements," *Emory International Law Review* (2006): 293.

97 For example, see Vienna Convention on Diplomatic Relations, April 18, 1961, 500 UNTS 95: art 41(1).

98 Sari, "Status of Forces," 81.

99 See Egan, "The Future of Criminal Jurisdiction."

100 Colonel Richard J. Erikson, USAF (Ret'd). "Status of Forces Agreements: A Sharing of Sovereign Perogative," *Air Force Law Review* (1994): 140.

101 "Framework Convention," Art 4(a)(5); "Tampere Convention," Art 5(1); "UNITAR Model for Disaster Relief Operations Rules," Annex A, Rule 14.

102 "Tampere Convention," Art 5(1)(a).

103 "Tampere Convention," Art 5(1).

The Framework Convention on Civil Defence Assistance, which promotes cooperation between civil defense services in the event of disaster, provides that the affected state will "grant all privileges, immunities, and facilities necessary for carrying out the assistance."[104] The UN Institute for Training and Research Model for Disaster Relief Operations provides that the disaster-affected state will "extend to the designated relief personnel the necessary facilities with a view to securing the expeditious performance of relief functions."[105] These three examples demonstrate the different levels of privileges and immunities that may be considered appropriate in disaster relief situations.

The ILC drafting committee adopted a draft article requiring the affected state to take the necessary measures to facilitate external assistance, in particular, privileges and immunities.[106] While some states expressed the view that formulating a model instrument for humanitarian relief operations patterned on a Status of Forces Agreement that could be annexed to the draft articles would serve a practical purpose, others have expressed doubt about the feasibility of that proposal.[107] Accordingly, it is not yet clear whether the ILC's work on disasters will advance the law beyond its current state in relation to privileges and immunities.

States have previously accepted legal regimes that codify privileges and immunities in other contexts,[108] which suggests there is scope for the acceptance of such a regime in relation to disaster relief personnel. Setting out privileges and immunities in a generally applicable legal framework on disaster relief, either through the ILC consideration of disaster assistance or another process, could make a positive contribution to facilitating disaster assistance and protecting the interests of both the assisting state and the disaster-affected state. This would assist in striking a balance between protecting the personnel of a mission from undue domestic interference while they are carrying out their duties to help facilitate their work, and respecting the sovereignty of the disaster-affected state. As demonstrated, a failure to strike this balance has the potential to jeopardize the provision of disaster relief assistance. Establishing a set of privileges and immunities that apply generally in disaster situations would provide a neutral basis on which to establish a relationship between the assisting state and the disaster-affected state.

CONCLUSION

The legal basis for the deployment of personnel to a disaster-affected state and the privileges and immunities such personnel will be accorded to carry out their

[104] "Framework Convention," Art 4(a)(5).

[105] "UNITAR Model for Disaster Relief Operations Rules," Annex A, Rule 14.

[106] Report of the International Law Commission. GAOR 63rd Sessions, UN Doc A/66/10 (2011): Article 14.

[107] Report of the International Law Commission, GAOR 64th Session, UN Doc A/67/10. 2012, 88 [65].

[108] Vienna Convention on Diplomatic Relations; "Convention on the Privileges and Immunities of the United Nations." December 15, 1946, 1 UNTS 15.

functions is generally dealt with bilaterally on a case-by-case basis. The need to establish such arrangements in relation to the deployment of disaster relief personnel has the potential to raise regulatory barriers to the provision of disaster assistance as a result of the interaction between these legal arrangements and the sovereignty of the disaster-affected state. These barriers could hinder seriously the provision of timely assistance to victims of large-scale disasters.

In the aftermath of a natural disaster, when there is likely to be widespread suffering and loss of life, it may not be possible to put in place legal arrangements between the disaster-affected state and a state that is offering assistance in a way that respects the sovereign rights and duties of both states. Such a time-pressured environment is not the ideal time to be addressing legal barriers to the provision of assistance. Institutionalizing the key legal elements of a deployment in a "standing" legal framework that generally applies to the provision of disaster assistance may be a positive contribution to the law in this field. Ultimately, the appetite for states to consent to such a framework will depend on a range of factors, including the protection of their personnel, the costs involved, and the views of their populace. The success of any such project will depend on the necessary balance being struck between the rights of sovereign states, and their concomitant duties to their populaces.

PART V

Disaster Prevention and Relief: Anticipatory Responses from NGOs

FIGURE 8 **Red Cross loading food containers for distribution in Galveston, Texas after Hurricane Ike (2008).** Source: Greg Henshall, Federal Emergency Management Agency (FEMA).

15

The Role of International Organizations in Disaster Response: A Case Study of Recent Earthquakes in Japan

Kentaro Nishimoto

This chapter considers the legal and policy reactions of the Japanese government to the international offers of assistance received in the wake of the 2011 earthquake and tsunami that struck Japan, and the earlier offers of aid in the wake of the 1995 Kobe earthquake. The Japanese responses were incongruent – the earlier response generally being negative, the later response generally being positive or receptive. Not only does this chapter explore the reasons behind such incongruency; it also charts the progression of international disaster response law (IDRL) from the period of the League of Nations to the present day and assesses the relative impact of both completed treaties and exercises undertaken to draft new laws throughout the twentieth century and into the twenty-first century.

INTRODUCTION

On March 11, 2011, a 9.0 magnitude earthquake struck the east coast of Honshu, Japan. The earthquake was followed by a tsunami that washed across the Pacific coast of eastern Japan, inflicting devastating damage on the Tohoku region. This was the largest earthquake recorded in the history of Japan, which resulted in more than 15,000 deaths, 92.4 percent of which was the immediate result of the tsunami.[1] The earthquake and tsunami also led to a nuclear crisis at the Fukushima Daiichi NPP. As a combined result of the earthquake and the nuclear crisis, approximately 470,000 people were forced to evacuate in the immediate aftermath of the disaster. In this unprecedented situation, 163 states and regions and forty-three international organizations offered assistance for the relief efforts.[2] The Japanese government accepted rescue teams and experts from twenty-four states and regions and five

[1] Cabinet Office, Government of Japan, *White Paper on Disaster Management 2011: Executive Summary* (Provisional Translation), www.bousai.go.jp/hakusho/WPDM2011_Summary.pdf.

[2] Ministry of Foreign Affairs of Japan, "Emergency Assistance from Overseas," www.mofa.go.jp/j_info/visit/incidents/index2.html#assistance.

international organizations, and aid supplies and financial assistance from 126 states, regions, and international organizations.[3] Japan benefited greatly from such warm support from the international community.

In light of these developments, this chapter examines the legal aspects of accepting assistance from the international community in the immediate aftermath of a disaster.[4] Although international initiatives against disasters have increasingly come to stress disaster risk reduction (DRR), this chapter focuses on immediate responses to disasters, since the Japanese experience has shown that even the more prepared states may be temporarily overwhelmed by large disasters. The purpose of this inquiry is twofold. First is to examine the existing international framework for disaster response, and its interaction with sovereign states. Second is to give an account of the international assistance that was received by Japan immediately after the 2011 earthquake, and some of the legal issues that were involved in accepting such assistance. Through an examination of the international framework in general and the experience of Japan in particular, this chapter also touches upon the underlying foundations of law and institutions concerning international disaster response.

The issue of disaster response has been neglected in international law scholarship, and the attention given to it has been disproportionate to its practical importance.[5] This has led some to argue in favor of strengthening the legal framework for an International Disaster Response Law (IDRL),[6] while others are skeptical of a direct role for international law in this field.[7] There is much merit in the argument that since natural disasters are viewed as short-lived episodic events by states, there is little prospect for states to limit their sovereignty and develop rules of international law with respect to disaster relief.[8] Although this chapter does not

3 Ibid.

4 The term "disaster" has only recently been discussed as a term of art in international law. Since 1996, the International Law Commission has been working on the topic "protection of persons in the event of disasters," and has adopted the following definition: "a serious disruption of the functioning of society, posing a significant, widespread threat to human life, health, property or the environment, whether caused by accident, nature or human activity, and whether developing suddenly or as the result of complex, long-term processes" (Article 3). For discussions on this concept in the ILC and Sixth Committee, see Hiroki Horimi, "An Introductory Study of the Concept of "Disaster" under International Law" (in Japanese), in *International Law in the Globalized Age*, ed. Toshiya Ueki (Tokyo: Shinzansya, 2012), 159–200.

5 Peter Macalister-Smith, "Disaster Relief: Reflections on the Role of International Law," *Zeitschrift für ausländisches öffentliches Recht und Völkerrecht* (1985): 25–26; Naoya Okuwaki, "Natural Disasters and International Cooperation – Hyogo Declaration and the Contribution of Japan," *Juristo* 1321 (2006): 66–72.

6 Jenny R. Hernandez and Anne D. Johnson, "A Call to Respond: The International Community's Obligation to Mitigate the Impact of Natural Disasters," *Emory International Law Review* 25 (2011): 1087–1096; Elizabeth Ferris and Kei Hakata, "Beyond the Disaster: A Call for Japanese Leadership in International Disaster Response Law," *Brookings*, March 22, 2011, www.brookings.edu/research/opinions/2011/03/22-japan-leadership-ferris.

7 David P. Fidler, "Disaster Relief and Governance after the Indian Ocean Tsunami: What Role for International Law?" *Melbourne Journal of International Law* 6 (2005): 458–473.

8 Ibid., 461.

try to refute this thesis entirely, it will show the possibility that such a perception by states on this matter may be subject to gradual change over time; it will also argue that while disaster response has been predominantly discussed in terms of humanitarian assistance on the basis of solidarity of the international community, disaster response as "functional cooperation" for states to further their interests may provide a complementary model to understanding the international law of disasters.

To this end, we begin with a discussion of how international institutions and frameworks have been utilized in disaster response. Special attention must be given to assistance from international intergovernmental organizations (IGOs), especially within the United Nations system. Although international assistance from bilateral frameworks, regional frameworks, and non-governmental organizations (NGOs) are not of any lesser importance in international disaster response, the focus will be on IGOs as a measure of organized efforts of the international community as a whole. This discussion of the international frameworks will be followed by a discussion of the international assistance Japan received in the immediate aftermath of the 2011 earthquake. This part will first present an overview of the Japanese legal system that serves as an interface between the international frameworks in accepting assistance, and then discuss the involvement of international organizations with respect to the disaster situation in general, as well as the nuclear disaster at Fukushima. The conclusion offers a discussion of some of the implications for international law and institutions that may be drawn from the Japanese experience.

The experience of Japan demonstrates that international assistance matters even when the territorial state is exercising its primary responsibility in responding to disasters. As Hurricane Katrina in New Orleans has likewise shown, disasters may cause significant difficulties even for states with substantial resources for disaster preparedness and response.[9]

INTERNATIONAL FRAMEWORKS FOR DISASTER RESPONSE

Early Efforts at Organized Disaster Response

Most literature in international law on coordinated efforts by the international community in response to disasters start their discussion with the International Relief Union (IRU), the first IGO focused on disaster relief. The organization originated a proposal in 1921 by the president of the Italian Red Cross, G. Ciraolo, after the Messina earthquake in 1908. Originally envisaged as a mutual insurance against disasters to realize the right of disaster-stricken people to international assistance, the proposal was significantly watered down through discussions at the

[9] Katrina Lessons Learned Review Group, *The Federal Response to Hurricane Katrina: Lessons Learned*, Washington, D.C., 2006.

League of Nations.[10] After much political compromise, a diplomatic conference in 1927 adopted the Convention Establishing the International Relief Union, which entered into force in 1932[11]

The IRU Convention, in its preamble, states the desire of states parties "to render aid to each other in disasters, to encourage international relief by a methodical co-ordination of available resources, and to further the progress of international law in this field."[12] Article 2(1) of the convention establishes the purpose of the IRU: "In the event of any disaster due to force majeure, the exceptional gravity of which exceeds the limits of the powers and resources of the stricken people, to furnish to the suffering protection first aid and to assemble for this purpose funds, resources and assistance in all kinds." The IRU was to play a role in the coordination of the efforts of other relief organizations (Article 2(2)). As such, the IRU would take actions for the benefit of all stricken people, but only with the consent of the government of the affected state (Articles 3 and 4). A novel feature of the convention was that it also envisaged cooperation with the Red Cross and similar organizations, and the actual relief work was to be carried out on behalf of the IRU under overall direction of the Executive Committee. The Red Cross, in turn, was formally integrated into the structure of the IRU, and was allowed to participate in the decision-making process in a consultative capacity.[13]

Unfortunately, this first and unique attempt at a response to disasters by the international community could not achieve much success. It was only able to respond to two disasters in 1934 and 1935, and had effectively ceased to function by the mid-1930s.[14] The failure of the IRU is explained by the lack of financial resources, which ultimately resulted from the lack of political will among the member states. It should be noted, however, that although the IRU was advocated in terms of international solidarity, it was premised not only on humanitarian goodwill among member states, but also on the idea of mutual assistance.[15] The artificial nature of this reciprocity with regard to disaster assistance, however, is attested by the British stance toward the creation of the Union in 1924: "... a country such as England, not visited by earthquakes or similar disasters, would have to put into the common fund without the prospect of anything in return."[16]

[10] For a historical account of its creation, see John F. Hutchinson, "Disasters and the International Order – II: The International Relief Union," *International History Review* 23 (2001): 253–298.

[11] Peter Macalister-Smith, "The International Relief Union: Reflections on the Convention Establishing an International Relief Union of July 12, 1927," *Tijdschrift voor Rechtsgeschiedenis* 54 (1986): 365.

[12] League of Nations, *Convention Establishing an International Relief Union* (No. 3115), ser. League of Nations Treaties, 1927, vol. 135.

[13] Ibid., 267. [14] Macalister-Smith, "The International Relief Union," 370–372.

[15] Hutchinson, "Disasters and the International Order," 262. [16] Quoted in ibid., 254.

Development of the International Humanitarian System

The United Nations is the successor organization to the League of Nations, and various disaster-related activities have taken place within the UN system. Although without explicit reference, response to disasters has been considered to fall under one of the purposes of the UN to "achieve international co-operation in solving international problems of an economic, social, cultural, or *humanitarian character* ...[17]" Operational agencies within the UN system were soon engaged in disaster response related activities. As the involvement of various UN agencies in both natural and man-made disasters grew, so did recognition of the need to coordinate humanitarian relief. The UN has struggled to achieve a working coordination mechanism on disaster relief, and has repeatedly remodeled the institutional architecture of its humanitarian assistance coordination arrangements.[18] The growth of NGOs involved in international humanitarian assistance has also added to the importance of developing the coordinating function of the UN.

The first major attempt at the creation of a coordination mechanism for disaster response was the establishment of the Office of the UN Disaster Relief Coordinator (UNDRO) in 1971.[19] The mandate of the Disaster Relief Coordinator (DRC) was, among other things, to "mobilize, direct and co-ordinate the relief activities of the various organizations of the United Nations system in response to a request for disaster assistance from a stricken State," and to coordinate this assistance with assistance given by IGOs and NGOs such as the International Red Cross.[20] It was also to serve as a clearinghouse for collecting and disseminating information. UNDRO made an attempt to develop a legal framework for facilitating disaster relief.[21] Although never adopted, a Draft Convention on Expediting Delivery of Emergency Assistance to expedite the delivery of assistance following disasters was prepared in 1984.[22]

[17] Charter of the United Nations, Art.1(3) (emphasis added).

[18] Laura M. E. Sheridan, "Institutional Arrangements for the Coordination of Humanitarian Assistance in Complex Emergencies of Forced Migration," *Georgetown Immigration Law Review*, 14 (2001): 944.

[19] See United Nations General Assembly, 26th Session. *Resolution 2816 (XXVI) [Assistance in cases of natural disaster and other disaster situations]*, December 14, 1971, 85. In *Resolutions adopted on the reports of the Third Committee*

[20] Ibid.

[21] David Fisher, "The Law of International Disaster Response: Overview and Ramifications for Military Actors," in *International Law Studies – Volume 83: Global Legal Challenges: Command of the Commons, Strategic Communications, and Natural Disasters*, ed. Michael D. Carsten (Newport: Naval War College, 2007): 295.

[22] United Nations General Assembly and Economic and Social Council, Office of the United Nations Disaster Relief Coordinator, *Report of the Secretary-General*. A/39/267/Add.2-E/1984/96/Add.2. June 18, 1984. The draft convention set out certain basic principles to be applied in providing emergency assistance after disasters, such as respect for the sovereignty of the receiving states (Article 3(1)(a)), and provided rules for the technical aspects of providing assistance, such as entry of personnel, communications, customs, costs, and liabilities.

UNDRO was a direct response to the growing recognition that the UN was lacking in its capability to respond to disasters, after several major natural disasters. Its mandate focused on natural disasters, and on disaster relief.[23] The focus on natural disasters meant that so-called man-made disasters, and especially armed conflict, were regarded to be outside its scope. Although General Assembly Resolution 2816 (XXVI), which established UNDRO, was entitled "Assistance in case of natural disasters and other disaster situations," the additional term "other disaster situations" did not in practice function to enlarge the role of UNDRO in any significant way. The focus on disaster relief meant that relief coordination was considered to be the primary role of UNDRO. However, this focus would call into question how pre-disaster planning and post-disaster activities were to fit within the overall agenda against disasters.

During the time of UNDRO's operation, the adequacy and efficiency of the UN system to provide coordinated relief and assistance came to be questioned, and were made the subject of official studies conducted within the UN with a view to strengthen UN relief coordination.[24] UNDRO's approach to disaster relief was considered deficient, and a single agency empowered to coordinate the provision of both humanitarian relief and development, in both natural and man-made disasters was sought. Although there were different views on how to effectuate this reform, calls for a strengthened UN relief action culminated in General Assembly Resolution 46/182, which established new arrangements for relief and assistance.

This resolution on the "Strengthening of the coordination of humanitarian emergency assistance of the United Nations" defines the guiding principles for humanitarian assistance activities in the UN system and provides the basis for the current system of UN bodies associated with humanitarian emergency activities. The DRC was transformed into the Emergency Relief Coordinator, responsible for the Department of Humanitarian Affairs (DHA), which was subsequently reorganized into the Office for the Coordination of Humanitarian Affairs (OCHA).[25] The DRC's limited role in response to natural disasters was expanded in the mandate of the Emergency Relief Coordinator to cover complex emergencies, and some of the operational roles of the DHA were transferred to other agencies, focusing on the role of OCHA on coordinating relief efforts.[26]

The current international humanitarian system has developed various multilateral mechanisms that may be utilized in disaster response, and OCHA also serves as

[23] Peter Macalister-Smith, *International Humanitarian Assistance: Disaster Relief Actions in International Law and Organizations* (Martinus Nijhoff Publishers, 1985), 137–138.

[24] Ibid., 138–140; Sheridan, "Institutional Arrangements," 953–956.

[25] United Nations Office for the Coordination of Humanitarian Affairs, *Coordination to Save Lives: History and Emerging Challenges*, (United Nations: 2012), 4–5.

[26] Ibid., 5. Notably, disaster preparedness mitigation activities were transferred to the UNDP.

a Secretariat for a number of these mechanisms.[27] The Inter-Agency Standing Committee (IASC), under the leadership of the Emergency Relief Coordinator, is assigned the role of a "primary mechanism for inter-agency coordination"[28] between major humanitarian actors both within and without the UN system, such as the UN operational agencies and the relevant NGOs such as the International Committee for the Red Cross (ICRC) and the International Federation of Red Cross and Red Crescent Societies (IFRC). Its task is to meet as soon as possible in response to emergencies, to assess needs and structure an efficient response.

The United Nations Disaster Assessment and Coordination (UNDAC) system is a designed for training disaster managers with a uniform methodology for disaster relief coordination and deploying a team of managers to coordinate incoming international assistance immediately after the occurrence of a disaster. The International Search and Rescue Advisory Group (INSARAG) is a network of the world's urban search and rescue providers, engaged in setting standards in rescue capabilities and facilitating operational coordination through the UNDAC system. Numerous standards and guidelines have been developed within this system as well. While varied in their degree of acceptance by the actors comprising the international humanitarian assistance system, some of the guidelines and standards have come to be perceived by the relevant actors as norms to be complied with in providing assistance.[29]

Although a full assessment of the development of the current system is beyond the scope of this chapter, it is possible to identify a shift in the context within which the organizational responses to disasters are situated. The establishment of the IRU was motivated by a sense of international solidarity and even altruism for the victims of those affected by the immediate consequence of natural disasters, but it was still framed as mutual insurance between states in response to a specific type of event. In contrast, the current regime for coordinating *humanitarian* relief based on UNGA Resolution 46/182 has shifted the focus of attention considerably from an episodic event that causes large-scale suffering, to the suffering of the affected victims. This shift in focus was a natural move in ensuring efficiency for efforts against human suffering, and is also faithful to the objectives of the UN as provided in its Charter, "solving international problems of … *humanitarian* character" (Article 1(3)). However, it is significant that this shift has caused the normative foundation of international disaster response to be increasingly understood in terms

[27] For activities of these mechanisms, see Giovanni De Siervo, "Actors, Activities and Coordination in Emergencies," in *International Disaster Response Law*, ed. Andrea d Guttry et al. (The Hague: TMC Asser, 2012), 485–515.

[28] United Nations General Assembly, 48th Session. *Resolution 57 [Strengthening of the coordination of humanitarian emergency assistance of the United Nations]*, A/Res/48/57, January 31, 1994.

[29] Dug Cubie, "An Analysis of Soft Law Applicable to Humanitarian Assistance: Relative Normativity in Action?"*International Humanitarian Legal Studies* 2 (2011): 177–215.

of international solidarity.[30] This focus on affected victims can also be considered as forming the foundations for a rights-based approach to disaster and humanitarian relief. In this context, international disaster response is no longer only about reciprocal cooperation between states to counter extraordinary situations.

Specialized International Organizations in Disaster Response

The involvement of the international community in disaster response, however, is not limited to the context of humanitarian assistance. Many specialized international organizations include certain aspects of disaster response in their mandates.[31] Since such international organizations are often established to address the issues that affect mutual interests of states, disasters become their concern when they hamper international cooperation in their specific fields. Treaties developed in close connection with these organizations also contain provisions specifically aimed at responding to specific types of disasters or facilitating disaster relief.

An example of the former is the 1990 International Convention on Oil Pollution Preparedness, Response and Co-operation (OPRC), which was adopted under the auspices of the International Maritime Organization (IMO) in response to the *Exxon Valdez* accident in 1989.[32] The OPRC provides for surveillance and reporting (Article 4), assessment of the situation and notification to affected States (Article 5), and international cooperation in oil pollution response (Article 7). The OPRC was further extended by the 2000 Protocol on Preparedness, Response and Co-operation to pollution Incidents by Hazardous and Noxious Substances (OPRC-HNS Protocol). The IMO assumes a role in implementing both the convention and the protocol. An example of the latter is included in the 1944 Convention on International Civil Aviation. The convention requires contracting states to facilitate entry, departure, and transit of relief flights "undertaken in response to natural and man-made disasters which seriously endanger human health or environment.[33]"

30 An example of such understanding within the UN system may be found in UN documents: *e.g.*, New International Humanitarian Order: Humanitarian assistance to victims of natural disasters and similar emergency situations. *Report of the Secretary-General*, (UN Doc. A/45/587), par.5. October 24, 1990.

31 For a full discussion on the role of international organization in disasters, see Eulalia W. Petit, "Les actions d'urgence dans les catastrophes : évolution des activités des organisations interétatiques et des organisations non gouvernementales," in *The International Aspects of Natural and Industrial Catastrophes*, ed. David D. Caron and Charles Leben (The Hague: Nijhoff, 2001): 539–589; see also Akiho Shibata, "Creating An International Urgent Assistance Mechanism In Case Of Natural And Industrial Catastrophes," ibid., 457–535.

32 Jean Claude Sainlos, "The International Maritime Organization and the Protection of the Marine Environment," in *The World Ocean in Globalisation: climate change, sustainable fisheries, biodiversity, shipping, regional issues*, ed. David Vidas et al. (Leiden: Martinus Nijhoff, 2011), 339–340.

33 International Civil Aviation Organization, *Annex 9 to the Convention on International Civil Aviation: Standard* 8.8, (2005).

Similarly, the 1965 Convention on Facilitation of International Maritime Traffic, adopted at the IMO, provides that "public authorities shall facilitate the arrival and departure of ships engaged in disaster relief work."[34]

A detailed treaty regime for a specific type of disaster, which came to be of special relevance to the nuclear disaster at Fukushima that followed the earthquake and tsunami are the two multilateral conventions adopted by the IAEA General Conference in the aftermath of the Chernobyl nuclear accident in 1986. The two conventions, The Convention on Early Notification of a Nuclear Accident (hereinafter "the Notification Convention")[35] and the Convention on Assistance in the Case of a Nuclear Accident or Radiological Emergency ("the Assistance Convention")[36] both not only creates rights and obligations between states but also assign a significant role to the IAEA in their operation.

The Notification Convention creates an obligation on the state parties to notify other states of a nuclear accident. Under Article 1, the Notification Convention applies to "any accident involving facilities or activities of a State Party or of persons or legal entities under its jurisdiction or control … from which a release of radioactive material occurs or is likely to occur and which has resulted or may result in an international trans-boundary release that could be of radiological safety significance for another State." In the event of a nuclear accident covered by Article 1, the state shall "forthwith notify," directly or through the IAEA, those states that are or may be physically affected and the IAEA of the nuclear accident, its nature, the time of its occurrence, and its exact location where appropriate (Article 2). In addition, states are to provide information relevant to minimizing the radiological consequences in those states (Article 5), and may also voluntarily notify nuclear accidents not covered by Article 1 with a view to minimizing the radiological consequences (Article 3). Such voluntary notifications are relevant where it is not clear whether Article 1 is applicable to certain situations.[37] For both obligatory and voluntary notifications, the IAEA assumes a role under the Notification Convention as a possible focal point for collecting and disseminating information.

The Assistance Convention sets out an international framework for cooperation among states parties and with the IAEA to facilitate prompt assistance and support in the event of nuclear accidents or radiological emergencies. It requires states to notify the IAEA of their available experts, equipment, and other materials for providing assistance. In case of a request, each state party decides whether it can render the requested assistance as well as its scope and terms. Assistance may be

[34] International Maritime Organization, *Annex to Convention on Facilitation of International Maritime Traffic: Standard* 7.8, (1965).

[35] International Atomic Energy Agency, *Convention on Early Notification of a Nuclear Accident*. 27 October 1986, 1439 UNTS 275.

[36] International Atomic Energy Agency, *Convention on Assistance in the Case of a Nuclear Accident or Radiological Emergency*. 26 February 1987, 1457 UNTS 133.

[37] Andrea Gioia, "Nuclear Accidents and International Law," in *International Disaster Response Law*, ed. A. de Guttry et al. (The Hague: T. M. C. Asser, 2012), 85–102.

offered without costs, taking into account *inter alia* the needs of developing countries and the particular needs of countries without nuclear facilities. Within this framework, the IAEA is requested to use its best endeavors to promote, facilitate, and support the cooperation between states parties provided for in the Assistance Convention (Article 1).

It plays a role in channeling information on assistance between states (Article 2), and the IAEA itself is required to respond to requests for assistance by making available appropriate resources, transmitting promptly such requests to other states and international organizations, and by coordinating the assistance at the international level if requested by the requesting state (Article 2(6)). Other functions of the IAEA include: (a) to collect and disseminate information relating to the types of assistance that could be made available in the event of nuclear accidents or radiological emergencies; (b) to assist a state party or a member state, when requested, in the area of emergency preparedness; (c) to make available resources for conducting an initial assessment of the accident or emergency; (d) to offer its good offices in the event of an accident or emergency; and (e) to establish and maintain liaison with relevant international organizations for obtaining and exchanging relevant information and data (Article 5).

In fulfilling its role and responsibilities under the Notification and Assistance Conventions, the IAEA has developed mechanisms and tools to fulfill its role as a clearinghouse and facilitator for information and assistance among the state parties. In 2005 the IAEA established the Incident and Emergency Centre (IEC) to serve as a global focal point for preparedness and response to nuclear and radiological incidents.[38] The IEC operates on a twenty-four-hour basis to ensure that prompt response may be taken for notifications and request for assistance under the conventions. The IEC manages the global Response and Assistance Network (RANET), which coordinates disaster response capacity of states in case of emergency. It also manages a web-based information exchange platform, the Unified System for Information Exchange in Incidents and Emergencies (USIE), which was established in 2011 to provide a secure platform for communication and coordination between states.[39]

ACCEPTING INTERNATIONAL ASSISTANCE: THE GREAT EAST JAPAN EARTHQUAKE

Before moving onto an analysis of the case study concerning assistance Japan received from the international community through IGOs, it should first be emphasized that assistance from these organizations was not necessarily the

[38] International Atomic Energy Agency, *Incident and Emergency Centre*, www-ns.iaea.org/tech-areas/emergency/incident-emergency-centre.asp.

[39] International Atomic Energy Agency, *Easier Information Exchange Strengthens Emergency Response*, www.iaea.org/newscenter/news/2011/easierinfo.html.

primary or only international assistance that was received.[40] A major contribution in terms of international assistance for immediate disaster response was bilateral assistance from the United States involving 24,500 U.S service members, 189 aircraft, and 24 naval ships. The United States provided assistance in search and rescue, and in delivering goods to the affected region. The U.S. armed forces were especially helpful in quickly restoring the infrastructure at the Sendai Airport and Miyako and Hachinohe ports, which were essential in terms of the logistics for the relief efforts. Bilateral assistance in the forms of search and rescue teams was accepted from a number of other states as well. International NGOs were also active in the relief efforts immediately after the earthquake and tsunami.

Domestic Framework for Accepting International Assistance

An important turning point in Japanese law and administrative practices with respect to receiving international assistance was the Great Hanshin Awaji earthquake (Kobe earthquake) in 1995. At that time, the Japanese government initially declined offers of international assistance in the form of emergency rescue teams and medical teams.[41] Although this reluctance was said to have stemmed from factors such as a strong sense of self-sufficiency, bureaucratic gridlock, and unwillingness of officials to relax domestic rules,[42] the largest reason seems to be simply that the central and local authorities in charge of disaster response had not anticipated accepting international assistance.[43] Due to this lack of preparation, accepting international assistance was perceived to be more burdensome than beneficial.

After the earthquake, efforts were made to rectify the situation through amendments to the domestic legal framework that is relevant to accepting international assistance. The Basic Act on Disaster Control Measures, which sets out the general legal framework for disaster situations was amended to include matters "relating to the acceptance of disaster prevention assistance from overseas" (Article 8(2)(15)) among the list of responsibilities of the state and local government with respect to disasters. The same amendment also created emergency powers for the cabinet to enact an ordinance in order to take necessary steps to accept assistance from overseas, when the Diet is unable to function (Article 109–2). However, the legal framework only provides general provisions that recognize acceptance of international assistance as functions to be fulfilled by the state and the local government. Although some aspects of practical issues such as customs, immigration, and quarantine procedures with respect to incoming international assistance are

40 "Emergency Assistance from Overseas."

41 The Cabinet Office, Government of Japan, *Materials on Lessons Learned from the Great Hanshin Awaji Earthquake* (in Japanese), 1-4-4, www.bousai.go.jp/kyoiku/kyokun/hanshin_awaji/data/detail/pdf/1-4-4.pdf.

42 John Hannigan, *Disasters without Borders* (Cambridge: Polity Press, 2012), 101.

43 "Emergency Assistance from Overseas."

addressed by legal rules and regulations, others are left to be addressed at the administrative level through inter-ministerial understandings.[44]

With respect to how the matter is addressed at the administrative level, the Basic Disaster Management Plan, laid down by the Central Disaster Management Council in accordance with the Basic Act on Disaster Control Measures contains some guidance on this matter.[45] The plan requires for the state to collect and accumulate information on available overseas assistance, to review the possible areas were such assistance may be beneficial, and to establish in advance the necessary measures for accepting assistance. In accordance with the plan, an arrangement has been created in 1998 by the "Liaison Conference among Relevant Ministries on the Acceptance of Assistance from Overseas."[46] This arrangement identifies the possible areas for international assistance, defines the ministries in charge of each area, and provides for the measures to be taken by the ministries concerned. Further, the "Headquarters for Emergency Disaster Control Manual" provides that a team for coordinating incoming international assistance (the C7 team) composed of officials from the relevant ministries.[47]

This was the framework under which international assistance was accepted at the time of the earthquake in 2011. Although customs, immigration, and quarantine procedures are reported to have operated smoothly, considerable difficulties were met with in matching the incoming international assistance with the specific needs of the disaster stricken areas, and with respect to costs and methods for actually delivering the assistance.[48] Although these reflect the difficulties in information gathering and transportation in the immediate aftermath of the disaster and the need to avoid burdening afflicted areas with the unnecessary aid, there have also been suggestions that the government should have relied more extensively on coordination mechanisms under the UN system.[49]

[44] For example, Article 4(5) of the Ordinance for the Import Quarantine of Dogs, etc. provides that dogs may be removed from quarantine facilities in special circumstances such as when necessary for disaster relief. Likewise, Article 15(1)(3) of the Customs Tariff Act on exemption from customs duty for "supplies donated for charity or relief and articles," although not confined to disasters, applies to incoming international assistance.

[45] Central Disaster Management Council, *Basic Disaster Management Plan*, www.bousai.go.jp/taisaku/keikaku/pdf/20111227_basic_plan.pdf (in Japanese) 2012, 31.

[46] *Arrangement by the Liaison Conference among Relevant Ministries on Acceptance of Assistance from Overseas*. To the author's knowledge, this document has not been publicized, although its existence has been mentioned in official documents and in articles written by public officials.

[47] The operation of this system is described in a material prepared by the Cabinet Office. The Cabinet Office, Government of Japan, *The Activities of the C7 Team (Coordination Team for Coordinating Incoming International Assistance) within the Headquarters for Emergency Disaster Control (Special Headquarters for Assisting the Lives of Disaster Victims)* (in Japanese), www.bousai.go.jp/oukyu/higashinihon/6/pdf/naikakufu1.pdf, 2–7.

[48] Ibid., 8–12.

[49] Yutaka Katayama, "International Emergency Assistance during the Great East Japan Earthquake and the Ministry of Foreign Affairs," (in Japanese) *Journal of International Cooperation Studies* 20 (2013): 60–62.

Another issue was the adequacy of the domestic laws regarding legal issues arising in connection with the activities of foreign rescue and medical teams. For example, the issue of qualification to practice under the Medical Practioners' Act was addressed in a notice dated March 14, 2011 from the Ministry of Health, Labor, and Welfare addressed to the three prefectural governments (Iwate, Miyagi, and Fukushima).[50] Whether such ad hoc arrangements suffice in creating a stable legal environment for foreign personnel should be reexamined;[51] similar issues exist for how damages caused by foreign personnel during their relief efforts should be handled as a matter of law.

Consequently, there still remain improvements to be made to the Japanese legal system with respect to accepting international assistance. Although some discussions for reform are underway, it seems that some aspects of the domestic framework for accepting international assistance in Japan will continue to be governed by flexible discretional measures to be taken by the relevant ministries as an emergency measure. However, what is important is that the domestic framework for accepting international assistance has significantly evolved from where it was in 1995. The experience of Japan with international assistance to major disasters has changed how the assistance is perceived; accepting international assistance is now firmly included in the inventory of domestic measures to be taken in the event of a major disaster in Japan.

Utilizing the International Humanitarian Assistance System

In receiving international assistance immediately after the 2011 earthquake, the coordinating mechanism of the international humanitarian assistance system was also utilized. The Japanese government requested the UNDAC team to be sent to Japan. This was the first time that a full UNDAC team was deployed to a developed country for disaster mitigation and response.[52] The team arrived in Japan on March 13, two days after the earthquake.[53] The team then immediately set up a coordination cell in Tokyo, which provided information overview of the crisis through daily "situation reports" disseminated through "ReliefWeb" operated by OCHA.[54]

[50] *On the Treatment of Medical Practice by Persons with a Foreign Medical License*, stating that "... since the Medical Practitioners' Act does not envisage emergency times such as the present, engaging in essentially necessary medical practice for disaster victims would constitute justifiable cause for noncompliance with the law under Article 35 of the Penal Code."

[51] For example, the IFRC's *Guidelines for the Domestic Facilitation and Regulation of International Disaster Relief and Initial Recovery Assistance* calls upon affected states to "establish expedited procedures" for temporary recognition of professional qualifications of foreign medical personnel (Art. 16(1)(c)).

[52] Yosuke Okita, "The UNDAC Team in Response to the Tohoku-Pacific Ocean Earthquake," (in Japanese) *Journal of Japan Society for Natural Disaster Science* (2011): 279.

[53] Ministry of Foreign Affairs of Japan, "Arrival of a team of UNDAC," www.mofa.go.jp/announce/announce/2011/3/0314_06.html.

[54] Okita, "The UNDAC Team in Response," 283.

These reports proved to be a valuable source of reliable information for international assistance efforts, especially since most of the information released by the Japanese government was written in Japanese. The team also advised the Japanese government on coordination of international offers of humanitarian assistance, and provided support for urban search-and-rescue efforts in the Tohoku region.[55] The UNDAC team continued its activities until March 23 when its operation was taken over by OCHA, which continued activities in Japan until April 2.

In total, Japan accepted international assistance in the form of twenty-four teams from foreign states, and five teams from international organizations.[56] The majority of the former were search-and-rescue teams that arrived in Japan one to three days after the earthquake and engaged in rescue operations for a short period, and the rest were medical support teams that arrived later on and stayed for longer periods. The assistance from international organizations, in addition to the UNDAC and OCHA team referred to earlier, included a team from the World Food Programme (WFP), a team of experts from the IAEA at the request of Japan, and the Food and Agriculture Organization/IAEA team comprised of experts on food monitoring.[57] In dealing with the disaster at hand, the Japanese government requested assistance for specific areas where it saw international organizations as better-equipped, such as advice on the coordination of incoming international assistance, disseminating information, and ensuring transparency in the eyes of the international community.

The fact that the UNDAC team was requested by Japan indicates that accepting international assistance could be a difficult task even for developed states.[58] The operation of the UNDAC team in Japan was an exceptional one in that the team only provided advice to the Japanese government, and the coordination activities itself were conducted by the government. However, the Japanese government was eager to get involved with the international humanitarian assistance system, which was, in itself, a marked change from Tokyo's policy at the time of the Great Hanshin Awaji earthquake in 1995. This reflects the fact that the international humanitarian assistance system, with its coordination mechanisms and standards, has come to be regarded as an important asset in the overall efforts in immediate disaster response.

International Assistance in Responding to the Nuclear Disaster

The response to the situation after the 2011 earthquake was made difficult by the fact that it was a "complex disaster:" – a combination of a natural disaster followed

[55] United Nations, Office for the Coordination of Humanitarian Affairs, *OCHA Annual Report 2011* (2011): 3.

[56] "Emergency Assistance from Overseas." [57] Ibid.

[58] For the situation in the United States after the Hurricane Katrina in 2005, see for example, Alejandra de Urioste, "When Will Help be on the Way? The Status of International Disaster Response Law," *Tulane Journal of International and Comparative Law* (2006): 198.

by a nuclear disaster. The IAEA played a substantial role with respect to the responses to the nuclear disaster at the Fukushima Daiichi NPP. In carrying out its functions under the Notification and Assistance Conventions, the IAEA served a valuable role as a facilitator for emergency information and assistance, both disseminating information with respect to the situation in Fukushima to member states, and providing information on available assistance to the Japanese government.

Since the earthquake, the Japanese authorities have kept close contact with the IAEA's Incident and Emergency Centre (IEC). In response, IAEA had offered its "good offices" to Japan even before the grave situation at the nuclear power plants came to be known.[59] On March 16, 2011, the Japanese government requested information on remotely controlled robots, ground vehicles, and aerial vehicles owned by contracting parties to the Assistance Convention through the IAEA.[60] The IAEA, through its IEC forwarded the request for assistance to contact points in the United States and France.[61] Responses were received after March 17, and the Japanese government received necessary equipment. These roles played by the IAEA are based on Article 5 of the Assistance Convention, which sets out the functions of the IAEA with respect to the convention.[62]

However, while the Japanese government has continued to work closely with the IAEA in discharging these functions, it has been reluctant to regard its interactions with the international organization as obligatory under the Notification Convention. Compliance with the Notification Convention became an issue with respect to the intentional release of low-level radioactive water to secure storage space for the large amount of high-level radioactive water that had resulted from emergency operations at the nuclear plant.[63] The measure itself, and the fact that the Japanese government was belated in informing neighboring states about the measure, aroused some concerns with respect to the lawfulness of the measure.[64] As for the

59 International Atomic Energy Agency, *IAEA Activities in Response to the Fukushima Accident*, (GOV/INF/2011/8), 5.

60 Investigation Committee on the Accident at the Fukushima Nuclear Power Stations of Tokyo Electric Company, *Interim Report* (in Japanese), 2011, 425.

61 International Atomic Energy Agency, *Response to the Fukushima Accident*,.20.

62 International Atomic Energy Agency, *Convention on Assistance*. Collecting and disseminating information on experts, equipment, and material is provided under paragraph (a)(i) of the article, and offering "Good Offices" to states in the event of a nuclear accident is provided in paragraph (d).

63 See Tokyo Electric Power Company, "Press Release (Apr 04,2011): Discharge of low level radioactive accumulated water in the Fukushima Daiichi Nuclear Power Station to the sea (2nd Release)," www.tepco.co.jp/en/press/corp-com/release/11040404-e.html.

64 For a critical assessment of the situation under law, see Yen-Chiang Chang and Yue Zhao, "The Fukushima Nuclear Power Station incident and marine pollution," *Marine Pollution Bulletin* 64 (2012): 897–901. Although a full discussion of whether the measures taken were consistent with international law is beyond the scope of this paper, it should at least be noted that one of Chang and Zhao's conclusions that the measure was in violation of the London Convention is unfounded since release of material from land does not fall under the definition of "dumping," which is the subject of regulation under the Convention.

notification scheme under the Notification Convention, the Japanese government has taken the position that the obligation under the Notification Convention to inform other states did not apply to the case, and that the notification made to the IAEA shortly before the discharge and to neighboring states slightly after the discharge was a voluntary notification pursuant to Article 3 of the Notification Convention.

This was based on the interpretation that the measure taken was not a "release of radioactive material ... that could be of radiological safety significance for another State" (Article 1(1)) to which the obligation to notify under Article 2 applies. Similarly, obligation under Article 198 of the UN Convention on the Law of the Sea (LOSC), which provides for immediate notification to other states and to competent international organizations in cases "in which the marine environment is in imminent danger of being damaged or has been damaged by pollution," was also considered inapplicable.[65] With regard to the incident however, Foreign Minister Matsumoto stated in the Diet that the measure was "not in line with the object of the LOSC, which places a general obligation to prevent pollution of the marine environment from any source."[66] Moreover, the report by the official Accident Investigation and Review Committee concluded that the "even if there was no obligation to notify under the convention, it was necessary by reason to inform the neighboring countries concerned."

While discussions in Japan with respect to the Notification Convention have focused on the intentional discharge incident, little attention has been given to the legal status of the notifications made in the course of contacts with the IAEA that had begun immediately after the earthquake. However, when asked whether the government had ever officially made a notification under Article 2 of the Notification Convention, Matsumoto stated "the important thing is that we have continued to provide notification corresponding to those that would be required under Article 2."[67] This answer seems to imply that the official position of the Japanese government is that while all notifications to IAEA were intended as voluntary notifications under Article 3 of the Notification Convention, they would nevertheless be also valid as notifications under Article 2 in case it was considered that the provision would apply. Also with respect to the Assistance Convention, Matsumoto stated that "although we have received various assistance from foreign States, we do not understand it as assistance based on the

[65] Investigation Committee on the Accident at the Fukushima Nuclear Power Stations of Tokyo Electric Company, *Interim Report*, 420. The International Maritime Organization has assumed its role as "competent international organization" under Article 198 of the Convention. Secretariat of the IMO, *Implications of the United Nations Convention on the Law of the Sea for the International Maritime Organization*, LEG/MISC.6, Annex, 2009, 12.

[66] Records of the Foreign Affairs Committee of the House of Representatives (in Japanese), 177th Session, 2011, Vol.6, 17.

[67] Ibid., Vol.7, 21.

Assistance Convention," when asked in the parliamentary discussions whether the Assistance Convention was activated.[68]

Apart from the function exercised by the IAEA, another area where a specialized international organization may have had a facilitative role under an international convention was the emergency release of low-level radioactive water into the sea. Although the discharge from the nuclear plant itself is outside the scope of the London Convention on dumping, both the original 1972 Convention and the 1996 Protocol provide an exception in emergencies, and assigns a role for the IMO in this regard.[69] Article 8(2) of the Protocol provides that exceptional permits may be issued by contracting parties for "emergencies posing an unacceptable threat to human health, safety, or the marine environment and admitting of no other feasible solution." To invoke this provision, the contracting parties are to consult countries that may be affected and the IMO. The IMO, after consultations, will recommend the most appropriate procedures to adopt, and the contracting party is to "follow these recommendations to the maximum extent feasible consistent with the time within which action must be taken and with the general obligation to avoid damage to the marine environment." In the actual situation, however, there are no signs that the provision was considered as a policy option. This may be explained by the urgency of the situation, and the fact that the discharge directly from the facilities on land was considered to be the most feasible measure.

CONCLUSION

The experience of Japan in receiving assistance from the international community has mixed implications with respect to the relationship between the international community and the state in times of disaster. The Japanese government, although faced with much difficulty from the enormous disaster damage, had in the end effectively dealt with the disaster at hand. It is recognized that, after all, response to disasters occurring within their territory are the primary responsibility of the territorial state; a state with sufficient capabilities will seek to assume full control in dealing with disasters within their territory. This is partly reflected in the fact that the Japanese government has attempted to distance itself, wherever legitimately possible, from the legal implications of the international framework, even though it has maintained a close working relationship with relevant international organizations. This attitude may have stemmed from the view that obligations under international agreements limit the discretion of the state in how to most effectively

[68] Records of the Committee on Foreign Affairs and Defense of the House of Councilors (in Japanese), 2011, 177th session, Vol.7, 12.

[69] 1996 Protocol to the Convention on the Prevention of Marine Pollution by Dumping of Wastes and Other Matter, 1972, 36 *ILM* 7. A similar provision appears in Article 5(2) of the 1972 London Convention.

respond to the disaster situation, and only place additional unnecessary burdens on an already dire situation.

However, the experience of Japan also sheds light on the possibility of mechanisms for international assistance to be perceived as reciprocal instruments to further state interests.[70] It was seen that through experience with incoming international assistance, the perception of the Japanese government to international disaster assistance has significantly changed; this is evidenced by the evolution of the domestic framework for accepting aid and its willingness to receive whatever assistance it requires from the international community. Japan has traditionally regarded itself as a leader in disaster management, and it has regarded its primary role in the international community as that of a contributor to international cooperation against disasters. There is now increasing recognition that that accepting international assistance in times of disaster is of importance even if Japan is better prepared for major disasters; this in turn has led to an increased sense of reciprocity.

Of course, there are significant limits to what may be drawn from this case study. The 2011 earthquake was a situation where, generally speaking, the territorial state had full capabilities and was entirely willing to exercise its responsibilities. The complementary role to be played by assistance from the international community through international organizations was not necessarily large in proportion to the overall disaster relief efforts. In this sense, the earthquake in Japan did not involve the sharp conflict of interests that constitute the most contested themes in the role of the international community and international law in times of disaster: when the disaster-affected state is unable, or especially unwilling, to respond to a major disaster within their territory.[71] Moreover, not all states are as prone to disasters as Japan is, especially with respect to earthquakes. Since regional disparities in the occurrence of natural hazards create unequal distribution of relative gains as a result of international cooperation, the experience of Japan may not be generalized to better-situated states. As Great Britain indicated with respect to the IRU in the early twentieth century, states that are less affected by disasters may continue to conceive of international disaster response only in terms of *giving* assistance to other disaster-stricken states.

[70] See also Watabe Masaki, "'Development of the International Humanitarian System and the East Japan Great Earthquake," (in Japanese) *Yearbook of World Law*, 32 (2013): 211–212, discussing a trend toward humanitarian assistance and disaster aid as mechanisms for mutual help.

[71] For a discussion of whether the concept of Responsibility to Protect (R2P) should apply to natural disasters, see for example, T. R. Saechao, "Natural Disasters and the Responsibility to Protect: From Chaos to Clarity," *Brooklyn Journal of International Law* 32 (2007): 663–707; J. Benton Heath, "Disasters, Relief and Neglect: The Duty to Accept Humanitarian Assistance and the Work of the International Law Commission," *New York University Journal of International Law and Politics* (2011): 419–477.

Although the preceding considerations indicate the significant limits of the 2011 earthquake and disaster in Japan as a case sufficient to study the state of international law with respect to disasters, the example does provide some insights on the foundations of so-called International Disaster Response Law. Although development of this body of law and the role of international community under the law have increasingly been associated with notions such as humanity and solidarity,[72] humanitarian assistance is not the only context from which the international framework may be conceptualized. The disparity among states in their capacity to respond to disasters has tended to divide states into the usual givers and usual receivers. This asymmetrical character of the humanitarian context has in some ways exacerbated the tension between the territorial sovereignty of the affected state and the effective delivery of aid to the affected peoples.

In contrast, for certain types of disasters, especially with transnational effects, states have entered into agreements to mutually limit their sovereignty to attain ends they cannot by themselves; in this context, more "classical" functional cooperation through international organizations have played a prominent role. More generally, states may in the long run come to perceive disasters as threats to their national security and enter into cooperation, entrusting some roles to the international community. While the international community will undoubtedly continue to face hard cases in the humanitarian context, more roles for international assistance and international law in disaster situations would be anticipated in a world where more states regard the framework for international assistance as valuable assets in exercising their response against disasters.

[72] Abdul G. Koroma, "Solidarity: Evidence of an Emerging International Legal Princple," in *Coexistence, Cooperation and Solidarity: Liber Amicorum Rüdiger Wolfrum*, ed. Holger P. Hestermeyer et al. (Leiden: Martinus Nijhoff, 2011), 123; Boniface Okere and Ernest M. Makawa, "Global Solidarity And The International Response To Disasters," in *The International Aspects of Natural and Industrial Catastrophes*, ed. David D. Caron and Charles Leben (The Hague: Nijhoff, 2001), 429–456.

16

International Investment Law and Disasters: Necessity, Peoples, and the Burden of (Economic) Emergencies

Ibironke T. Odumosu-Ayanu

Disasters – natural and economic – are not new but international law's response to disasters has been limited. International investment law is a part of this limited response system. The debate in international investment law about the scope of the plea of necessity at customary international law, relevant emergency clauses in investment treaties, and their application and limitations in the event of economic emergencies, and the suffering that follows such emergencies, is the focus of this chapter. Necessity's component requirements, and related considerations like temporal limitations and compensation that impact the necessity plea, do not permit a clear application to economic emergencies and disasters. Given its nature, necessity appears unable to capture the impossible choices that economic emergencies present. Balancing national interests and economic emergencies against honoring commitments to foreign investors does not present an easy decision-making process. Juxtaposed against the commitment to foreign investors, emergencies present difficult choices for decision makers in international investment law. Hence, in order to adopt a comprehensive approach to economic emergencies, it is necessary to take the interests and suffering of foreign investors, states, and peoples seriously. In order to take suffering seriously, and not defer difficult questions to applications of the plea of necessity, the allocation of risk of economic emergencies must be considered in specialized regimes including applicable investment arrangements. At a time when the world's economy is exposed to emergencies, it is essential to craft clearly articulated responses to emergencies and disasters of serious magnitude before they occur.

INTRODUCTION

At a time when the world faces serious natural disasters and economic emergencies, the theme of this volume, international law and disasters, is appropriately timed. While the incidence of these disasters is readily acknowledged as a cause for concern, international law's limited ability to respond adequately is somewhat alarming. In light of the socioeconomic emergencies that states have faced in recent times, the attendant losses that investors encounter, and the suffering that peoples are confronted with, international investment law is becoming familiar with economic emergencies. This encounter between international investment law and economic disasters has brought the limitations of international law's ability to respond to these disasters to the fore.

The plea of necessity in international law and investment treaty provisions that attempt to provide some responses to economic and other emergencies have been the subject of intense debate.[1] The focus of analyses on this subject has been on states *qua* governments and foreign investors. This chapter analyzes the restrictive nature of the necessity plea and the limited interpretation of emergency provisions in investment treaties in light of peoples' suffering in emergencies.

International law is not a stranger to human suffering. From international human rights law to international humanitarian law, international law sometimes engages difficult questions of suffering and seeks to regulate the activities of states (and other actors) in order to mitigate such suffering. In this regard, Professor Upendra Baxi has written about "taking suffering seriously."[2] In fact, notions of suffering inform Baxi's analysis of human rights.[3] He trains his lenses on human suffering in articulating the future of human rights. While some international human rights scholars and activists are able to readily account for the need to take suffering seriously, similar assumptions may not necessarily hold true for other areas of international law like international investment law. International investment law mostly defines the obligations of states toward foreign investors. Hence, considerations of broader issues of public importance often involve difficult conversations regarding state obligations and public interest.[4] However, when emergencies occur (human) suffering becomes more readily apparent. From investors losing large sums of money to citizens of host states facing untold hardship due to a natural or economic emergency, suffering is very real. In order to adopt a holistic view of the impacts of emergencies, international investment law must be able to take suffering seriously. And to "take suffering seriously we must address both individual and group oppression."[5]

1 On necessity in international law, see generally, Andrea Bjorklund, "Emergency Exceptions: State of Necessity and 'Force Majeure'," in *The Oxford Handbook of International Investment Law*, ed. Peter Muchlinski, Federico Ortino, and Christoph Schreuer (Oxford University Press, 2008), 459.

2 Upendra Baxi, "Taking Suffereing Seriously: Social Action Litigation in the Supreme Court of India," *Third World Legal Studies* (1985): 107.

3 Upendra Baxi, *The Future of Human Rights* 3rd ed. (New Delhi: Oxford University Press, 2008).

4 See generally, Gus van Harten, *Investment Treaty Arbitration and Public Law* (Oxford University Press, 2007).

5 William F. Felice, "The Case for Collective Human Rights: The Reality of Group Suffering," *Ethics and International Affairs* (1996): 48.

Making difficult choices is one of the challenges of contemporary interpretations of necessity as a viable defense to states' failure to fulfill their international obligations.[6] Even if these kinds of choices were not traditionally part of the defense of necessity as Robert Sloane contends,[7] adjudicating serious economic emergencies involves balancing diverse, perhaps even irreconcilable interests – the interests of investors to whom the state owes an international obligation and local peoples who also suffer the consequences of the disasters. It is therefore not surprising that international law's treatment of disasters has been somewhat hesitant as the questions that disasters raise are inherently difficult ones that defy easy responses. In fact, international investment arbitration tribunals' treatment of economic crises has invited the comment that states' long-term willingness to participate in the system and the system's very legitimacy may be threatened.[8]

The responses to economic emergencies in international investment law have been inadequate in comprehensively addressing the challenges that states, foreign investors, local peoples, and adjudicators face. The plea of necessity, as articulated in Article 25 of the International Law Commission's (ILC) Draft Articles on Responsibility of States for Internationally Wrongful Acts 2001 (ILC State Responsibility Draft Articles),[9] is restrictive in its application and economic emergencies do not seem to be its forte. Investment treaty law does not also provide conclusive answers, for non-precluded measures (NPM) clauses vary from treaty to treaty and are completely omitted in some.[10] Article XI of the Treaty between the United States of America and the Argentine Republic Concerning the Reciprocal Encouragement and Protection of Investments (US-Argentina BIT), an example of the NPM clause, is particularly relevant in light of this chapter's analysis of arbitral decisions on Argentina's economic crisis.[11] Even with cases that have relied on the

6 See Robert Sloane, "On the Use and Abuse of Necessity in the Law of State Responsibility," *American Journal of International Law* (2012): 447.

7 Ibid.

8 William W. Burke-White, "The Argentine Financial Crisis: State Liability under BITs and the Legitimacy of the ICSID System," *Asian Journal of WTO & International Law and Policy* (2008): 210.

9 United Nations, "Draft Articles on Responsibility of States for Internationally Wrongful Acts." *Report of the International Law Commission on the Work of its Fifty-Third Session*, UN GAOR 56th Session Supplement No. 10, UN Doc. A/56/10. 2001. www.un.org/documents/ga/docs/56/a5610.pdf. For commentaries on the draft articles, see "Draft Articles on Responsibility of States for Internationally Wrongful Acts, with Commentaries, 2001," http://untreaty.un.org/ilc/texts/instruments/english/commentaries/9_6_2001.pdf.

10 The non-precluded measures language is adopted following authors like Burke-White and von Staden. See for example, William W. Burke-White, and Andreas von Staden, "Investment Protection in Extraordinary Times: The Interpretation and Application on Non-Precluded Provisions in Bilateral Investment Treaties," *Virginia Journal of International Law* 48 (2008): 307.

11 *Treaty between the United States of America and the Argentine Republic Concerning the Reciprocal Encouragement and Protection of Investment* (14 November, 1991, entered into force 20, October 1994), (1992) 31 I.L.M. 128 [*US-Argentina BIT*]. Article XI provides: "This Treaty shall not preclude the application by either Party of measures necessary for the

US-Argentina BIT where arbitral tribunals have been presented with identical treaty provisions, tribunals have adopted conflicting interpretations.[12]

Essentially, this area of the law and its application to economic emergencies requires dedicated attention, especially if international law is to take human suffering seriously. Clarification of investment treaty obligations in cases of emergencies is necessary.[13] Such clearer definition may be informed by the challenges that interpretations of Article 25 and the NPM clauses in investment treaties have posed. Even if Article 25 did not pose challenges in terms of scope and interpretation, Article 25 is not well suited to economic emergencies. Hence it is necessary to pre-negotiate interests, obligations, risk, and liabilities in investment treaties and contracts. These agreements, especially the treaties, have not been without controversy. But since states and investors continue to insist on adopting investment treaties, it is necessary to ensure that they account for the interests of all relevant stakeholders.

ECONOMIC EMERGENCIES AND INTERNATIONAL INVESTMENT LAW

Investment Arbitration

Argentina's crisis at the turn of the twenty-first century, which forms the basis of this chapter's analysis, has been the subject of much scholarly analysis.[14] As a result, the cause of the crisis and the events that occurred will not be repeated here. However, it is necessary to note that in addition to the losses that investors suffered, the people of Argentina also suffered untold hardship. There was intense poverty, chronic unemployment, hunger, a decrease in social welfare, accelerated social inequality, protests, and several deaths. The crisis, which has been labeled "financial," was in fact, a socioeconomic crisis. Indeed the crisis also had political dimensions because

maintenance of public order, the fulfilment of its obligations with respect to the maintenance or restoration of international peace or security, or the protection of its own essential security interests."

12 See generally, Diane A. Desierto, "Necessity and 'Supplementary Means of Interpretation' for Non-Precluded Measures in Bilateral Investment Treaties," *University of Pennsylvania Journal of International Law* 31 (2010): 827.

13 The United Nations Conference on Trade and Development has considered national security measures in investment treaties. See UNCTAD. *The Protection of National Security in IIAs*. New York, Geneva: United Nations, 2009. See generally, UNCTAD. *World Investment Report 2012: Towards a New Generation of Investment Policies* (New York, Geneva: United Nations, 2012).

14 There have been several renditions of Argentina's socioeconomic crisis at the turn of the century, including one by the present author. See Paul Blustein, *And the Money Kept Rolling in (And Out): Wall Street, the IMF and the Bankrupting of Argentina* (New York: Public Affairs, 2005), 1; Ibironke Tinuola Odumosu, *ICSID, Third World Peoples and the Re-Construction of the Investment Dispute Settlement System* (Doctoral Dissertation: University of British Columbia, 2010): 212–20.

it became so severe that Argentina had five presidents within two weeks. The crisis clearly engendered serious suffering in Argentina.

This suffering, investors' losses, and the applicability of the plea of necessity and/or NPM clauses in treaties have been difficult for arbitral tribunals to reconcile. Investment arbitration tribunals have encountered some fundamental questions regarding economic emergencies. This section discusses five of those questions: the legal/extra-legal nature of emergency measures; applicability of necessity and NPM clauses to economic emergencies; whether Article 25 of the ILC State Responsibility Draft Articles is reflective of customary international law; the relationship between Article 25 and NPM clauses; and the (non)self-judging nature of NPM clauses.

Parties may raise defenses to economic emergencies under multiple regimes.[15] In recent arbitral jurisprudence, economic emergencies, measures taken to address these emergencies, and the arbitrability of these measures, have been debated in cases initiated against Argentina following its crisis at the turn of the century. As a defense to these claims, Argentina has raised *inter alia* three pleas – necessity under Argentine law, treaty provisions that exempt liability including NPM clauses, and necessity under international law. Necessity, as codified in Article 25 of the ILC Draft Articles, has emerged as a major plea in defense of measures taken to address economic emergencies that would otherwise be in breach of obligations to foreign investors. (See Appendix)

The ILC State Responsibility Draft Articles define necessity "as those exceptional cases where the only way a State can safeguard an essential interest threatened by a grave and imminent peril is, for the time being, not to perform some other international obligation of lesser weight or urgency."[16] The plea of necessity precludes wrongfulness.[17] Necessity has been used in various circumstances including military, humanitarian, environmental, and economic circumstances.[18] Its use in economic circumstances has been recently publicized by Argentina's crisis. Argentina's crisis, its International Centre for Settlement of Investment Disputes (ICSID) cases, and academic literature on these issues and the ILC State Responsibility Draft Articles have contributed to the five questions discussed here.

[15] See August Reinisch, "Necessity in Investment Arbitration," *Netherlands Yearbook of International Law*, 2010: 148–152.

[16] ILC Draft Articles, Commentaries, 80.

[17] Ibid. On necessity as an excuse or justification, see Sloane, "On the Use and Abuse," 472 (noting that Article 25 "recasts necessity as a justification rather than an excuse"). For a more detailed discussion, see ibid., 482–486. See generally, Vaughan Lowe, "Precluding Wrongfulness or Responsibility: A Plea for Excuses," *European Journal of International Law* 10 (1999): 405.

[18] Sarah F. Hill, "The 'Necessity Defense' and the Emerging Arbitral Conflict in its Application to the U.S.-Argentina Bilateral Investment Treaty," *Law and Business Review of the Americas*, 2007: 551–557.

First, the legal or extra-legal nature of emergencies remains a fundamental debate.[19] These questions are often debated as part of jurisdictional challenges before international arbitral tribunals.[20] ICSID tribunals often reject these jurisdictional objections by classifying aspects of emergency measures as impacting on foreign investment. The argument is that while the general measures that states adopt in response to crises may not be subject to arbitral jurisdiction, the impacts of these measures on the investments of foreign investors are within the jurisdiction of arbitral tribunals.

In the decision on jurisdiction in *CMS Gas Transmission Company v. The Republic of Argentina* (*CMS*), the tribunal expressed the view that the general measures that Argentina adopted were not the subject of the proceedings but the extent to which they violated "specific commitments" that had been given to investors.[21] Although tribunals correctly note that a country's "economic choices" are not arbitrable,[22] one cannot separate the impacts of these choices on foreign investors from the economic measures themselves.

Hence, investment arbitration tribunals are charged with deciding questions that are highly significant not only for foreign investors and states but also for peoples of the state facing the emergency. In the same breath, tribunals exclude measures adopted to alleviate peoples' suffering from their decision making while making decisions about the impacts on foreign investors of these same measures adopted to mitigate peoples' suffering. Tribunals inevitably place themselves in the province of measures that impact on human suffering but refuse to pronounce on such suffering because international investment law does not empower them to do so.

Second, until Argentina's investment arbitration cases, it was not entirely clear whether the plea of necessity applies to economic emergencies.[23] Although there were arguments akin to necessity in the *Russian Indemnity Case*,[24] the *Rainbow Warrior* arbitration cast doubts on the plea of necessity as recently as 1990.[25] Even though the application of necessity to economic emergencies has now been firmly

[19] Reinisch, "Necessity," 140.

[20] Ibid., 141. Reinisch refers to Argentina's challenges to ICSID jurisdiction on the basis that the emergency questions did not qualify as a legal dispute and to decisions of the International Court of Justice in the Nuclear Weapons Case (*Legality of the Threat or Use of Nuclear Weapons*), *Advisory Opinion*. (International Court of Justice, 1996) and the *Legal Consequences of the Construction of a Wall in the Occupied Palestinian Territory*, *Advisory Opinion*. (International Court of Justice, July 9, 2004). [*Wall Case*].

[21] *CMS Gas Transmission Company v. The Republic of Argentina (Decision of the Tribunal on Objections to Jurisdiction)*. ICSID Case No. ARB/01/8 (July 17, 2003).

[22] Ibid., at par. 29. [23] Reinisch, "Necessity," 145.

[24] *Russian Claim for Interest on Indemnities (Damages Claimed by Russia for Delay in Payment of Compensation Owed to Russians Injured During the War of 1877–1878) Russia v. Turkey*. (November 11, 1912).

[25] *Case Concerning the Differences between New Zealand and France Concerning the Interpretation of Application of Two Agreements, Concluded on July 9, 1986 between the Two States and which Related to the Problems Arising from the Rainbow Warrior Affair*. (April 30, 1990).

established, some commentators support the view that necessity is ill-suited to financial crises.[26] Argentina's arbitral cases found that the plea of necessity may apply to economic emergencies although most tribunals were of the view that Argentina's crisis did not satisfy all the cumulative requirements for invoking the necessity plea.

While some may contend that an economic emergency does not rise to the level that it could engender the application of a plea such as necessity, Argentina's crisis suggests otherwise. During the course of Argentina's emergency and the period before and after the emergency, the impact of the crisis on the people of that country was immense and demonstrated that economic emergencies hold the potential of causing serious devastation. At the same time, necessity as currently construed may not be the plea of choice as the decisions analyzed in this chapter demonstrate.

Indeed, necessity appears ill-suited to an emergency of the nature that occurred in Argentina, for Argentina's cases demonstrate that in order to take suffering seriously, it is necessary to take deliberate steps to accommodate socioeconomic emergencies within a legal framework that deliberately contemplates the multiple layers of suffering that occur as a result of emergencies. For economic emergencies, the plea of necessity does not appear able to contemplate these multiple layers of suffering.

Third, the status of Article 25 as customary international law is debated. While it is readily acknowledged in judicial and quasi-judicial decisions that Article 25 is reflective of customary international law,[27] Daniel Bodansky and John Crook observe that the article appears to be self-professing.[28] They note that the "confident, direct quality" of the ILC State Responsibility Draft Articles "adds to their seeming authority and certainty ... their seeming clarity and formal presentation may lead readers to take the articles too much at face value, believing that they indeed state 'the law.'"[29] According to the authors, the impression could be deceptive as the State Responsibility Draft Articles sometimes present elements of "progressive development" or the ILC's judgments regarding the state of existing law.[30]

[26] Michael Waibel, "Two Worlds of Necessity in ICSID Arbitration LCMS and LG&E," *Leiden Journal of International Law*, 2007: 637.

[27] On Article 25 of the ILC Draft Articles and customary international law, see *Case Concerning the Gabcikovo-Nagymaros Project (Hungary/Slovakia)*. (International Court of Justice, September 25, 1997); *Wall Case*. par. 140. Arbitral tribunals also share a similar view. See for example, *CMS Gas Transmission Company v. The Argentine Republic*. ICSDI Case No. ARB/01/08 (May 12, 2005). (Noting that "Article 25 of the Articles on State Responsibility adequately reflects the state of customary international law on the question of necessity.")

[28] Daniel Bodansky and John R. Crook, "Symposium: The ILC's State Responsibility Articles – Introduction and Overview," *American Journal of International Law*, 2002: 773.

[29] Ibid., 787. [30] Ibid.

Bodansky and Crook adopt Article 25 in their illustration arguing that it "had a bootstrapping quality, helping to shape the law to match the draft."[31] They argue that the International Court of Justice (ICJ) gave necessity a "powerful boost" in the *Case Concerning the Gabcikovo-Nagymaros Project (Hungary/Slovakia)*.[32] Referencing the ICJ's judgment, Bodansky and Crook note that "legal development had a circular quality: the ILC's draft helped produce that Judgment. Article 25 thus played an important role in its own validation."[33]

Sloane argues that "[b]efore an earlier draft of Article 25 led the ICJ to analyze necessity in its explicit terms, few, if any, decision makers had understood necessity – as a defense in the law of state responsibility – in the terms set forth in that article."[34] For Sloane, from necessity's "origins in the classical law of nations to the last decade of the twentieth century, international custom recognized a narrow defense of necessity, which was based on the supposedly natural right of states to preserve themselves."[35] Necessity was a "highly circumscribed defense in the law of state responsibility" that focused on the "very existence" of states rather than any of the state's essential interests.[36] From the perspective of a state's peoples suffering the impact of a socioeconomic catastrophe, necessity as presented under Article 25 might have been relaxed in terms of its core application to essential interests, but it still includes many questions that are not definitively answered.

Fourth, the relationship between Article 25 and treaty NPM clauses is debated in investment arbitration. It appears that the requirements of Article 25 of the ILC State Responsibility Draft Articles are more stringent than the NPM clause in the US-Argentina BIT. Of the first four ICSID decisions on Argentina's plea of necessity, *CMS*, *Enron Corporation and Ponderosa Assets, L.P. v. Argentine Republic* (*Enron*),[37] and *Sempra Energy International v. Argentine Republic* (*Sempra*)[38] rejected the defense.[39] In *Suez, Sociedad General de Aguas de Barcelona S.A., and InterAgua Servicios Intergrales del Agua S.A. v. The Argentine Republic*, the tribunal also rejected the defense.[40] However, the tribunal in *LG&E Energy Corp., LG&E*

[31] Ibid., 788.

[32] *Gabcikovo-Nagymaros Project Case.*; Bodansky and Crook, "Symposium," 788.

[33] Ibid. [34] Sloane, "On the Use and Abuse," 470. [35] Ibid., 470–471. [36] Ibid. at 471.

[37] *Enron Corporation and Ponderosa Assets, L.P. v. Argentine Republic.* ICSID Case No. ARB/01/3 (May 22, 2007). http://italaw.com/documents/Enron-Award.pdf.

[38] *Sempra Energy International v. Argentine Republic (Award).* (ICSID Case No. ARB/02/16) (September 28, 2007) https://icsid.worldbank.org/ICSID/FrontServlet?requestType=CasesRH&actionVal=showDoc&docId=DC694_En&caseId=C8 ["*Sempra*, Award"].

[39] For a detailed analysis of the decisions in *CMS*, *LG&E*, *Enron*, and *Sempra*, see Jose E. Alvarez and Kathryn Khamsi, "The Argentine Crisis and Foreign Investors: A Glimpse into the Heart of the Investment Regime," in *The Yearbook on International Investment Law and Policy 2008/2009*, ed. Karl P. Sauvant (New York: Oxford University Press, 2009), 379.

[40] *Suez Sociedad General de Aguas de Barcelona S.A. and InterAgua Servicios Intergrales del Agua S.A. v. The Argentine Republic.* ICSID Case No. ARB/03/17 (July 30, 2010) http://italaw.com/documents/SuezInterAguaDecisiononLiability.pdf. At par. 243, the tribunal noted that it "denies Argentina's plea of the defense of necessity against the Claimant's claims of BIT violations, because Argentina's and the Province's measures in violation of the BITs were not

Capital Corp., LG&E International Inc. v. Argentine Republic (*LG&E*) accepted the defense.[41]

William Burke-White argues that the tribunals that rejected the defense read the requirements of necessity under customary international law into the NPM provision in the US-Argentina BIT while the *LG&E* tribunal considered the BIT and customary international law differently.[42] He notes that: "For the *LG&E* Tribunal, the NPM clause is a separate risk allocation device and an explicit part of the bargain in the U.S.-Argentina BIT, providing the state parties greater protections than would have been available in customary law."[43]

Jurgen Kurtz has also analyzed the relationship between NPM clauses in bilateral investment treaties (BITs) and the customary international law on necessity. He notes that three methods are potentially applicable – "confluence," "*lex specialis*," and "primary-secondary applications."[44] The confluence method has been prominent and was adopted by the initial tribunals in *CMS*, *Enron*, and *Sempra*.[45] Kurtz expresses support for the primary and secondary applications but notes that the "choice . . . comes down to an election between" *lex specialis* and the primary and secondary applications.[46] In addressing the relationship between treaty provisions and Article 25, August Reinisch, following the decision of the Annulment Committee in *CMS*,[47] notes that tribunals are to determine, first, whether states have acted in accordance with their investment treaty obligations including those clauses that permit "emergency measures."[48] Where these clauses are unavailable or where they do not avail the state a defense, then "the secondary rules of State responsibility including Article 25 will come into play."[49] Depending on how they are drafted, the NPM clauses in BITs (may) provide a more lenient standard for state measures than the cumulative requirements of Article 25. Essentially, a state does not need to fulfill all the requirements under Article 25 in order to meet the required standards under a BIT.[50]

Articulating the requirements for necessity, Article 25 of the ILC State Responsibility Draft Articles takes the view that there must be a "grave and imminent peril" and serious impairment of the state's "essential interest."[51] Some tribunals analyze

the only means to satisfy its essential interests and because Argentina itself contributed to the emergency situation it was facing in 2001–2003."

41 *LG&E Energy Corp., LG&E Capital Corp., LG&E International Inc. v. Argentine Republic. (Decision on Liability)*. ICSID Case No. ARB/02/1 (October 3, 2006).

42 Burke-White, "State Liability & Legitimacy," 212. 43 Ibid., 212–213.

44 Jurgen Kurtz, "Adjudging the Exceptional at International Investment Law: Security, Public Order and Financial Crisis," *International Comparative Law Quarterly*, (2010): 325.

45 Ibid., 341. 46 Ibid., 371.

47 *CMS Gas Transmission Company v. The Republic of Argentina. (Decision of the Ad Hoc Committee on the Application for Annulment of the Argentine Republic)*. ICSID Case No. ARB/01/8 (September 25, 2007 https://icsid.worldbank.org/ICSID/FrontServlet?requestType=CasesRH&actionVal=showDoc&docId=DC687_En&caseId=C4.

48 Reinisch, "Necessity," 156. 49 Ibid. 50 Ibid.

51 United Nations, "Draft Articles on Responsibility."

these terms even when the applicable treaties include NPM clauses. In addressing whether Argentina faced a grave and imminent peril, the *CMS* tribunal expressed the view that "the situation was difficult enough to justify the government taking action to prevent a worsening of the situation and the danger of total economic collapse. But neither does the relative effect of the crisis allow here for a finding in terms of preclusion of wrongfulness."[52]

The tribunal in *Enron* also considered the plea of necessity under customary international law.[53] In determining whether Argentina's crisis "qualifies as affecting an essential interest of the State," the tribunal noted that there was indeed a severe crisis in Argentina.[54] However, it rejected the view that the "situation compromised the very existence of the State and its independence so as to qualify as involving an essential interest of the State" or that "the events were out of control or had become unmanageable."[55] The *Enron* tribunal's reading of essential interest as a situation that compromises the existence of a state is reminiscent of the pre-Article 25 enunciation of necessity.[56] Under this formula the necessity defense would be unavailable except the state was facing imminent collapse.

In *LG&E* the tribunal adopted a different view.[57] It rejected the view that the NPM clause in the US-Argentina BIT only applies to military action and war. For the tribunal:

> [T]he conditions in Argentina in December 2001 called for immediate decisive action to restore civil order and stop the economic decline. To conclude that such a severe economic crisis could not constitute an essential security interest is to diminish the havoc that the economy can wreak on the lives of an entire population and the ability of the Government to lead. When a State's economic foundation is under siege, the severity of the problem can equal that of any military invasion.[58]

In *Continental Casualty Company v. Argentine Republic* (*Continental Casualty*) the tribunal also noted that a "severe economic crisis may thus qualify" under the

52 *CMS* (Award), par. 322.
53 *Enron* (Award).
54 Ibid., pars. 305–6.
55 Ibid., pars. 306–7.
56 See generally, Sloane, "On the Use and Abuse."
57 *LG&E* (Liability). The LG&E decision is the subject of pending annulment proceedings. In *Sempra*, the tribunal declined to follow *LG&E*, choosing to adopt the reasoning in *CMS* and *Enron*. See especially par. 346. See also *BG Group Plc. v. The Republic of Argentina. (Final Award)*. (UNICTRAL, November 3, 2008). http://italaw.com/documents/BG-award_000.pdf. (Finding at par. 381 that there was no state of emergency or state of necessity defense available to Argentina under the U.K.-Argentina BIT. The tribunal also noted at par. 385 that "the Argentina-U.K. BIT does not include a national security exception analogous to Article XI of the Argentina-U.S. BIT." According to par. 407, Argentina's defense based on Article 25 of the ILC Draft Articles also failed.); *National Grid Plc. v. Argentine Republic. (Award)*. (UNCITRAL, November 3, 2008) http://italaw.com/documents/NGvArgentina.pdf. (Finding at par. 262 that "the Respondent's own evidence shows that its contribution to the crisis was substantial.")
58 *LG&E*, (Liability). Ibid., par. 238.

BIT "as affecting an essential security interest."[59] Analyzing the protection of essential security interests under the US-Argentina BIT, the tribunal noted that:

> It is impossible to deny . . . that a crisis that brought about the sudden and chaotic abandonment of the cardinal tenet of the country's economic life . . .; the near collapse of the domestic economy; the soaring inflation; the leap in unemployment; the social hardships bringing down more than half of the population below the poverty line; the immediate threats to the health of young children, the sick and the most vulnerable members of the population, the widespread unrest and disorders; the real risk of insurrection and extreme political disturbances, the abrupt resignations of successive Presidents and the collapse of the Government, together with a partial breakdown of the political institutions and an extended vacuum of power; the resort to emergency legislation granting extraordinary legislative powers to the executive branch, that all of this, taken together, does not qualify as a situation where the maintenance of public order and the protection of essential security interest of Argentina as a state and as a country was vitally at stake . . . The protection of essential security interests . . . does not require that "total collapse" of the country or that a "catastrophic situation" has already occurred before responsible national authorities may have recourse to its protection . . . There is no point in having such protection if there is nothing left to protect.[60]

The quotes from *LG&E* and *Continental Casualty* indicate tribunals' willingness to take suffering into account in their decision making and to adopt interpretive approaches that analyze investment disputes in light of the socioeconomic upheaval that generated the disputes. However, one cannot ignore the opposite approach that other tribunals adopted based on the same facts. Hence, in order to take suffering seriously pre-emergency, and in order to afford arbitral tribunals the opportunity to take suffering seriously post-emergency, the enunciation of clear rules before emergencies occur is necessary. It is clear that the present state of the law is unsatisfactory.

Fifth, it has been necessary to determine whether NPM clauses are self-judging. Even if possible or appropriate to determine and define beforehand, situations that constitute an emergency, the question remains whether the state involved has the capacity to singularly determine the necessity of the measures it adopts or whether these measures can be subject to judicial and arbitral scrutiny and review. In Argentina's cases, the arbitral tribunals shared the view that the necessity plea and Article XI of the US-Argentina BIT are not self-judging.[61] As expressed in

59 *Continental Casualty Company v. Argentine Republic. (Award)*. ICSID Case No. ARB/03/09 (September 5, 2008). italaw.com/documents/ContinentalCasualtyAward.pdf.

60 Ibid., par. 180 (footnotes omitted).

61 Reinisch, "Necessity," 143–144. See for example, *LG&E* (Liability), par. 212; *Enron* (Award), par. 339 (noting that "Article XI [of the *US-Argentina BIT*] is not self-judging and that judicial review in its respect is not limited to an examination of whether its invocation or the measures adopted were taken in good faith.") Discussing the international law on necessity, the ICJ was

Continental Casualty, if the provision were self-judging, tribunals are effectively prevented from inquiring into the legality of these provisions if they have concluded that the crisis qualified under the NPM provision in the treaty.[62]

The tribunal in *Continental Casualty* noted that there must be "a significant margin of appreciation for the State applying the particular measure" for "a time of grave crisis is not the time for nice judgments, particularly when examined by others with the advantage of hindsight."[63] While not self-judging, some margin of appreciation is thought to be apposite. The 2012 United States Model BIT has inserted language that seeks to clarify the self-judging character of its NPM clause.[64]

Argentina's cases have clarified some of the points outlined in the foregoing discussion. First, tribunals have consistently asserted that they have jurisdiction to consider the challenges that economic emergencies present for foreign investors. Second, there is some agreement that the necessity plea is applicable to economic emergencies. However, the requirements for determining whether an economic emergency has met the requisite threshold for being so classified seem to fluctuate from moderate to severe. Third, while ICSID tribunals consider Article 25 to be reflective of customary international law, literature suggests that the debate is not entirely settled. Fourth, the relationship between Article 25 and NPM clauses is perhaps, the most unsettled area of the economic emergency debate in investment arbitration. Fifth, the tribunals are also consistent in holding that NPM clauses and necessity pleas are not self-judging.

However, this conclusion might be modified in light of a newer generation of treaties that insert language that suggest that these clauses are self-judging. Even though these five (formerly) contentious points are being clarified, the decisions in Argentina's cases demonstrate that it remains difficult to take suffering seriously; difficult to offer a balanced view of peoples' suffering and investors' losses with the current regime that prevails.

of the view that "the state of necessity can only be invoked under certain strictly defined conditions which must be cumulatively satisfied; and the State is not the sole judge of whether those conditions have been met." *Hungary/Slovakia*, par. 51.

62 *Continental Casualty* (Award), par. 182. See Burke-White and von Staden, "Investment Protection in Extraordinary Times," 368–86. (Discussing *inter alia* self-judging and non-self-judging non-precluded measure clauses in BITs.).

63 *Continental Casualty* (Award). Ibid., par. 181. For the margin of appreciation, see generally, Yuval Shany, "Toward a General Margin of Appreciation Doctrine in International Law?" *European Journal of International Law*, 2005: 907; Janneke Gerards, "Pluralism, Deferrence and the Margine of Appreciation Doctrine," *European Law Journal*, 2011: 80. But see Eyal Benvenisti, "Margin of Appreciation, Consensus, and Universal Standards," *New York University Journal of International Law and Policy* (1999): 843.

64 Article 18(2) of the 2012 United States Model Bilateral Investment Treaty provides that: "Nothing in this Treaty shall be construed to preclude a Party from applying measures that it considers necessary for the fulfilment of its obligations with respect to the maintenance or restorations of international peace or security, or the protection of its own essential security interests." www.ustr.gov/sites/default/files/BIT%20text%20for%20ACIEP%20Meeting.pdf.

NECESSITY AND COMPETING INTERESTS

> As is many times the case in international affairs and international law, situations of this kind are not given in black and white but in many shades of grey.[65]
>
> Be that as it may, "shades of grey" are omnipresent in international life and do not justify glossing over Argentina's plea of necessity.[66]

Even though necessity arguments are now common in investment arbitration cases addressing economic emergencies, the availability of the necessity defense is difficult to establish. For arbitrators, judges, and parties to disputes, the plea presents difficulties that result in often unsatisfactory results. The inconsistent decisions in Argentina's socioeconomic crisis present an apt example. For the people affected by these decisions, the consequences are greater. It has been estimated that "Argentina's total potential liability" in the cases that were initiated following this crisis "is several dozen billion US dollars."[67] The ultimate bearer of that liability is the Argentine people. At the same time, investors suffer significant losses. Hence, the question regarding whether foreign investors should bear the burdens of financial and other crises is a question that admits of no easy answers.[68] It is also difficult to establish a single rule that applies in all cases that suggests that the state and inevitably its people are the bearers of the risk in all cases.

The necessity plea is not a panacea for economic emergencies. The potential for abuse is one of the reasons for the stringent rules on necessity. As the ILC notes, "necessity will only rarely be available to excuse non-performance of an obligation and ... is subject to strict limitations to guard against possible abuse."[69] In order "to emphasize the exceptional nature of necessity and concerns about its possible abuse, article 25 [of the ILC Articles] is cast in negative language ('Necessity must not be invoked ... unless')."[70] While safeguarding necessity against abuse is essential, the other extreme – an inability to take suffering seriously – is also a very real possibility.

At least four of the requirements in Article 25 of the ILC State Responsibility Draft Articles, or related to Article 25, make a successful necessity argument difficult in economic emergencies.[71] These requirements – the "only way" requirement, temporal limitation, state contribution, and compensation – are discussed in

65 *CMS* (Award), par. 320. 66 Waibel, "Two Worlds," 645.

67 Ibid., 637, footnote 1. Burke-White draws from commentators that estimate that "Argentina's potential liability from these cases alone could be greater than eight billion U.S. dollars, more than the entire financial reserves of the Argentine government in 2002." Burke-White, "State Liability & Legitimacy," 204.

68 In *LG&E* (Liability), par. 264., the tribunal concluded that the investor would bear the burden of the measures that Argentina adopted during the state of necessity.

69 "ILC Draft Articles, Commentaries," 80. 70 Ibid., 83.

71 Michael Waibel identifies three unsatisfactory requirements of the international law on necessity that make it ill-suited for financial crises. Waibel, "Two Worlds," 641–3.

turn. Discussion of the temporal limitation and state contribution requirements rely in part on Michael Waibel's comments on these points.

First, Article 25(1)(a) of the ILC State Responsibility Articles includes the requirement that the measures taken must have been the "only way" to " safeguard an essential interest against a grave and imminent peril." In the *Hungary/Slovakia* case the ICJ expressed the view that "Hungary had means available to it, other than the suspension and abandonment of the works, of responding to that situation."[72] This remains the case even though the other measures involved more costly techniques. In concluding that the construction of a wall in the occupied Palestinian territory is contrary to international law in the Advisory Opinion in the case of the *Legal Consequences of the Construction of a Wall in the Occupied Palestinian Territory*, the ICJ noted that it was not convinced that "the construction of the wall along the route chosen was the only means to safeguard the interests of Israel against the peril which it has invoked as justification for that construction."[73] As the Argentine cases demonstrate, showing that a measure taken is the only way to safeguard an essential interest may be a more difficult requirement to satisfy in economic emergencies.

Second, there is a "temporal limitation" on necessity.[74] Article 27(a) of the ILC State Responsiblity Draft Articles requires "[c]ompliance with the obligation in question, if and to the extent that the circumstance precluding wrongfulness no longer exists." Here the necessity plea and the NPM clause in the US-Argentina BIT appear similar. The ICJ recognized this principle in the *Hungary/Slovakia* case.[75] In *LG&E*, the tribunal concluded that "Argentina is excused under Article XI [of the US-Argentina BIT] from liability for any breaches of the Treaty between 1 December 2001 and 26 April 2003," which were in the tribunal's view the operative dates of the state of necessity.[76] The tribunal marked "these dates as the beginning and end of the period of extreme crisis in view of the notorious events that occurred during this period."[77]

All Argentina's measures that violated the provisions of the BIT before and after the period that the tribunal identified were to be compensated.[78] The tribunal later issued an award for damages for violations that occurred outside the period of the emergency.[79] Waibel expresses the view that while the temporal limitation on the observation of obligations works well in cases of use of force, the same does not apply well to economic emergencies.[80] In emergencies that affect the state's ability to pay its debt, resumption of the obligation following the necessity period, "could

72 *Hungary/Slovakia*, par. 55. 73 *Wall Case*, par. 140. 74 Waibel, "Two Worlds," 641.

75 *Hungary/Slovakia*, par. 101. 76 *LG&E* (Liability), par. 229.

77 Ibid., par. 230. 78 Ibid., par. 263.

79 *LG&E Energy Corp., LG&E Capital Corp., LG&E International, Inc. v. Argentine Republic*. (Award). ICSID Case No. ARB/02/1 (July 25, 2007) https://icsid.worldbank.org/ICSID/FrontServlet?requestType=CasesRH&actionVal=showDoc&docId=DC786_En&caseId=C208.

80 Waibel, "Two Worlds," 641–2.

unnecessarily prolong the country's financial woes."[81] And by prolonging the country's "financial woes," it inevitably prolongs the peoples' suffering.

Third, by Article 25(2)(b) of the ILC State Responsibility Draft Articles, the state must not have contributed to the state of necessity. In addition, of significance to a successful plea of necessity, albeit somewhat reduced, is the requirement that the international obligation in question must not exclude the possibility of invoking necessity, which the tribunal alluded to in *BG Group Plc. v. The Republic of Argentina* (*BG Group*).[82] Waibel identifies the exclusion of the plea in cases where the state contributed to the necessity as a characteristic of necessity that makes it ill-suited to financial crises.[83] The ILC clarifies that the contribution must be substantial and not peripheral or incidental.[84] In *National Grid Plc. v. Argentine Republic* (*National Grid*), the rejection of the plea of necessity was based on the tribunal's conclusion that Argentina's "contribution to the crisis was substantial," thereby resulting in failure to meet one of the cumulative tests for finding necessity under customary international law.[85] In Argentina's cases, tribunals have found it difficult to navigate this requirement. The nature of the crisis lends credence to the assertion that "[n]o ICSID tribunal will ever be able to disentangle the exogenous and endogenous causes of such crises" as "[m]acroeconomists will still be arguing about the causes of and solutions for Argentina's financial crisis in twenty years' time."[86] In Argentina's case, there has been enough blame to go around, with some being allocated to the International Monetary Fund.[87]

Fourth, like other international law defenses that preclude wrongfulness, necessity is not an absolute defense. Article 27(b) of the ILC Draft Articles supports this view by reserving "[t]he question of compensation for any material loss caused by the act in question." The possibility that compensation may (not) be required is perhaps the most challenging requirement for accounting for the positions of all the actors involved. Essentially, even if necessity as articulated in Article 25 is available, it does not necessarily absolve the state of the requirement to compensate investors. This raises questions: Who bears the risk of an economic emergency – the state that could have partially contributed to the crisis and ultimately its peoples, or a foreign investor that invests in a country with promises of protection but with recognition that business comes with attendant risks? Is there a possibility for shared financial burden and costs?

In the *CMS* award and similar cases, as well as the *CMS* annulment ruling, Argentina was held liable for damages.[88] These tribunals ultimately allocated a significant portion of the risk of the emergency, irrespective of the applicability of necessity, to Argentina. Tribunals also appear to be aware of the challenges involved in allocating risk, liability, and financial burdens in economic

81 Ibid., 642. 82 *BG Group*, par. 409. 83 Waibel, "Two Worlds," 642.
84 "ILC Draft Articles, Commentaries," 84. 85 *National Grid*, note 57 at par. 262.
86 Waibel, "Two Worlds," 643. 87 Hill, "The 'Necessity Defense'," 548.
88 *CMS* (Award), pars. 388–390.

emergencies. While noting that Argentina's crisis had a "severe impact on the Claimant's business," in *CMS*, the tribunal was of the view that the "impact must to some extent be attributed to the business risk the Claimant took on when investing in Argentina."[89] The tribunal alluded to both parties "sharing some of the costs of the crisis in a reasonable manner" and agreeing with Argentina that the tribunal's decision should not "amount to an insurance policy against business risk."[90] Ultimately however, the tribunal did not favor the view that the burden should be placed on the claimant alone.[91] The *CMS* tribunal ordered Argentina to pay $133.2 million in compensation to the claimant. In *Sempra*, the tribunal also noted that it would take Argentina's crisis into account in determining compensation due to the claimant.[92]

The *LG&E* and *Continental Casualty* tribunals took a different approach. In *LG&E*, the tribunal stated that the investor should bear the damages it suffered during the state of necessity.[93] These tribunals laid the weight of their decisions on the NPM clause in the treaty. With an emphasis on treaty, the decision regarding whether compensation would be awarded in any given case would depend on the language of the treaty in question and on the principles of treaty interpretation. For example, *BG Group* and *National Grid* are often analyzed separately because they are not based on the US-Argentina BIT but on the Agreement between the Government of the United Kingdom of Great Britain and Northern Ireland and the Government of the Argentine Republic for the Promotion and Protection of Investments (UK-Argentina BIT).[94]

The ILC notes that Article 27 refers to the "possibility of compensation" in some cases "because it is not possible to specify in general terms when compensation is payable."[95] It clarifies that compensation does not refer to compensation as reparation for wrongful conduct.[96] Rather, Article 27 addresses situations where, in spite of the plea of a circumstance precluding wrongfulness, a state should "make good any material loss suffered by any State directly affected."[97] The ILC recognizes that the range of possibilities that could be covered by circumstances precluding wrongfulness is wide and as a result, it does not outline situations where compensation should be payable.[98]

Ultimately, the conclusion regarding compensation might depend on the standards being applied – treaty standards or customary international law standards. Under some treaties, the application of NPM or similar clauses absolves the state of liability as well as the duty to pay compensation. However, the position under

89 Ibid., par. 248. 90 Ibid. 91 Ibid.

92 *Sempra* (Award), par. 397. 93 *LG&E* (Liability), par. 264.

94 *Agreement between the Government of the United Kingdom of Great Britain and Northern Ireland and the Government of the Argentine Republic for the Promotion and Protection of Investments* (December 11, 1990, entered into force on February 19, 1993).

95 "ILC Draft Articles, Commentaries," 86. 96 On reparation, see "ILC Articles," art. 34.

97 "ILC Draft Articles, Commentaries," 86. 98 Ibid.

customary international law remains unclear. In fact, the international law in this area would not necessarily be the first recourse for clear statements on this issue. However, it is arguable that the ambiguity surrounding compensation leaves open the question of the party that bears the financial costs of investors' losses resulting from emergencies and provides some potential for positive experimentation.

In offering a response to the dilemmas that crises present for the plea of necessity Sarah Hill argues that:

> [A] fairer proposal seems to be a "balancing test" or "necessity discount" approach that entails a weighing of the social, political, and economic interests of Argentina and the risks assumed by the investor against LG&E's lost profits in crisis, then calculating the total amount of the damages for which Argentina would have been responsible in the absence of crisis and discounting them for the urgency of the necessity measures involved.[99]

Ultimately, it is difficult to find much fault in an approach that seeks to balance the interests of parties affected by a crisis. A balancing approach, which does not necessarily have to follow Hill's admittedly "circumstantial and speculative approach,"[100] is one of the viable options for the continued adoption of necessity as a response to claims arising out of or as a result of measures taken during economic emergencies. However, the necessity plea does not provide recourse for taking the suffering of diverse actors seriously and doing so in a holistic manner. Given that it is perhaps the major plea in the event of disasters and emergencies, with *force majeure* being a distant second, the plea of necessity demonstrates international law's limited ability to adequately respond to economic emergencies.[101]

If the inconsistent decisions in the Argentina economic crisis cases have demonstrated anything, they have shown that international law has not clearly articulated the components of necessity and other measures for dealing with economic emergencies in investment arbitration. Substantial inconsistency remains in the area. However, Reinisch argues that the annulment decision in *CMS* appears to have provided an analytical framework for investment arbitration that other tribunals are following; that is, to first consider the BIT provisions before extending the analysis to the necessity plea in international law.[102] Until the decision in *Continental Casualty*, it appeared that most arbitral tribunals were leaning in favor of conclusions similar to the award in *CMS* (which held Argentina liable for compensation in spite of the defenses it raised) but the *Continental Casualty* tribunal reached

[99] Hill, "The 'Necessity Defense'," 565. [100] Ibid., 565–566.

[101] The commentaries on the ILC's Draft Articles on State Responsibility offer brief comparisons between necessity and the following: consent, self-defense, countermeasures, *force majeure*, and distress. "ILC Draft Articles, Commentaries," 80.

[102] Reinisch. "Necessity," 156.

conclusions similar to the conclusion in *LG&E* (which did not require Argentina to compensate the claimant for losses suffered during the period of the emergency).

MULTIPLE ACTORS, NECESSITY, AND ECONOMIC EMERGENCIES

As noted earlier, ICSID tribunals have reached conflicting decisions on the application of necessity to Argentina's economic crisis. The initial awards in *CMS*, *Enron*, and *Sempra* do not tell the complete story of the necessity defense in ICSID arbitration. In fact, a complete story has probably not yet unfolded. Subsequent annulment decisions have taken alternative views of the issues. As noted, although the annulment decision in *CMS* heavily criticized the initial award it did not annul the finding on damages.[103] However, in *Enron*,[104] the ad hoc committee annulled the award to the extent that Argentina was precluded from relying on the NPM clause in the US-Argentina BIT and on the necessity plea. Consequently, it also annulled the decision that Argentina should pay $106.2 million in compensation and the related interest payments. The ad hoc committee particularly criticized the failure to apply the tests for determining necessity under international law.[105]

Given the restricted scope of ICSID annulment proceedings, the committee's conclusion does not suggest that Argentina is entitled to rely on necessity or on the NPM clause in the BIT. In the words of the committee, it "cannot go further and make its own findings as to whether or not Argentina is entitled to rely on the principle of necessity under customary international law or on Article XI of the BIT or as to whether or not Argentina is responsible for breaches of its obligations *vis-à-vis* the Claimants..."[106] In *Sempra*,[107] the ad hoc annulment committee also annulled the award, this time in its entirety. Following the tribunal's reliance on Article 25 of the ILC State Responsibility Draft Articles, the committee relied *inter alia* on the failure to apply the NPM measures in the BIT.[108] It noted "the Award must be annulled in its entirety on the basis of manifest excess of powers (Article 52 (1)(b) of the ICSID Convention) in respect of failure to apply Article XI of the BIT."[109]

The inconsistences in arbitral jurisprudence on economic emergencies and the plea of necessity is only one example of the challenges that affect international law's treatment of the multiple interests that are impacted by economic

[103] *CMS* (Annulment).

[104] *Enron Creditors Recovery Corp. Ponderosa Assets L.P. v. The Argentina Republic. (Annulment Decision)*. ICSID Case No. ARB/01/3 (July 30, 2012). http://italaw.com/documents/EnronAnnulmentDecision.pdf.

[105] Ibid., pars. 368–395.

[106] Ibid., par. 408.

[107] *Sempra Energy International v. Argentine Republic. (Annulment Decision)*. ICSID Case No. ARB/02/16 (September 28, 2007). https://icsid.worldbank.org/ICSID/FrontServlet?requestType=CasesRH&actionVal=showDoc&docId=DC1550_En&caseId=C8.

[108] Ibid., pars. 196–209.

[109] Ibid., par. 159.

emergencies and other disasters. Even the status of the rules applied (as customary international law or not) is contested.[110] In assessing the "paradoxical relationship between form and authority" in the ILC State Responsibility Draft Articles, David D. Caron expresses the view that the ILC Draft Articles are not a "source" of law; rather, they are "evidence of a source of law."[111] As a result, arbitrators should be careful not to "defer too easily and uncritically to them."[112] Caron does not, however, disagree that the ILC State Responsibility Draft Articles "represent (at best) a restatement of the customary international law of the secondary principles of state responsibility."[113]

Crafting the nature of the ILC State Responsibility Draft Articles in this manner, and as recognized by the ILC itself in Article 55, allows derogations based on *lex specialis* rules enunciated in treaties.[114] In addition, Caron notes that the "articles are a mix of codification and progressive development" – with some difficulty sometimes in telling which is a larger component in a particular article.[115] He urges a welcoming attitude toward these articles that have now had significant influence on the work of tribunals interpreting Argentina's economic crisis in light of the ILC State Responsibility Draft Articles and in light of the US-Argentina BIT. More importantly, Caron recognizes that "international law must be able to evolve with the needs of the international community."[116]

Following the theme of international law evolving with the needs of the international community, some factors necessary for taking multiple actors' perspectives seriously in economic emergencies may be deduced from the foregoing discussion. These factors, encapsulated in the recourse to specifically defined treaty rules, incorporate pre-emergency decision making in treaties and post-emergency judicial and quasi-judicial determinations (where necessary) to adjudicate the competing interests presented when an emergency occurs. In making these decisions pre- and post-emergency, it is important to recall that "states are not themselves human individuals. In the last resort, states are not the bearers of ultimate value. They exist for the sake of human individuals."[117]

While it would have been easier to cast necessity in a state-state light to the exclusion of other actors, in the twenty-first century, direct considerations of the interests of non-state actors (including the peoples of the state and foreign investors) cannot be ignored. As Jeremy Waldron notes, the well-being of the "men, women, communities, and businesses who are committed" to

[110] See Bodansky and Crook, "Symposium."

[111] David D. Caron, "The ILC Articles on State Responsiblity: The Paradoxical Relationship between Form and Authority," *American Journal of International Law* (2002): 867.

[112] Ibid., 867. [113] Ibid., 872. [114] Ibid. [115] Ibid., 873. [116] Ibid.

[117] Jeremy Waldron, "The Rule of International Law," *Harvard Journal of Law and Public Policy* (2006–2007): 24. For a discussion of community interest in international investment law, see Christoph Schreuer and Ursula Kriebaum, "From Individual to Community Interest in International Investment Law" in *From Bilateralism to Community Interest: Essays in Honour of Judge Bruno Simma*, ed. Ulrich Fastenrath et al. (Oxford University Press, 2011): 1079.

the "care" of states, and "not the well-being of sovereign nation-states, is the ultimate end of international law."[118]

State sovereignty is a complex phenomenon, and investment protection and the powers of the state do not necessarily exist in binaries. In an examination of more expansive restrictions in more recent investment treaties, Jose Alvarez examines what he calls "the return of the state" or "the return of some states."[119] Allowing states more flexibility in investment regulation is an exercise in state sovereignty that some could argue would reduce investment protection. However, international investment law has not devised how else to ensure the protection of peoples *qua* peoples outside the framework of the state.[120] Pending international investment law's articulation of a strategy for protecting peoples without much recourse to states, these debates will continue.

Investment treaties may provide an opportunity for states to specifically respond to challenges and disasters that international economic law often presents and faces. Argentina's crisis aptly demonstrates this need. While the tribunals were not always clear in their analysis, decisions made based on the US-Argentina BIT made a special treaty regime relevant to the discussion, whereas other cases with investors from other states relying on other BITs were different. For example, in *National Grid*, the tribunal took the view that since there was no NPM clause in the UK-Argentina BIT,[121] it had to turn to customary international law for guidance.[122]

Treaties may impact the application of international law rules on necessity and other pleas potentially applicable to economic emergencies. In the annulment decision in *CMS*, the ad hoc annulment committee adopted a primary rule and secondary rule of international law dichotomy. The committee was of the view that "the excuse based on customary international law could only be subsidiary to the exclusion based on Article XI."[123] As well "the requirements under Article XI are not the same as those under customary international law as codified by Article 25 [of the ILC State Responsibility Draft Articles]."[124]

The *Continental Casualty* tribunal emphasized this difference.[125] The difference also has a significant impact on compensation. If Article XI of the BIT applied, compensation would not be payable whereas such compensation may be payable under customary international law.[126] According to the *CMS* annulment committee, if the initial tribunal was satisfied with the arguments based on Article XI of the

118 Waldron. Ibid., 24–5.

119 Jose E. Alvarez, "The Return of the State," *Minnesota Journal of International Law* (2011): 260.

120 Ibironke Odumosu, "The Law and Politics of Engaging Resistance in Investment Dispute Settlement," *Penn State International Law Review* (2007): 251.

121 U.K.-Argentina BIT.

122 *National Grid*, pars. 254–255.

123 *CMS* (Annulment), par. 132.

124 Ibid., par. 130.

125 *Continental Casualty* (Award), pars. 162–168.

126 *CMS* (Annulment), pars. 144–147. Compensation *may* be payable for as the Committee noted at par. 127, article 27 of the ILC Articles "is a 'without prejudice' clause, not a stipulation"

US-Argentina BIT, there would have been no breach of the BIT. In this case "Article XI [of the BIT] and Article 25 [of the ILC Articles] thus construed would cover the same field and the Tribunal should have applied Article XI as the *lex specialis* governing the matter and not Article 25."[127]

If necessity is construed as a "secondary rule of international law" that goes to the "issue of responsibility," the tribunal's initial obligation would be to consider whether there has been a breach of the treaty and whether Article XI excluded the breach.[128] If the tribunal concludes that there was conduct that did not conform to the treaty it would consider whether responsibility is precluded wholly or in part under customary international law.[129] Hence, the treaty is initially applicable. In the event that wrongfulness is found under treaty and is not excluded by any NPM clause, tribunals may have recourse to customary international law pleas that, in the words of the ILC State Responsibility Draft Articles, preclude wrongfulness.

Carefully worded treaty language that follows deliberate considerations of the impacts of economic emergencies is one of the means by which states may address the challenges that these emergencies present. With the benefit of hindsight with regard to the Argentine cases, states are properly positioned to weigh all relevant interests in crafting treaty language in this area. Such language would carefully consider investment protection as investment treaties have done in the past but be more aware of the possibilities of disasters that significantly impact states and their peoples. Specifically addressing the question of compensation is an important area that treaty drafters need to turn their attention toward.

Taking suffering seriously will involve treaty drafting that is different from the current crop of investment treaties that are being concluded. Contemporary investment treaties are investment protection instruments. They are designed to protect the investment of foreign investors from the sovereign powers of the state without much focus on the peoples of these states. They are state- and investor-centric. While investor and investment protection continues to be necessary, recent history demonstrates that investment protection and the treaties that ensure such protection cannot be divorced from broader socioeconomic issues that impact upon states and their peoples. Argentina's cases are being presented as investment protection disputes but they are clearly bigger than investment protection.

This realization should inform the conclusion of investment treaties in the future. Treaties need to account for more robust interactions, which are the reality of the international economic order. Analyzing measures for re-crafting investment treaties to generally take peoples' suffering seriously is beyond the scope of this chapter. However, with regard to economic emergencies, two courses of action are necessary with regard to treaty drafting. First, all investment treaties need to include

and it "does not attempt to specify in which circumstances compensation could be due, notwithstanding the state of necessity."

[127] Ibid., par. 133. [128] Ibid., par. 134. [129] Ibid.

NPM clauses that are drafted after appropriate considerations of the multiple actors and interests that the treaties' provisions could impact. Second, and more importantly, treaty drafters need to specifically define emergencies to include socioeconomic emergencies; specify whether the emergency clauses are self-judging or not; clearly identify the relationship between the clause and other areas of international law, especially the plea of necessity; and clarify questions of compensation in the event of an emergency.

Some attempts at clarification, for example by some arbitral tribunals, are already occurring. But given the inconsistent decisions that have plagued interpretation of the law on the plea of necessity and NPM clauses in investment treaties, especially in the context of Argentina's crisis, deliberate rethinking of the law is necessary. In seeking to clarify the position of the law on compensation during emergencies, the annulment committee in *CMS* suggests that if Article XI of the US-Argentina BIT applied, compensation is not payable. It notes that "Article XI, if and for so long as it applied, excluded the operation of the substantive provisions of the BIT. That being so, there could be no possibility of compensation being payable during that period."[130] Given that the doctrine of precedent does not apply to investment arbitration, one cannot rely on the *CMS* annulment committee to clarify the law especially as it relates to accounting not only for the losses of investors but also the suffering of the peoples of the state involved. Given that the investment treaty model is pervasive under contemporary international investment law, clarification under the current regime lies in clearer and more deliberate treaty drafting that takes people's suffering seriously.

Some may argue that investment treaties may offer less protection to investors if NPM clauses are included in all treaties and if the kind of clarification suggested in this chapter occurs. This is, however, not necessarily the case, for as William Burke-White and Andreas von Staden aptly note with regard to BIT NPM clauses: "In exchange for granting investors greater protections than would have been available in customary law, the states also sought to preserve for themselves greater freedom of action through the NPM clause than would have been available in customary international law."[131]

While treaty language is being reformulated, the plea of necessity may continue to influence the development of this area of international law. The debate surrounding the customary international law status of Article 25 of the ILC State Responsibility Draft Articles contributes significantly to the impact of necessity in international law. As drafted, Article 25 is defined in negative and restrictive terms in order to prevent a floodgate of frivolous reliance on the defense. The other extreme of this often necessarily restrictive definition is its inability to respond to serious economic emergencies that impact upon the peoples of a particular state.

130 *CMS*, (Annulment), par. 146.
131 Burke-White and von Staden, "Investment Protection in Extraordinary Times," 396.

However, if Article 25 is not in fact reflective of customary international law, then the restrictive nature of the defense does not automatically apply to serious emergencies in the absence of specific treaty provisions. Again, the other side of this position is that frivolous reliance on the defense could be rife.

Representing particular international law principles as having customary international law status does not preclude dynamism in interpretation and development. The essence of custom is that it is subject to change; change sometimes that is a response to the needs of the relevant community. Hence, while the necessity plea has developed in a restrictive manner in order to prevent abuse, it has the potential to continue to develop in a manner that is cognizant of the challenges that have been addressed in this chapter. Admittedly, this would involve a very slow process. However, as long as the process generates results that have been schooled by the interaction of the multiple actors and interests impacted by economic emergencies, the potential for a robust framework would be realized.

In addition to treaty drafting and the development of customary international law, there is room for judicial and arbitral interpretation in developing a multi-actor sensitive approach to addressing economic emergencies that impact not only on governments and investors but also on local peoples. Baxi's work that adopts the "taking suffering seriously" title, analyzes the proactive nature of the work of the Supreme Court of India in a bid to make the judiciary more accessible and responsive to the poor.[132] In terms of interpretation, to say that arbitral tribunals have experimented with diverse interpretative techniques might be an understatement. In *Continental Casualty*, for example, the tribunal referred *inter alia* to GATT and WTO case law in its analysis of necessity.[133]

The extent to which such forays into areas like international trade is sustainable is open to debate.[134] Given the prominence that arbitral tribunals have assumed in interpreting the applicable laws in this area, their contributions are important to the development of the system. This prominence does not however preclude considerations of other forums that may be more appropriate for addressing public interest-impacting disputes that the law currently regards as investment disputes. Sometimes these disputes which are regarded as investment disputes are broader given the cause of the disputes and the diversity of actors that are impacted by such disputes. They need to be analyzed as such – as disputes that have significant impacts beyond the disputing investors and states.

[132] Baxi, "Taking Suffering Seriously."

[133] See *Continental Casualty* (Award) par. 192. For analysis of *Continental Casualty*, see Jose E. Alvarez and Tegan Brink, "Revisiting the Necessity Defense: Continental Casualty v. Argentina," *Yearbook on International Investment Law and Policy*, 2010–2011: 315.

[134] See Alvarez and Brink, "Revisiting the Necessity."

CONCLUSION

In light of the financial crises that impacted most countries since 2007, it is not conjecture to suggest that there might be other crises of Argentina's magnitude or greater in the making. It is the responsibility of states to take people's suffering seriously. In order not to only respond to investor concerns but also to direct the course of international investment law to a more disaster-sensitive mode, states must take deliberate and concerted steps.

Necessity in international law is restrictive and difficult to apply to economic crises. Necessity was developed during a different era for purposes other than investment disputes and economic crises. Hence the restrictive interpretations of the plea of necessity might be inevitable and international actors may need to accept these restrictions pending changes to international custom. Investment treaties have also been subject to severe criticisms due to their often lopsided nature. But in light of states' continued recourse to the investment treaty mechanism, it is essential to tailor these instruments to respond to the needs of multiple actors.

States have crafted legal rules in order to take investment protection seriously and even in times of economic crises, states are usually held to these obligations. Argentina's socioeconomic crisis has reminded the international community that crises impact on multiple actors. In fact, the decisions of the tribunals are replete with several renditions of the impacts of the crises on multiple stakeholders. However, the law does not seem to have the language to adequately articulate the interests of these other stakeholders, especially the people of the state involved in an investment dispute. This chapter is a call to consciously have this multi-stakeholder conversation while the necessity debate is still vibrant.

17

Clarifying the *Acquis Humanitaire*: A Transnational Legal Perspective on the Internalization of Humanitarian Norms

Dug Cubie

Although the international obligations and institutional frameworks for disaster response are not yet settled, the diverse source and nature of drafting initiatives suggests that the international community is engaged in a process of norm creation, elaboration, and interpretation, reflecting a desire for legal clarity in humanitarian operations. Situated within the framework of transnational law, this chapter argues that an *acquis humanitaire*, based on the principle of humanity, encapsulates the evolving body of law and practice specifically relating to the protection of persons in times of humanitarian crisis in both armed conflicts and natural or human-made disasters. Reflecting the nontraditional, non-statist, dynamic, and normative basis of transnational legal process, as elaborated by Harold Koh, the constant flow of ideas and principles between the national, regional, and international spheres provides an analytical framework for the ongoing transnational dialogues on the social, political, and legal internalization of humanitarian norms. Drawing on the internalization of humanitarian norms within the United Kingdom, this chapter concludes that as the international community examines the codification of a universal legal framework for the protection of persons in the event of disasters it is necessary to understand the transnational process of interpretation and internalization of humanitarian norms, and how this may vary across different regions and countries.

INTRODUCTION

While major natural disasters such as the 2004 Indian Ocean tsunami and Hurricane Katrina, or protracted and bloody armed conflicts such as those in Iraq, Sudan, or Syria, can provide a public focus for the need to provide humanitarian assistance to communities suffering from the effects of war or catastrophe,[1] the past

[1] See for example: Victoria Bannon and David Fisher, "Legal Lessons in Disaster Relief from the Tsunami, the Pakistan Earthquake and Hurricane Katrina," *ASIL Insights* 10, no. 6 (2006).

twenty-five years have also seen a growing momentum regarding the importance of legal frameworks to support humanitarian relief operations.[2] Yet many countries still lack contingency plans for offers and acceptance of international humanitarian assistance should the need arise, and it is recognized that national authorities (intentionally or unintentionally) often create bureaucratic barriers to the entry and deployment of international relief personnel and equipment, while simultaneously providing insufficient oversight of international relief efforts.[3] Indeed, it has been argued that international disaster relief is a "swashbuckler's paradise" with a scattered and underused international legal regime and scant applicable law at the national level.[4]

Nevertheless, a range of divergent provisions does exist at the national, regional, and international levels regarding the general principles for and operational delivery of humanitarian assistance; and both binding international law and nonbinding normative standards[5] aim to regulate international mechanisms for prevention, preparation, and response to humanitarian crises. Additionally, there is extensive state practice in the provision of humanitarian assistance to alleviate suffering in times of war and disaster, as well as huge amounts of public and private donations for emergency appeals by humanitarian actors.[6] Yet legal frameworks for such activities remain fragmented and incomplete. Furthermore, although too much legislation and bureaucracy at the national level can result in delayed entry and

2 For early examinations of international legal frameworks for disaster response, see: Peter Macalister-Smith, *International Humanitarian Assistance: Disaster Relief Actions in International Law and Organization* (Martinus Nijhoff Publishers, 1985); Peter Walker, "Victims of Natural Disaster and the Right to Humanitarian Assistance: A Pracitioner's View," *International Review of the Red Cross* (December 1998): 611–617. Rohan Hardcastle and Adrian Chua, "Humanitarian Assistance: Towards assistance: towards a Right of Access to Victims of Natural Disasters," *International Review of the Red Cross* (December 1998): 589–609.

3 *The Right Aid at the Right Time: Progress Report on the Guidelines for the Domestic Facilitation and Regulation of International Disaster Relief and Initial Recovery Assistance.* Geneva, 2009. 1–2. For analysis of the importance of "bottom up" approaches to the development of national disaster response frameworks, see: Claire Clement, "International Disaster Response Laws, Rules and Principles: A Pragmatic Approach to Strengthening International Disaster Response Mechanisms." See Chapter 5 in this volume.

4 David Fisher, "International Disaster Relief: A Growing Regulatory Dilemma," Proceedings of the Annual Meeting (ASIL) (2007): 114–118.

5 Reflecting conceptual and terminological difficulties surrounding the phrase "soft law," the phrase "non-binding normative standards" refers to guidelines, recommendations and minimum standards developed at operational, policy and political levels by states, humanitarian agencies and international organizations. See: Dug Cubie, "An Analysis of Soft Law Applicable to Humanitarian Assistance: Relative Normativity in Action?" *Journal of International Humanitarian Legal Studies* (2012): 177–215.

6 The Global Humanitarian Assistance Report estimates an average of US $16 billion-17.2 billion humanitarian assistance was provided annually (2007–2011) by donor governments and private donations for conflict and non-conflict humanitarian crises. *Global Humanitarian Assistance Report 2012–2013*, Development Initiatives (Somerset, 2012–2013): 511. C.f. the UNOCHA Financial Tracking Service reports an average of only US $1.6 billion annually (2000–2012) in response to specific appeals following natural disasters. http://fts.unocha.org/.

delivery of humanitarian assistance, a lack of legislation and policies can lead to poor quality and uncoordinated recovery operations.[7]

Irrespective of whether or not a general right to humanitarian assistance currently exists in international law,[8] this chapter argues that we have reached a stage in the process of our evolving understanding of humanitarian norms in times of war and disaster where one can talk of an emerging *acquis humanitaire*. In other words, a defined body of law and practice specifically relating to the protection of persons in times of humanitarian crisis that includes provisions in general and customary international law, as well as a broad range of nonbinding sources that provide operational and normative guidance for humanitarian preparations and responses to both armed conflicts and natural or human-made disasters. As explained in the following sections, the evolution of this emerging *acquis humanitaire* is best understood through the mechanism of transnational legal processes, described by Koh as being nontraditional, non-statist, dynamic, and normative.[9]

The multiple flows of ideas, principles, and regulations between national, regional, and international frameworks encapsulated by the *acquis humanitaire* presents an important mechanism for the internalization of international and regional humanitarian norms at the domestic level, reflecting Koh's argument that transnational legal process is pivotal to our understanding of why nations obey international law.[10] This chapter therefore examines various methods of internalization of humanitarian norms into national law and policy, in particular examining the social internalization of such norms in the United Kingdom. The concept of an *acquis humanitaire* based on the principle of humanity provides important insights into the ongoing dialogue between transnational actors as to the norms and standards applicable in times of humanitarian crisis.

THE EMERGING 'ACQUIS HUMANITAIRE'

In the 1980s, Steven Marks noted that generally before one can postulate a "right to . . ." there is the development of a "law of . . ." He argued this was because "a new body of legal norms or a revision of legal thinking on a given problem provides the conceptual framework for identifying first the legal implications of the problem, then the human rights implications, and finally the reformulation of the whole

[7] See: IFRC. Introduction to the Guidelines for the Domestic Facilitation and Regulation of International Disaster Relief and Intitial Recovery Assistance. Geneva, 2008., See: known as the "IDRL Guidelines," 3–5.

[8] For conflicting views, see: Yoram Dinstein, "The Right to Humanitarian Assistance," *Naval War College Review* (2000): 77–92; and David Fisher, "The Right to Humanitarian Assistance," in *Incorporating the Guiding Principles on Internal Displacement into Domestic Law: Issues and Challenges*, ed. Walter Kälin, et al. (Washington DC: ASIL, 2010), 51–2.

[9] Harold Koh, "Transnational Legal Process," *Nebraska Law Review*, 1996: 181–207.

[10] Harold Koh, "Review Essay: Why Do Nations Obey International Law," *Yale Law Journal*, 1997: 2599–2659.

problem in terms of a new human right."[11] So it is instructive to consider whether the disparate laws, rules, regulations, and practice for responding to humanitarian crises can be classified as a distinct body of law. It is well recognized that "international humanitarian law" refers exclusively to the protection and support of specific groups of people within international and non-international armed conflicts, rather than all "humanitarian" crises.[12]

Nevertheless, the consistent focus on humanitarian assistance by the UN General Assembly, particularly following natural and human-made disasters, highlights the importance governments place on the effective delivery of aid.[13] In 2001, the International Federation of Red Cross and Red Crescent Societies (IFRC) coined the phrase "International Disaster Response Laws" or IDRL, which was quickly amended to include "Rules and Principles," in recognition of the nonbinding sources of many humanitarian norms.[14] The IFRC's activities were recently changed to simply "Disaster Laws" to reflect the difficulty of separating the regulation of international disaster relief from domestic responses, and to move beyond emergency responses to include areas such as disaster risk reduction and regulatory barriers to post-disaster shelter.[15] Meanwhile, the International Law Commission's (ILC) Special Rapporteur on the Protection of Persons in the Event of Disasters, Eduardo Valencia-Ospina, recently acknowledged the existence of "the *acquis* of the international law of disaster response."[16]

Martha Finnemore and Kathryn Sikkink have argued there is an identifiable process whereby norms emerge, develop, and then achieve international acceptance: *norm entrepreneurs* develop the initial understanding of an emerging norm, which is taken up by other individuals and groups as *norm sponsors* who work

11 Steven Marks, "Emerging Human Rights: A New Generation for the 1980s?" *Rutgers Law Review* (1980–1981): 442. Indeed, Marks discusses a potential right to humanitarian assistance in non-conflict settings as one of six emerging human rights, at 449–450.

12 While the concept of an *acquis humanitaire* sketched in this paper includes the provisions of international humanitarian law relating specifically to the delivery of humanitarian assistance to civilians in times of war, the primary focus of this paper is on the role of the *acquis humanitaire* in understanding the evolving legal and policy frameworks at the international, regional and national level regarding preparations for and responses to natural and human-made disasters.

13 Arjun Katoch, "International Natural Disaster Response and the United Nations," in *International Disaster Response Laws, Principles and Practice: Reflections, Prospects and Challenges* (Geneva: IFRC, 2003), 47–60.

14 See generally: Horst Fischer, *International Disaster Response Law: A Preliminary Overview and Analysis of Existing* (Geneva: IFRC, 2003). . and International Federation of the Red Cross and Red Crescent Societies, *Law and Legal Issues in International Disaster Response: A Desk Study*. (Geneva: IFRC, 2007).

15 IFRC, "The IFRC's IDRL Programme is Becoming the Disaster Law Programme," February 2, 2012, www.ifrc.org/PageFiles/53419/name%20change%20announcement_FINAL%20LATEST%2002.02.2012.pdf; and IFRC, "*Disaster Law Programme: Summary Plan 2013*," www.ifrc.org/PageFiles/53419/Disaster%20Law%20Summary%20Plan%202013.pdf.

16 Eduardo Valencia-Ospina, *Fourth Report on the Protection of Persons in the Event of Disasters*, May 11, 2011, UN Doc: A/CN.4/643, par. 97.

toward internalization of the norm domestically. Agreement between a critical mass of actors around an emergent norm can then create a tipping point after which agreement (or consensus) becomes widespread.[17] The key role played by the IFRC and the broader Red Cross Movement as a norm entrepreneur and sponsor for disaster laws should not be underestimated. Their work in identifying and analyzing the legal implications of technical and operational aspects of major disasters[18] paved the way for the current work of the ILC on the Protection of Persons in the Event of Disasters and the increasing awareness of the need for legal clarity for humanitarian actors in disaster settings. For despite previous attempts to codify international responses to natural and human-made disasters, most notably the 1927 Convention establishing an International Relief Union,[19] the 1982 UNITAR Model Rules for Disaster Relief Operations,[20] and the 1984 draft UN Convention on Expediting the Delivery of Emergency Assistance,[21] the legal implications of disasters are only now being comprehensively analyzed both conceptually and operationally.

Notwithstanding the failure of these early initiatives to produce a universal legal framework for disaster responses, general international law does contain express rules for the provision of humanitarian assistance, while under customary international law civilians in times of armed conflict are entitled to specific humanitarian support.[22]

[17] Martha Finnemore and Kathryn Sikkink, "Norm Dynamics and Political Change," *International Organization* 52 (1998): 887–917, in particular 892–893. Koh similarly argues key agents in transnational legal processes include transnational norm entrepreneurs, governmental norm sponsors, transnational issue networks, law-declaring fora, and bureaucratic compliance procedures: Harold Koh, "Bringing International Law Home," *Houston Law Review* (1998–1999): 623–681 at 646–655.

[18] For example Georg Potyka and Katrien Beeckman, *The Regulatory Framework for Disaster Response Established within the European Union: A Focus on Humanitarian Aid and Civil Protection* (Geneva: IFRC,2005); IFRC, *Analysis of Law in the EU and a Selection of Member States pertaining to Cross-Border Disaster Relief: Synthesis Report and Recommendations* (Geneva: IFRC, 2010); and British Institute of International and Comparative Law (BIICL), *Analysis of Law in the United Kingdom pertaining to Cross-Border Disaster Relief* (London: British Red Cross, June 2010).

[19] League of Nations. Treaty Series. Vol. 135: 247. See: Peter MacAlister-Smith, "The International Relief Union: Reflections on the Convetion Establishing an International Relief Union of July 12, 1927," *Tijdschrift voor Rechtgeschiedenis* (1986): 363–374; John Hutchinson, "Disasters and the International Order I: Earthquakes, Humanitarians, and the Ciraolo Project," *International History Review* (June 2001): 1–36.; and John Hutchinson, "Disasters and the International Order II: The International Relief Union," *International History Review* (2001): 253–298.

[20] Mohammed el Baradei, *Model Rules for Disaster Relief Operations. Policy and Efficacy Studies* No. 8, UN Institute for Training and Research (UNITAR), 1982.

[21] A/39/267/Add.2-E/1984/96/Add.2, June 18, 1984. See also: Peter MacAlister-Smith, "Disaster Relief: Reflections on the Role of International Law," *Heidelberg Journal of International Law* (1985): 25–43.

[22] For discussion of customary humanitarian norms, see: Jean-Marie Henckaerts "Customary International Humanitarian Law: Taking Stock of the ICRC Study," *Nordic Journal of International Law* (2010): 435–468.

Furthermore, useful analogies and substantive provisions of the *acquis humanitaire* can be deduced from other branches of international law,[23] reflecting the acknowledged interconnections, overlaps, and gaps between traditionally distinct aspects of international law.[24]

The fundamental importance of non-governmental sources in international law making have long been recognized within transnational law,[25] so any *acquis humanitaire* will include a wide range of nonbinding texts developed by intergovernmental and non-governmental humanitarian actors. Relevant sources include the influential 1994 Code of Conduct for the International Red Cross and Red Crescent Movement and NGOs in Disaster Relief, which elaborated the "humanitarian imperative" to alleviate human suffering without discrimination wherever it may be,[26] while more recent Red Cross guidance for governments in respect of their national legislation for disaster-preparedness and response includes the IDRL Guidelines,[27] and the draft Model Act,[28] and draft Model Emergency Decree for the Facilitation and Regulation of International Disaster Relief and Initial Recovery Assistance.[29] Other influential nonbinding texts for state and non-state

[23] See: David Fisher, "Domestic Regulation of International Humanitarian Relief in Disasters and Armed Conflict: A Comparative Analysis," *International Review of the Red Cross* (June 2007): 345–372. See Rebecca Barber, "Facilitating Humanitarian Assistance in International Humantarian and Human Rights Law," *International Review of the Red Cross* (June 2009): 371–397.; and Ruth Stoffels, "Legal Regulation of Humanitarian Assistance in Armed Conflict: Achievements and Gaps," *International Review of the Red Cross* (September 2004): 515–546.

[24] See: Theodor Meron, "On the Inadequate Reach of Humanitarian and Human Rights Law and the Need for a New Instrument," *American Journal of International Law* (July 1983): 589–606. See: Rachel Brett and Eve Lester, "Refugee Law and International Humanitarian Law: Parllels, Lessons and Looking Ahead," *International Review of the Red Cross*, (September 2001): 713–726.; and Stephane Jaquemet, "The Cross-Fertilization of International Humanitarian Law and International Refugee Law," *International Review of the Red Cross* (September 2001): 651–674

[25] Michael Reimann, "From the Law of Nations to Transnational Law: Why We Need a New Basic Course for the International Curriculum,'" *Penn State International Law Review* 22 (2003-2004): 397–415.

[26] See: Dorothea Hilhorst, "A living document? The Code of Conduct of the Red Cross and Red Crescent Movement and NGOs in Disaster Relief," research paper for conference *Ten Years Code of Conduct: Principles in Practice*, Disaster Studies Wageningen University (2004); and *Code of Conduct for the International Red Cross and Red Crescent Movement and NGOs in Disaster Relief*, Relief and Rehabilitation Network Paper 7, Overseas Development Institute (September 1994).

[27] IFRC. "*IDRL Guidelines.*"

[28] The IFRC has worked with the UN Office for the Coordination of Humanitarian Affairs (UNOCHA) and Inter-Parliamentary Union to produce detailed legislative guidance for governments. A pilot version of the Model Act was presented at the 31st International Conference of the Red Cross Red Crescent Movement in November 2011, and the final version was launched at the General Assembly of the Inter-Parliamentary Union in March 2013 in www.ifrc.org/en/what-we-do/idrl/model-act-on-idrl.

[29] The IFRC and UNOCHA are currently developing a draft national Model Emergency Decree for situations where a clear legal framework is not in place before the onset of a large-scale disaster, or where the adoption of more comprehensive legislation is not practicable. www.ifrc.org/what-we-do/idrl/draft-model-emergency-decree.

responses to humanitarian crises include the Sphere Handbook,[30] IASC Operational Guidelines,[31] UN Guiding Principles on Internal Displacement,[32] and the Guiding Principles of UN General Assembly Resolution 46/182.[33] However, as with other developing bodies of law,[34] in addition to promoting a systematic framework for the emerging legal order by identifying positive legal rights and obligations applicable in humanitarian crises, such binding and nonbinding sources of the *acquis humanitaire* also provide practical and operational guidance to humanitarian agencies. Yet considering the wide divergence of cultural, religious, and ideological backgrounds of the multitude of humanitarian actors likely to respond to a major humanitarian crisis, can one identify a fundamental or underlying basis for the proposed *acquis humanitaire*?

CONCEPTUAL UNDERPINNINGS OF THE 'ACQUIS HUMANITAIRE'

The concept of an "acquis" comes originally from the institutional and legal framework established by the European Union (EU). First introduced in the Treaty on European Union,[35] the *acquis communautaire* is not defined in the core texts of

[30] Published in 2000, and updated twice, the Sphere Project's "*Humanitarian Charter and Minimum Standards in Humanitarian Response*." (3rd ed,, 2011) provides a comprehensive set of standards and indicators for the provision of humanitarian assistance. The Humanitarian Charter centers around three overarching principles: the right to life with dignity; the right to receive humanitarian assistance; and the right to protection and security; while. Meanwhile, the Minimum Standards contain sector-specific indicators covering: water supply and sanitation; food security and nutrition; shelter and non-food items; and health.

[31] Inter-Agency Standing Committee, "Operational Guidelines on the Protection of Persons in Natural Disasters," *Brookings-Bern Project on Internal Displacement* (January 2011). The Operational Guidelines provide practical advice to humanitarian agencies for implementation of rights-based approaches to humanitarian assistance.

[32] UN Commission on Human Rights. "The Guiding Principles on Internal Displacement." E/CN, 4/1998/53/Add.2, 1998. The Guiding Principles were drafted by international legal experts convened by the Representative of the UN Secretary-General on Internally Displaced Persons. They have been endorsed by the Secretary-General, General Assembly and Security Council, plus specialized agencies including UNHCR, UNICEF, UNOCHA and UNHCHR. Walter Kälin, "Guiding Principles on Internal Displacement: Annotations." *ASIL and Brookings-Bern Project on Internal Displacement, Studies in Transnational Legal Policy* (Spring 2008): vii.

[33] United Nations, "*Strengthening of the Coordination of Humanitarian Emergency Assistance of the United Nations*." GA Res. 46/182, 1991., UNGA elaborated Guiding Principles covering the rationale and modalities for providing humanitarian assistance while also revising and strengthening the UN system for disaster response coordination and funding.

[34] For discussion of the evolving norms of international climate justice, see: Teresa Thorpe, "Climate Justice: A Constitutional Approach to Unify the *Lex Specialis* Principles of International Climate Law," *Utrecht Law Review* 8(3) (2012):7–37.

[35] *The Consolidated Version of the Treaty on European Union (TEU), OJ* C340, 10 *November* 1997. Article 2 TEU provides the EU must "maintain in full the *acquis communautaire* and build on it"; while Article 3 TEU states the EU must "attain its objectives while respecting and building upon the *acquis communautaire*."

the EU, although is generally held to be the "body of common rights and obligations which bind all the Member States together within the European Union."[36] The French expression "acquis" meaning "accepted" or "established"[37] has likewise been used by the Council of Europe to describe its normative activities in the fields of democracy, the rule of law, and fundamental human rights and freedoms,[38] and is generally understood to comprise a body of laws, rules, and regulations established within an institutional framework.[39]

Meanwhile, following the conclusion of the four Geneva Conventions in 1949, the modern conceptual understanding of "humanitarianism" can be traced to the 1965 International Conference of the Red Cross, which adopted the seven fundamental principles of the International Red Cross and Red Crescent Movement. The first principle is that of "humanity," which states:

> The International Red Cross and Red Crescent Movement, born of a desire to bring assistance without discrimination to the wounded on the battlefield, endeavors, in its international and national capacity, to prevent and alleviate human suffering *wherever it may be found* [emphasis added]. Its purpose is to protect life and health and to ensure respect for the human being. It promotes mutual understanding, friendship, cooperation and lasting peace amongst all peoples.[40]

However, several of the Red Cross principles (namely voluntary service, unity, and universality) relate specifically to the internal working and structure of the Red Cross Movement. So following the failure of the 1984 draft UN Convention on Expediting the Delivery of Emergency Assistance, a nonbinding but no less

[36] As defined in "Glossary of EU Legislation," http://europa.eu/legislation_summaries/glossary/community_acquis_en.htm.. Alternatively: "The *acquis* comprises all the EU's treaties and laws (directives, regulations, decisions), declarations and resolutions, international agreements and the judgments of the Court of Justice. It also includes action that EU governments take together in the area of Justice and Home Affairs (Freedom, Security and Justice) and the Common Foreign and Security Policy – but it excludes decisions to set up 'enhanced cooperation' arrangements." Vaughne Miller, "The EU's *Acquis Communautaire*," SN/IA/5944, 26 April 2011, Library of the UK House of Commons, p.2. See also: Amichai Magen, "The *Acquis Communautaire* as an Instrument of EU External Influence," *European Journal of Law Reform* 9(3) (2007): 361–392.

[37] Jean-Benoit Ormal-Grenon and Natalie Pomier (eds.), *Oxford Hachette French Dictionary*, 3rd ed. (Oxford University Press, 2001).

[38] Council of Europe, Resolution 1290 (2002): *Future of the co-operation between European institutions*, Assembly debate, June 26, 2002 (20th Sitting).

[39] See for example references to the GATT *acquis* in: WTO, *Japan – Taxes on Alcoholic Beverages*, Report of the Appellate Body (AB-1996–2): 14–15; and the UNCITRAL *acquis* in: Asa Markel, "American, English and Japanese Warranty Law Compared: Should the US Reconsider Her Article 95 Declaration to the CISG?" *Pace International Law Review* 21 (Winter 2009): 163–204 at 200.

[40] The seven fundamental principles of the Red Cross Red Crescent Movement are: humanity, impartiality, neutrality, independence, voluntary service, unity and universality. In *IFRC*, www.ifrc.org/en/who-we-are/vision-and-mission/the-seven-fundamental-principles.

influential examination of humanitarian principles took place in the UN General Assembly. The resulting Resolution 46/182 of 1991 on "Strengthening of the coordination of humanitarian emergency assistance of the United Nations" contained not just technical improvements for the working of UN humanitarian operations but a set of Guiding Principles for humanitarian action.[41] The Resolution stresses that "[h]umanitarian assistance must be provided in accordance with the principles of humanity, neutrality and impartiality." Although these humanitarian principles were left undefined, humanity is generally understood to mean that "human suffering must be addressed wherever it is found, with particular attention to the most vulnerable populations such as children, women, and the elderly. The dignity and rights of all victims must be respected and protected."[42]

The debate regarding the ethical and practical standing of the concepts of neutrality and impartiality, particularly in humanitarian actions in times of war, is well known.[43] Yet the applicability of "humanity" as the overriding motivation and foundation for humanitarian action in both armed conflicts and disasters, as confirmed by Resolution 46/182, is widely accepted in law and practice. The recent entry into force of the African Union's Convention for the Protection and Assistance of Internally Displaced Persons in Africa, has created regional binding obligations on signatory states to "respect and ensure respect for the principles of humanity and human dignity of internally displaced persons."[44]

Meanwhile, the International Court of Justice (ICJ) noted in the Nicaragua case that "[t]here can be no doubt that the provision of strictly humanitarian aid to persons or forces in another country ... cannot be regarded as unlawful intervention, or as in any other way contrary to international law."[45] The ICJ proceeded to

[41] For background to UNGA Resolution 46/182, see: Edward Tsui, *Analysis of Normative Developments in Humanitarian Resolutation since the Adoption of 46/182* (Geneva: UNOCHA, 2009).

[42] GA Res. 46/182, Annex, Guiding Principle 2. See for example: *Guiding Principles for Public-Private Collaboration for Humanitarian Action*, prepared by the World Economic Forum and the UN Office for the Coordination of Humanitarian Affairs (December 2007): 1; and *OCHA on Message: Humanitarian Principles* (Version 1, April 2010) http://ochanet.unocha.org/p/Documents/OOM_HumPrinciple_English.pdf.

[43] It is outside the scope of this paper to detail the evolution of the academic and practitioner debate surrounding humanitarian principles, such as the conceptual split between the Red Cross Movement and the *sans frontières* movement. For critical analysis see: Hugo Slim, "Relief Agencies and Moral Standing in War: Principles of Humanity, Neutrality, Impartiality and Solidarity," *Development in Practice*, 1997: 342–352. See also: David Kennedy, *The Dark Sides of Virtue: Reassessing International Humanitarianism*, (Princeton University Press, 2004). See also: Martin Barnett and Thomas Weiss (eds.), *Humanitarianism in Question: Politics, Power, Ethics*. (Cornell University Press, 2008).

[44] African Union, Convention for the Protection and Assistance of Internally Displaced Persons in Africa ("Kampala Convention") (adopted October 22, 2009, entered into force December 6, 2012), Art 3(1)(c).

[45] *Nicaragua v. United States of America*, par.242. *Military and Paramilitary Activities in and against Nicaragua* (Nicaragua v. United States of America), Merits, Judgment. I.C.J. Reports (1986): 14, par.242. As early as 1949, the ICJ affirmed in the *Corfu Channel Case* that

recall that humanitarian assistance must be provided based on "the purposes hallowed in the practice of the Red Cross, namely 'to prevent and alleviate human suffering,' and 'to protect life and health and to ensure respect for the human being'." Above all, humanitarian assistance must be given without discrimination to all in need.[46] Humanitarian agencies have likewise stressed a firm commitment to humanitarian principles. For example, the influential, albeit nonbinding, Sphere Humanitarian Charter explicitly reaffirms the humanitarian imperative and the need to act in accordance with the principles of humanity and impartiality.[47] The Charter notes: "[b]y this we mean … that all possible steps should be taken to prevent or alleviate human suffering arising out of conflict or calamity, and that civilians so affected have a right to protection and assistance."[48]

Most recently, the ILC Draft Articles on the protection of persons in the event of disasters explicitly contain the humanitarian principles of humanity, neutrality and impartiality, with the Commission noting that these are "core principles recognized as foundational to humanitarian assistance."[49] Although the Commission did not find it necessary to determine whether they were currently part of general international law, the Commentaries stress that "[t]he principle of humanity stands as the cornerstone of the protection of persons in international law."[50]

Furthermore, the Commission highlights that the principle of humanity is an element of both international humanitarian law and international human rights law, while also informing the development of law relating to the protection of persons in the event of disasters. In this, the ILC is drawing on the role that humanity plays as the foundation for actions undertaken by the Red Cross Movement to alleviate suffering. As argued by the ICRC, the principle of humanity is

elementary considerations of humanity were well-recognized principles of the international legal order "even more exacting in peace than in war." *United Kingdom of Great Britain and Northern Ireland v. Albania*, Merits, Judgment of 9 April 1949, I.C.J. Reports (1949): 22.

46 Ibid., *Nicaragua v. United States of America*, ibid., par. 243.

47 The concept of impartiality provides that humanitarian assistance must be provided without discriminating as to ethnic origin, gender, nationality, political opinions, race or religion. Relief of the suffering must be guided solely by needs and priority must be given to the most urgent cases of distress. See: International Committee of the Red Cross (ICRC), *The Fundamental Principles of the Red Cross and Red Crescent*." ICRC Publication, Ref: 0513 (Geneva: ICRC, 1996):: 4–5.

48 *Sphere Handbook*, 16.

49 For adopted draft articles, see: A/CN.4/L.758, A/CN.4/L.776, A/CN.4/L.794, and A/CN.4/L.812. For analysis of the on-going work of the ILC in this regard, see: Dug Cubie, "'An Enchanted Tool? Humanitarian Assistance and the ILC Draft Articles on the Protection of Persons in the Event of Disasters," Irish Yearbook of International Law IV–V (2009–2010): 119–151.; Arnold Pronto, "Consideration of the Protection of Persons in the Event of Disasters by the International Law Commission," *ILSA Journal of International and Comparative Law* (2008–2009): 449; and J. Benton Heath, "Disasters, Relief, and Neglect: The Duty to Accept Humanitarian Assistance and the Work of the International Law Commission," *NYU Journal of International Law and Politics* (2010–2011).

50 For Commentaries on draft articles 1–5, see: "Official Records of the General Assembly, Sixty-fifth Session, Supplement No. 10." A/65/10, Chapter VII, 322–330.

neither too vague nor general to form the basis of the Red Cross Movement's work as "the words used . . . – to prevent, to alleviate, to protect, to ensure respect – require very concrete efforts."[51]

The ILC Draft Articles uniquely include an operative article invoking human dignity. In hortatory language, the draft articles state that "[i]n responding to disasters, States, competent inter-governmental organizations and relevant non-governmental organizations shall respect and protect the inherent dignity of the human person."[52] By recognizing human dignity as the core principle that informs and underpins international human rights law, the draft articles connect the humanitarian principles of draft article 6 with the provisions on human rights in draft articles 7 and 8.[53] Such an approach reflects the argument advanced by Christopher McCrudden that although the concept of human dignity may be interpreted differently in domestic adjudication systems, it provides an important method for interpreting human rights.[54] In particular, McCrudden identifies the basic minimum content of human dignity across religious, cultural and judicial spectrums, and argues that the core of human dignity contains three key elements:

(1) every human being possesses an intrinsic worth, merely by being human;
(2) this intrinsic worth should be recognized and respected by others; and
(3) the intrinsic worth of the individual requires that the state should be seen to exist for the sake of the individual human being, and not vice versa.[55]

[51] ICRC, "*Fundamental Principles*," 3.3. For discussion of the universality of the concept of humanity in international humanitarian law and human rights law see: Coupland, Robin. "Humanity: What is it and How Does it Influence International Law?" International Review of the Red Cross, December 2001: 969–989.

[52] For Commentaries on draft articles 6–11, see: United Nations. "Official Records of the General Assembly, Sixty-sixth Session, Supplement No. 10." A/66/10, Chaper IX, 255–270.

[53] Draft Article 8 states: "Persons affected by disasters are entitled to respect for their human rights." As noted by Christopher McCrudden, many judicial decisions based on human dignity view the concept not as a synonym for human rights but rather as "expressing a value unique to itself, on which human rights are built." See: Christopher McCrudden, "Human Dignity and Judicial Interpretation of Human Rights," *European Journal of International Law*, (2008): 655–724 at 680–681.

[54] For specific discussion of the concept of human dignity in humanitarian and human rights instruments, see ibid., 667–671.

[55] Ibid., 679–680. Ibid. at 679–680. There is a growing body of literature on the concept of human dignity in both domestic and international law, for example: Stéphanie Hennette-Vauchez, "A Human *Dignitas*? Remnants of the Ancient Legal Concept in Contemporary Dignity Jurisprudence," *International Journal of Constitutional Law* 9(1) (2001): 32–57; Conor O'Mahony, "There Is No Such Thing As a Right to Dignity," *International Journal of Constitutional Law* 10(2) (2012): 551–574; and Emily Kidd White, "There Is No Such Thing As a Right to Dignity: A Reply to Conor O'Mahony," *International Journal of Constitutional Law* 10(2) (2012): 575–584.

Despite the lack of commonality or universality exhibited in domestic judicial interpretations of the concept of human dignity, the existence of a "minimum core" relating to the intrinsic worth of each individual ties directly into the motivation and foundations of humanitarian action: namely assisting people in times of extreme distress to ensure a certain minimum of protection and respect is accorded to them despite the challenging circumstances they find themselves in.[56]

Likewise, the ILC's examination of humanitarian principles and the legal conception of human dignity as it relates to international human rights law provides an authoritative recognition of the desire to alleviate suffering in times of war and disaster through humanitarian action. In recognition of the operational and moral debates regarding the application of the principles of neutrality and impartiality, especially but not exclusively in times of armed conflict,[57] the principle of humanity can therefore be considered as the overarching umbrella for the laws and rules of humanitarian action contained within the *acquis humanitaire*. An approach that focuses specifically on the concept of humanity explains the underlying rationale and motivation for humanitarian rules and practice in both armed conflicts and disasters, without imposing constraints on the content or status of norms within the *acquis humanitaire*, thereby providing leeway for the development of bottom up domestic mechanisms. Furthermore, it reflects the reality of the work undertaken by many humanitarian actors to alleviate suffering in both conflict and disaster settings, while recognizing their divergent operational modalities.[58]

DEMARCATING THE 'ACQUIS HUMANITAIRE'

Although one cannot situate the evolving body of humanitarian norms within a single institutional framework, in contrast to the EU's *acquis communautaire*, it is possible to identify a variety of framework documents that establish specialized institutional regimes for component parts of the *acquis humanitaire*. For example, the Rome Statute of the International Criminal Court provides the key foundation for individual criminal responsibility for internationally accepted crimes, including the deliberate targeting of humanitarian personnel and the willful denial of

[56] For discussion of the minimum core set out by McCrudden in reference to serious violations of human rights, see: Paolo Carozza, "Human Dignity and Judicial Interpretation of Human Rights: A Reply," *European Journal of International Law* (2008): 931–944 : "with regard to the most acute and urgent human rights problems worldwide, a judicial (among other actors) reinforcement and consolidation of the content of the minimum core of human dignity is well worth pursuing, even if it is limited to the bare outlines of the three interrelated claims McCrudden identifies."

[57] See: Slim, "Relief Agencies and Moral Standing in War."

[58] As noted in the 2010 Humanitarian Accountability Partnership's "*Standard in Accountability and Quality Management*": "The work of many organisations is not limited to providing humanitarian relief, but includes all types of assistance to people affected by or prone to disasters, conflict, poverty or other crises," 2.

humanitarian assistance, as well as international judicial structures for the prosecution of such offences.[59]

Similarly, the Geneva Conventions and Additional Protocols provide the overarching framework for the legal obligations of states and combatants in times of war, including requirements to provide protection and assistance to civilians, while also reaffirming the institutional mandates of the International Committee of the Red Cross and other national relief societies.[60] One can therefore envisage that the ILC Draft Articles for the protection of persons in the event of disasters could become the institutional and overarching framework for humanitarian actions relating to major natural and human-made disasters, including actions undertaken pre-disaster for risk reduction, prevention and mitigation purposes,[61] and those taken post-disaster for recovery and rehabilitation. Indeed, there is a certain level of support both within the ILC and from states for the draft articles to be concluded as a Framework Convention.[62]

In another key difference to the EU's *acquis communautaire*, the proposed *acquis humanitaire* does not imply that states are obliged to incorporate into their domestic law all the provisions contained within. Rather, the *acquis humanitaire* provides a mechanism to identify the sources that form the legal and policy framework of humanitarian actions in times of crisis. For despite the *acquis humanitaire* encompassing both highly developed systems of positive international law, such as the humanitarian law of armed conflict, and evolving systems such as international disaster laws, the common thread of humanity underpins the origins of these various sources. Yet as with the concept of transnational legal process itself,[63] the aim of the *acquis humanitaire* is not just to define the sources of humanitarian rules, but to explain the normativity of the process as international interactions become internalized at the domestic level. By incorporating relevant provisions of international human rights law including the concept of human dignity, in

59 See: Rome Statute of the International Criminal Court, 2187 UNTS (entered into force 1 July 2002), Article 6 (Genocide), Article 7 (Crimes Against Humanity) and Article 8 (War Crimes). See also: *Elements of Crimes*, Official Records of the Review Conference of the Rome Statute of the International Criminal Court, Kampala, 31 May-11 June 2010 (ICC, RC/11); and Dug Cubie, "Is the Wilful Denial of Humanitarian Assistance in Natural and Human-Made Disasters a Crime Against Humanity?" *ASIL Accountability Newsletter* (Summer 2012).

60 The *acquis humanitaire* only comprises aspects of general and customary international humanitarian law in armed conflicts relating specifically to the provision of humanitarian assistance and protection to civilian populations, so does not include rules relating to the conduct of hostilities in either international or non-international armed conflicts.

61 For example in the context of the UN International Strategy for Disaster Reduction. "Hyogo Framework for Action 2005–2015: Building the Resilience of Nations and Communities to Disasters." World Conference on Disaster Reduction. Kobe, Hyogo, Japan, January 18–22, 2005.

62 General Assembly Official Records, Sixty-first session, Supplement No.10 (A/61/10), Annex C, par.24; and UNGA Sixth Committee's 22nd to 25th meetings in 2008 (A/C.6/63/SR.22 – A/C.6/63/SR.25), and 20th to 22nd meetings in 2009 (A/C.6/64/SR.20 – A/C.6/64/SR.22).

63 Koh, "Transnational Legal Process," 184.

addition to technical regulations for the effective delivery of humanitarian assistance,[64] the *acquis humanitaire* provides an inclusive framework for promoting and respecting the rights of individuals and groups affected by humanitarian crises, while simultaneously recognizing the fundamental role played by operational standards and frameworks for efficient prevention and response activities.

Therefore, supported by the work of international norm entrepreneurs and sponsors, it is undeniable that the international community is currently engaged in a process of creating and elaborating a specific body of law relating to humanitarian action in crisis settings. Interactions between transnational actors are stimulating a range of academic, judicial, practitioner, and state interpretations of the applicable humanitarian norms both internationally and at the domestic level. The resultant *acquis humanitaire* contains binding obligations from existing international law sources, including judicial decisions by international tribunals such as the International Court of Justice[65] and the European Court of Human Rights,[66] as well as authoritative nonbinding normative standards for humanitarian action, such as the Sphere Handbook and RCRC Code of Conduct. Taken together, the content and sources of these various aspects of prevention and response in conflict and non-conflict settings represent an emerging body of humanitarian laws and practice.

Furthermore, as explained below, the transnational legal process of multiple interactions leading to legal interpretations of emerging and existing humanitarian norms contained within this *acquis humanitaire* ultimately helps promote the internalization of such norms into domestic legal and policy frameworks. As such, the *acquis humanitaire* is not intended as a constrictive description of purely positive law, but a dynamic reflection of the sources that influence state and humanitarian actors in their preparations for and responses to humanitarian crises.

These various sources may overlap both in approach and content, but also reflect separate and distinct bodies of law and practice. So while the humanitarian motivations behind the *lex specialis* of the international humanitarian law of armed conflict share many similarities with those of international criminal law or disaster laws, in actual cases one must examine the specific humanitarian crisis to determine which binding or nonbinding aspects of the *acquis humanitaire* may be applicable. Although a conceptually broad and porous tent, the boundaries of the *acquis humanitaire* can nevertheless be demarcated by the principle of humanity in

64 Such as the Tampere Convention on the Provision of Telecommunication Resources for Disaster Mitigation and Relief Operations (1998), 2296 UNTS 5 (entered into force January 8, 2005).

65 For example, *Nicaragua v. United States of America*, 14, pars. 242–243.

66 For discussion of state responsibility to undertake preventative measures, see: *Öneryildiz v. Turkey*. Application 48939/99 (November 30, 2004). (methane explosion in shantytown built on rubbish dump); and *Budayeva & Others v. Russia*. Applications 15339/02, 2116602, 20058/02, 11673/02 and 1543/02 (March 20, 2008) (mudslide that destroyed a village).

times of crisis which provides a common thread running throughout the current transnational dialogues regarding applicable humanitarian norms and practice.

TRANSNATIONAL LEGAL APPROACHES TO DISASTER RESPONSE

While the international normative obligations and institutional frameworks for disaster response are not yet settled, the diverse source and nature of such initiatives suggests that the international community is engaged in a process of norm creation, elaboration and interpretation reflecting a desire for legal clarity in humanitarian operations. Situated within the framework of transnational law[67] and concomitant implications for the creation of binding and nonbinding normative standards, the constant flow of ideas and principles between the national, regional and international spheres regarding humanitarian norms can best be described and analyzed from a transnational legal process approach.

Philip Jessup defined "transnational law" as that which "regulates actions or events that transcend national frontiers. Both public and private international law are included, as are other rules which do not wholly fit into such standard categories."[68] He thereby provided the name for a concept already challenging the traditional view of international law as the regulation of relations between sovereign states. Reflecting Jessup's experience of both law and politics,[69] it is not surprising he promoted an understanding of international law which included the interactions of non-state entities, as well as inter-disciplinary approaches to the creation of rules, norms and customs.[70]

Yet despite Jessup's stature, the field of "transnational law" was initially slow to take off. Nevertheless, building on Jessup's concept of transnational law and expansive view of international legal sources and influences, plus the American legal process tradition,[71] Harold Koh developed his theory of "transnational legal

67 See generally: Reimann, "Transnational Law."

68 As quoted in: Covey Oliver, "Philip C. Jessup's Continuing Contribution to International Law," *Columbia Law Review* 62(7) 62(7) (November 1962):1132–1137. See also: Christian Tietje, Alan Brouder, and Karsten Nowrot (eds.), "Philip C. Jessup's *Transnational Law* Revisited on the Occasion of the 50th Anniversary of its Publication," *Essays in Transnational Economic Law* 50 (February 2006).

69 Jessup served as a US government official and diplomat to the newly created UN and subsequently as a Judge on the International Court of Justice. Jessup's humanitarian credentials include involvement in drafting the UN Charter in 1945 and role as Under-Secretary-General of the UN Relief and Rehabilitation Agency during WWII. See: Oscar Schachter, "Philip Jessup's Life and Ideas," *American Journal of International Law* (October 1986): 878–895.

70 Peer Zumbansen, "Transnational Law," in *Encylcopedia of Comparative Law*, ed. Jan M. Smits (Edward Elgar Publishing, 2006), 738–40.

71 The legal process approach was adopted by US legal scholars to help distinguish between objective law and subjective policy, and was applied to international law in Abram Chayes, Thomas Ehrlich, and Andreas F. Lowenfield, *International Legal Process: Materials for an Introductory Course*, (Boston: Little, Brown, 1968). As noted by Anne-Marie Slaughter, this was

process" which he defines as the theory and practice of how public and private actors, including states, international organizations, multinational enterprises, non-governmental organizations, and private individuals, interact in various public and private, domestic and international fora to make, interpret, enforce, and ultimately, internalize rules of transnational law.[72] Key to this conception are four aspects, namely that transnational legal processes are non-traditional, non-statist, dynamic, and normative.[73] Professor Koh proceeds to note that:

> [f]rom this process of interaction, new rules of law emerge, which are interpreted, internalized, and enforced, thus beginning the process all over again. Thus, the concept embraces not just the descriptive workings of a process, but the *normativity* of that process. It focuses not simply upon how international interaction among transnational actors shapes law, but also on how law shapes and guides future interactions: in short, how law influences why nations obey.[74]

Transnational legal process can therefore be used to explain how international norms are created and complied with based on multiple interactions at various levels by a multitude of actors. These interactions prompt legal interpretations which may ultimately lead to the internalization of a norm into either domestic, regional or international law thereby promoting obedience to the norm.[75]

The developing *acquis humanitaire* is undergoing such a process to determine which norms and standards should apply at the national, regional, or international levels. Key to this is the internalization of international and regional norms within

a pragmatic response to the realist challenge to the relevance of (international) law in the face of power politics, by concentrating on how international legal rules are created, interpreted and applied in practice, rather than providing a descriptive analysis of international laws' normative content. Anne-Marie Slaughter Burley, "International Law and International Relations Theory: A Dual Agenda," *American Journal of International Law* (1993): 205–239 at 213–214. See also: Mary Ellen O'Connell, "New International Legal Process," *American Journal of International Law* (1999):205–239. See also: Anne-Marie Slaughter, Andrew S. Tulumello, and Stepan Wood, "International Law and International Relations Theory: A New Generation of Interdisciplinary Scholarship," *American Journal of International Law* (1998): 367–397.

72 Koh, "Transnational Legal Process," 183–184.

73 Ibid., "Transnational legal process has four distinctive features. First, it is *non-traditional:* it breaks down two traditional dichotomies that have historically dominated the study of international law: between domestic and international, public and private. Second, it is *non-statist:* the actors in this process are not just, or even primarily, nation-states, but include non-state actors as well. Third, transnational legal process is *dynamic,* not static. Transnational law transforms, mutates, and percolates up and down, from the public to the private, from the domestic to the international level and back down again. Fourth and finally, it is *normative.*"

74 Ibid.

75 *C.f.* there is on-going debate regarding the factors that influence State behavior. For example, Oona Hathaway has criticized norm-based scholarship, such as transnational legal process, for not providing clear guidance regarding which laws will be internalized, or sufficient predictive evidence of what states will actually do. See: Oona Hathaway, "Between Power and Principle: An Integrated Theory of International Law," *University of Chicago Law Review* (2005): 468–536.

domestic legal structures.[76] Professor Koh has identified three distinct forms of internalization:

(1) "social internalization" when a norm acquires sufficient public legitimacy to gain widespread general adherence;
(2) "political internalization" when political elites accept an international norm and advocate its adoption as a matter of government policy; and
(3) "legal internalization" when an international norm is incorporated into the domestic legal system through the executive, legislature, or judiciary, or a combination of all three.[77]

Koh argues that the sequencing of internalization will vary from case to case, with social internalization occurring before political or legal internalization in some instances, while legal internalization may prompt social and political internalization in others.

INTERNALIZATION OF HUMANITARIAN NORMS

International and Regional Development of Humanitarian Norms: The multiplicity of sources of ideas and principles of disaster response, encompassing top down international and bottom up national approaches, in addition to regional mechanisms, reflects the symbiotic relationship between the identification of humanitarian norms and the process for determining rights and responsibilities in disaster settings. A transnational legal process approach recognizes these multiple flows of influence, whereby domestic law and practices can directly impact international understanding of relevant norms and vice versa.[78] Consequently, a wide range of traditional international law topics,[79] including international humanitarian law and human rights law, are engaged in the development of consensus of humanitarian norms, alongside an evolving body of nonbinding normative standards of humanitarian response.

These international laws and standards feed directly into regional and sub-regional structures, such as the European Union's internal and external civil protection mechanisms[80] or the ASEAN disaster management

[76] Few countries have developed comprehensive national legal frameworks for disaster responses, particularly covering offer and acceptance of international assistance by other states and humanitarian agencies. See: IFRC, "*IDRL Guidelines*," 3–5.

[77] Harold Koh, "How is International Human Rights Law Enforced?" *Indiana Law Journal* (1998–1999): 1397–1417.

[78] Koh, "Transnational Legal Process," 184.

[79] For the blurring of distinctions between private and public international law, and international and domestic law since the end of the WWII, see: Reimann, "Transnational Law," 406–408.

[80] Since 2010, the European Communities Humanitarian Office (ECHO) has encompassed both external humanitarian aid and internal civil protection. See: European Communities Humanitarian Office, "Council Decision Establihsing a Community Civil Protection

system,[81] and then into domestic law, policies and practices. Such approaches reflect Professor Koh's description of "layered cooperation" within legal diplomacy, whereby in any given area of international cooperation, the choice between international agreements and non-legal alternatives is not binary, and states may choose non-legal mechanisms to facilitate international cooperation.[82]

In the context of sovereignty as responsibility, Francis Deng and his colleagues similarly referred to "layers of assistance" whereby states exercise sovereignty at home, then turn to neighbors for assistance, and subsequently request assistance at a regional level before looking for a global response via the UN.[83] This layered approach to diplomacy and sovereignty reflects the reality of international humanitarian responses. If one takes Cyclone Nargis in 2008, it took a month until Western relief workers were allowed to assess the stricken areas of Myanmar, while their governments argued over the legally of a military humanitarian intervention.[84] However, Indian, Chinese, and other Asian neighbors were allowed earlier access to assist the Myanmar relief effort,[85] and there were numerous local initiatives despite government restrictions.[86]

Clearly, by delaying much-needed international assistance the Myanmar government failed in their responsibilities as a sovereign state toward persons on their territory. Nevertheless, the role played by regional bodies should not be underestimated. Since 2005 the ASEAN Agreement on Disaster Management and Emergency Response has regulated the regional responses to natural disasters, in a region well-known for its strong views on the importance of state sovereignty and non-intervention, and so it was through assistance from ASEAN states that international support first arrived in Myanmar.

Meanwhile, at the global level one can identify a range of top down approaches initiated by the UN or other global bodies including non-governmental actors.

Mechanism (recast)," 2007/779/EC, Euratom, OJ L314/9, November 8, 2007); and The European Consensus on Humanitarian Aid, "Joint Statement by the Council and Representatives of the Governments of Member States Meeting within the Council, the European Parliament and the European Commission," OJ C25, January 30, 2008.

81 See: IFRC, *Disasters in Asia: the Case for Legal Preparedness*, (Geneva: IFRC, 2010).

82 Harold Koh "Twenty-First Century International Lawmaking," (speech at Georgetown University Law Center, Washington DC, October 17, 2012), 10.

83 Francis Deng, et al., *Sovereignty as Responsibility: Conflict Management in Africa*. (Brookings Institution Press, 1996), xviii.

84 Stuart Ford, Stua "Is Failure to Respond Appropriately to a Natural Disaster a Crime Against Humanity? The Responsibility in the Aftermath of Cyclone Nargis," *Denver Journal of International Law and Policy*, 2009–2010: 227–276.

85 See also: Martin Jacques, "His Master's Voice," *The Guardian*, May 15, 2008. www.guardian.co.uk/commentisfree/2008/may/15/hismastersvoice1. See also: Tripartite Core Group of Representatives from the Government of Myanmar, ASEAN, and UN, "Post-Nargis Joint Assessment," July 2008.

86 "Comic Arrested for Burma Aid Runs," *BBC News*, June 5, 2008. http://news.bbc.co.uk/2/hi/7437251.stm.

Whatever the final format of the ILC Draft Articles on the protection of persons in the event of disasters, it is likely the views of the primary UN body for codification and progressive development of international law will have a major impact on our understanding of the norms and principles, as well as technical and operational modalities, of international humanitarian action.[87] Yet the development of consensus within the ILC or UN is not just influenced by transnational standards but simultaneously exerts its own influence on the operations and practice of state and non-state humanitarian agencies. For example, it is still unclear what, if any, influence the removal of a "right to humanitarian assistance" from the latest version of the UN IASC Operational Guidelines[88] will have when, conversely, the (non-governmental) 2011 Sphere Handbook[89] now bases its entire approach on the existence of such a right.

Alternatively, reflecting the dynamic nature of transnational processes, the complexities and diversity of domestic approaches to disasters may ultimately help in the development of an international consensus,[90] although the importance of sovereignty and governmental desire to control or restrict access of international humanitarian assistance is likely to remain high. While bottom up approaches may influence domestic structures, mechanisms, and standards for disaster response,[91] it is unclear to what extent any individual practice by particular states will have on the global understanding of humanitarian principles and practices. As David Caron notes, the haphazard bottom up response by states "mirrors the fortuity of the catastrophes humanity encounters."[92] But perhaps recognizing the evolving international body of law and practice contained within the *acquis humanitaire* will allow domestic actors the opportunity to apply the concept of humanity and its operational requirements to alleviate suffering more clearly within their national legal and policy structures. As argued by Christopher McCrudden, the principle of human dignity subject to its core minimum content allows the domestic context to

[87] Although as noted by David D. Caron in the context of the ILC Draft Articles on state responsibility: "It is essential to recognize that internationally, the freezing effect of a code may be greater [than national legal codes] because the system lacks both a legislative body to amend the codes and an authoritative judiciary with jurisdiction over enough cases to adapt them to new circumstances and needs." David D. Caron, "The ILC Articles on Responsibilty: The Paradoxical Relationship Between Form and Authoritary," *American Journal of International Law* (2002): 857–873 at 861.

[88] IASC, *Operational Guidelines*.

[89] *Sphere Handbook*.

[90] Koh, "Transnational Legal Process," 184.

[91] For discussion of human rights implications arising from responses to Hurricane Katrina, see: Susannah Sirkin, "The Debacle of Hurrican Katrina: A Human Rights Response," *Fletcher Forum of World Affairs* (2006): 223–227.; and Maria Isabel Medina, "Confronting the Rights Deficit at Home: Is the Nation Prepared in the Aftermath of Katrina? Confronting the Myth of Efficiency," *Califronia Western Law Review* (2006–2007): 9–20.

[92] David Caron, "Addressing Catastrophes: Conflicting Images of Solidarity and Separateness," in *The International Aspects of Natural and Industrial Catastrophes*, by Hague Academy of International Law, 3–4. (Martinus Nijhoff Publishers, 2001), 3–4.

be incorporated under the guise of a universal principle, so each jurisdiction can develop its own practice in the application of international human rights norms.[93]

Case Study—The Internalization of Humanitarian Norms within the United Kingdom: Inherent within the concept of transnational legal process is the fact that the development and elaboration of norms is by necessity an on-going and long-term process. Consequently, the widespread acceptance of nonbinding normative standards by non-governmental humanitarian agencies in the UK provides an interesting example of the social internalization of international norms for humanitarian responses, which can be contrasted with the limited political and legal internalization of such norms within UK domestic law and policy.

Of 515 humanitarian agencies globally to sign the RCRC Code of Conduct, 115 or 22% are UK-based. The next largest group of signatories based on country of origin is from the USA, yet at only 7% of the total.[94] Similarly, 29% of the 209 member organizations of the People in Aid Code of Good Practice in the Management and Support of Aid Personnel are UK-based,[95] as are 19% of signatories to the Humanitarian Accountability Partnership's 2010 HAP Standard in Accountability and Quality Management,[96] which suggests a marked tendency for both large and small UK-based NGOs to accede to international (nonbinding) standards regarding humanitarian responses.[97]

Moreover, the HAP Standard was originally derived from the 1997 Humanitarian Ombudsman Project initiated by a group of British humanitarian organizations, including the British Red Cross.[98] While there may be a variety of reasons behind the willingness of UK-based organizations to commit to such voluntary codes and standards, it is noticeable that such a high proportion of UK-based non-governmental agencies appear to support strengthening international norms regulating humanitarian responses, while the UK government has steadfastly rejected the concept of binding international obligations for disaster responses.[99]

93 McCrudden, "Human Dignity," 719–720.

94 List of Signatories, October 24, 2012. www.ifrc.org/Global/Publications/disasters/code-of-conduct//codeconduct_signatories.pdf.

95 People as Aid, "Members' Directory," www.peopleinaid.org/membership/directory.aspx. The next largest country group is from the USA, at 6 percent.

96 The Humanitarian Accountability Partnership calls the Standards the "most widely recognized humanitarian programme norm." Humanitarian Accountability Partnership, *Membership*. www.hapinternational.org/membership.aspx. The next largest country group is from Pakistan, at 17 percent.

97 These high numbers potentially also reflect the large number of such organizations within British civil society – which may be unusually large compared with other Western democracies.

98 See: David Peppiatt, "The Ombudsman Project: Pilot Project to Investigate the Concept of an Ombudsman for Humanitarian Assistance," *Humanitarian Exchange Magazine* (November 1997): 9. See David Peppiatt, "Update: Ombudsman for Humanitarian Accountability," *Forced Migration Review* (December 1999): 43.

99 Although as noted in the 2011 government-initiated review of the UK's humanitarian emergency response policies and capacity, the UK government has funded several of these standard

Indeed, although the UK signed the 1927 Convention establishing an International Relief Union, it did not extend the application of the Convention to its colonies or overseas territories, many of which faced higher risks of disaster than the UK itself. It was subsequently the first country to pull out of the Convention in 1948 on the establishment of the UN.[100] This reluctance is similarly manifest today in the debates in the UN General Assembly Sixth Committee on the ILC's work on the protection of persons in the event of disasters. Although most states have reserved their positions regarding the final form of the draft articles, the UK has clearly expressed their preference for nonbinding guidelines or a set of principles.[101] Furthermore, despite ranking highly in terms of the overall level of humanitarian assistance provided, the lack of independence and impartiality of the UK government's humanitarian aid has recently been noted.[102] Indeed, there is a gap between the policy and practice of many EU Member States, which has seen the increased politicization of humanitarian aid, particularly in conflict zones such as Afghanistan and Somalia where the objectives of humanitarian and military actors can be at variance.[103]

Yet a transnational legal process of political and legal internalization appears to be following the existing social internalization of nonbinding humanitarian norms within UK-based non-governmental organizations who have acted as norm sponsors. For example, it is notable that the UK Department for International Development has recently become a signatory to the HAP Standards, while the UK government worked closely with the British Red Cross study of UK disaster laws

setting initiatives. See: Chaired by Lord Ashdown, "Final Report of the Humanitarian Emergency Response Review," (March 28 2011): 26. However the resultant governmental Humanitarian Policy takes a limited position on the applicability of international norms and standards. While confirming its adherence to the principles of humanity, impartiality, neutrality and independence in humanitarian actions, the UK government policy does not specifically reference non-governmental standard setting initiatives such as the Sphere Handbook or HAP Standard. See: Department for International Development, "Saving Lives, Preventing Suffering and Building Resilience: The UK Government's Humanitarian Policy," (London, September 2011).

[100] One of the critiques made by the UK delegation before the 1927 diplomatic conference cynically stated that measures to encourage the study of preventive measures against disasters, "must not lead to the setting up of a laboratory for preparing pills against earthquakes." As quoted in: Hutchinson, "The IRU," 284. See also: Macalister-Smith, "The IRU," 365.

[101] See: United Nations, "Summary of 23rd Meeting, General Assembly Sixth Committee." A/C.6/64/SR.20, 2009, par. 40); and "Summary of 20th Meeting, General Assembly Sixth Committee." A/C.6/64/SR.20, 2009, par. 40.

[102] DARA. The Humanitarian Response Index 2010: The Problems of Politicisation. Madrid, November 2010. This is particularly important as 75 percent of UK humanitarian assistance was provided in conflict zones in 2009–2010: DFID, 4, *Humanitarian Policy*, 4.

[103] Caritas Europa. *Bridging the Gap Between Policy and Practice: The European Consensus on Humanitarian Aid and Humanitarian Principles*. Brussels, October 2011. See also: Naomi Baird and Anne Street, "Bridging the Gap between Policy and Practice: the European Consensus on Humanitarian Aid and Humanitarian Principles," *Humanitarian Exchange Magazine*, May 2012.

in 2009.[104] By assisting in grouping together the disparate laws, regulations, policies, and plans relating to disaster response within the administrative regions of the United Kingdom, the study highlights the importance of bottom up approaches emanating at the domestic level based on international and regional norms.[105] Furthermore, social internalization by UK-based NGOs appears to have occurred primarily within those undertaking external humanitarian actions. As noted in the British Red Cross study, decisions on the request or acceptance of international assistance within the UK appear to be based on *ad hoc* arrangements due to a lack of relevant operational rules.[106] So the transnational dialogue will continue, with the social internalization of norms that has already taken place potentially acting as the precursor to more developed political and legal internalization.

CONCLUSION

This chapter provides a brief overview of transnational approaches to humanitarian norms. Yet the concept of an emerging *acquis humanitaire* and the role played by transnational legal processes of norm creation and elaboration provide a mechanism to analyze how norms and practices can be internalized, and therefore operationalized, in preparations for and responses to humanitarian crises. Top down international initiatives, exemplified by the International Law Commission's draft articles, have already had far-reaching impacts on our understanding of the normative values and operational mechanisms for disaster prevention and response. But these norms and practices are in a constant state of change, and on-going work is required to improve international consensus on these issues. Meanwhile, although many countries provide large amounts of humanitarian assistance in the face of humanitarian crises, the divergence of bottom up approaches to humanitarian norms and responses to specific events highlights the need for further interactions on and interpretation of the overarching rationale and framework for international offers and acceptance of support to alleviate suffering.

Ultimately, the internalization of humanitarian norms will be influenced by regional and domestic factors, not least by local interpretations of what the "international consensus" on humanitarian norms and operational standards may be. Therefore, it is important to understand this process of interpretation and internalization, including the sequencing of social, political, and legal internalization, and how this may vary in different regions and countries.

The concept of an *acquis humanitaire* based on the principle of humanity provides an analytical framework for this on-going dialogue between transnational actors as to the binding and nonbinding norms and standards applicable in times of

[104] See: BIICL. "*Analysis of Law in the UK*." plus discussion of the study in this volume: Clement. Chapter 5.

[105] Clement, ibid. [106] BIICL. "*Analysis of Law in the UK*," 79.

crisis. As the International Law Commission works toward developing a comprehensive global legal framework for the protection of persons in the event of disasters, it is useful to also focus attention on regional initiatives and to question how interactions between international, regional, and national approaches to humanitarian crises can improve the delivery and effective coordination of humanitarian action by multiple actors in actual disaster settings.

Conclusion

Anastasia Telesetsky

INTERNATIONAL LAW OF DISASTER IN A NO-ANALOG FUTURE

Much scholarly attention has been given to scientists' assertion that we have entered a no-analog ecological future. Examining fossil pollen, paleoecologists inform us that if the current climate change models are correct then the environmental communities of today may not survive the climate of tomorrow and we will instead see the rise of unpredictable novel ecosystems.[1] Everything scientists have to say today about the uncertain survival of non-human communities in the face of unprecedented change should also resonate within our human communities as we plan for averting future disasters.

We are currently living in a space of unprecedented change. In human history, we have never had so many individuals crowded into so many environmentally vulnerable places. We have never had so much unfrozen global waters or warmest recorded years on record. We have never been able to travel as far as we do in so little time. We have never had as concerted of an effort to protect individual human rights as we do today. We have never had so many opportunities for transboundary state cooperation. Like the pollen communities in their no-analog ecological future, we have entered a no-analog legal future regarding how we manage our human relationships within a rapidly changing environment.

It is because of the coexistence of a no-analog ecological future and a no-analog legal future that the law of international disasters matters. When the environment changes rapidly, the law must change too to protect what we socially value the most – human and non-human lives. As the contributions to this book have underscored, there is a necessity to be proactive in thinking about both emergency responses and disaster risk reduction and how they interact. International law and policy play an increasingly important role in defining both national and

[1] Douglas Fox, "Back to the no-analog Future?," *Science* 316, no. 5826 (May 11, 2007): 823–825 (citing scientific research by John Williams, Stephen Jackson, and John Kutzbach).

international best practices in responding to disasters because we have access to new warning, search and rescue, and post-disaster rehabilitation technologies. While we are better prepared than we have been historically, we are not yet fully prepared to face a world where certain kinds of disasters are more prevalent and possibly more destructive.

Where does the international law of disaster go from here? A number of authors in this book have suggested that international legal efforts to prevent disaster and to respond to disaster have reached a critical governance stage. When reflecting on a no-analog future, a number of interlocking themes emerge including the relevance of sovereign boundaries; the challenges of long-term financing; and the role of non-state actors.

Boundaries

We know that sovereign boundaries make disaster relief efforts logistically difficult. There may be internal reluctance to waive custom and immigration laws coupled with political resistance to receiving outside assistance. In the no-analog legal future where states may also not have a responsibility to protect disaster victims from other states, sovereign boundaries are likely to remain a legal fixture for defining state-to-state relations. Albeit, territorial boundaries may become a future subject of even greater friction among states if the predictions regarding "environmental refugees" materialize and individuals facing environmental threats flee across borders without permission of a resisting recipient state. It is likely that a large percentage of future "environmental refugees" will also be future disaster victims either fleeing an immediate disaster event such as an earthquake, or more likely fleeing in a post-disaster phase when it becomes apparent that viable restoration such as desalinizing fields is not possible. Are boundaries permeable when individuals are responding to disaster? What is the relationship between human rights and territorial boundaries?

One might dispose of this impermeable legal boundaries issue simply by relabeling what states are obligated to do in this emerging field. Instead of an international law of disasters, why not pursue a law of international disasters? This might be the path of least resistance if we want to avoid tricky boundary issues. When we have a transboundary disaster that involves more than one state, states could give explicit or implicit consent to trigger international or regional disaster responses. Where one state has responsibility for creating conditions that led to the disaster, the law of state responsibility might be implicated. Neither of these responses predicated on well-developed doctrines of consent and responsibility is particularly controversial. Yet, focusing only on "international disasters" seems an unnecessary narrowing of the field. We have many more individual human tragedies, including earthquakes, floods, cyclones, and droughts that demand international attention because the scale of human suffering may exceed the capacity of a nation-state or a nation-state may simply refuse to act to provide relief to its own citizens. In this no-analog future, some territorial boundaries may simply disappear. The future of the international law of disasters requires more thought to be given to how permeable state boundaries can be or should be when disaster strikes.

Financing

When we don't know exactly what the future might bring in terms of the scope or intensity of disaster events, it is hard to budget for the consequences of disaster. Disasters probe a curious aspect of human psychology. We are loss-adverse in that we suffer more from a given loss than we benefit from a similar gain. Yet, we collectively do little risk mitigation and instead live in a world of delayed maintenance. We instead assign large discount rates to the possibilities of low probability but high impact events. Given the contradictory impulses associated with loss aversion and delayed investment, the international law of disasters may play an important stimulus role in improving state accountability.

As a number of contributors to this volume have suggested, states need to collectively invest resources today to ensure that there are resources for tomorrow even if disaster strikes. Financing is integral to ensuring that national boundaries do not become the subject of friction and conflict. All states must expend some basic resources to plan for unlikely but possible disaster events by updating communication networks, ensuring a supply of disaster relief materials, and training disaster relief professionals. This long-term planning and financing is a hallmark of good governance. Yet few states seem to be making these investments as part of their national expenditures.

The future of international law of disaster requires greater attention on the responsibilities of a state to its citizens in preparing for disaster by financing disaster prevention and disaster relief. Disasters are not entirely unpredictable since what makes a particular event a disaster is often the lack of sound infrastructure or preparation on the ground. Today, global disaster risk reduction financing seems to be a priority concern of only a few states. Fifty percent of the financing for disaster risk reduction has come since 1993 from Japan and the World Bank.[2] Most of the financing has gone to middle-income countries such as China and Indonesia.[3] Many of the poorest and more vulnerable countries receive little support in pursuing risk mitigation and instead must rely largely on emergency relief and post-disaster relief, which accounts for 96.5 percent of the financing spent on disasters.[4] This approach keeps states locked in a spinning wheel of poverty since

[2] Jan Kellett and Alice Caravani, *Financing Disaster Risk Reduction: A 20 Year Story of International Aid*, Global Facility for Disaster Reduction and Recovery at the World Bank and the Overseas Development Institute (2013): iv.

[3] Ibid., vi. [4] Ibid., 13 (Calculated for 1991–2010).

any destruction to national assets forces a state to start over in building its social, economic, and ecological infrastructure.

Greater consideration by the international community should be given to a number of creative disaster relief financing interventions by states including the use of catastrophe bonds and disaster relief insurance. For example, in the Caribbean, the Caribbean Catastrophe Risk Insurance Facility has been providing for affordable insurance products for sixteen states to insure against the risk of earthquakes and hurricanes.[5] In Mexico, the Mexican government has negotiated a catastrophe risk bond, MultiCat Mexico 2012, to provide $315 million coverage against earthquakes and hurricanes.[6]

In 2013, the Marshall Islands, Samoa, Solomon Islands, Tonga, and Vanuatu are participating in a two-year Pacific Catastrophe Risk Insurance Pilot to insure against the impacts of tsunamis.[7] A combination of disaster risk reduction and disaster relief financing holds the potential for greater self-sufficiency for states rather than ongoing dependency. Yet our current aid budgets do not prioritize risk reduction. In fact, in 2010, while $9.5 billion was spent by on peacekeeping and $4.2 billion on food aid by international aid donors, only $1.1 billion was spent on disaster risk reduction activities such as the restoring of wetlands.[8]

Non-state actors

As states critically evaluate the role of territorial borders in the face of a disaster and contemplate how to adequately finance disaster risk reduction and disaster relief, there remains a critical question about what role non-state actors can and should play in the delivery of disaster aid and risk reduction. As a number of the contributions articulate, non-state actors are crucial for the delivery of emergency relief. But for individual search and rescue teams operating often in a voluntary capacity and organizations such as the International Federation of Red Cross and Red Crescent Societies delivering supplies and supplying qualified relief specialists, there would have been an even greater death toll from recent disasters. Non-state actors have played an essential role in improving disaster relief policies with groups such as the Red Cross and Red Crescent Societies advising policymakers.

But what should the relationship be between States and non-state actors? Can states merely delegate responsibilities to non-state actors? Can states allow the market to resume duties of disaster relief and risk reduction? Is it possible that in

5 Caribbean Catastrophe Risk Insurance Facility, www.ccrif.org/

6 World Bank, "Mexico Launches Second Catastrophe Bond to Provide Coverage Against Earthquakes and Hurricanes," http://treasury.worldbank.org/bdm/htm/Financing_Noteworthy/Mexico_Launches_Second_Catatrophe_Bond_Oct2012.html

7 World Bank, "Helping Small Islands State Cope with the Aftermath of Natural Disasters," (January 2013) http://treasury.worldbank.org/web/documents/PacificIslands_PCRFIpilot.pdf

8 Ibid., 8.

the no-analog future that states can contract out any state responsibilities that might be triggered by human rights obligations to groups of "humanitarian mercenaries?" If so, what legal accountability measures are there to protect individuals who are the recipient of disaster relief services? Does international human rights law provide any guidance as to what the minimum legal standards of accountability might be?

We cannot avoid future disaster events. But what can be avoided is the intensity of a disaster and the level of community devastation by each state making a good faith effort to address the conditions of social, economic, and environmental disaster. We are not powerless and one of the few tools that we have available to address the impacts of disaster on a community is law and international law in particular. Global law is an imperfect tool but it has the power to create normative shifts that can protect individual rights when an emergency strikes. Given that the international law of disaster will be the field of law that may govern how sates and non-state actors will respond in the coming years to the unfettered forces of nature and to human-engineered accidents, states must focus greater attention on defining their obligations to their citizens and their obligations to each other.

Given the international successes of legal harmonization in the field of international trade law captured in the General Agreement on Trade and Tariffs and the General Agreement on Services, it is not inconceivable that states can reach a similar mutually beneficial legal consensus on how best to prevent disasters, deliver emergency services when disaster occurs, and ensure sufficient funding and oversight to rebuild. To reach a working consensus that is adequately funded will take time. With the uncertainty of when and where the next disaster might unfold, there is no better time than the present for states to create progressive law that will promote disaster-resilient communities and protect individual rights. No state should be disinterested in designing the future of disaster policies since no state can guarantee immunity from disaster.

We do not need to fear the no-analog future described in this conclusion. States simply need to be creative with the resources that they have and committed to leaving the world a little better than they inherited it. This will require some compromise, reciprocity, an investment in long-term hope, and a certain degree of altruism. But this should not be too much to ask of the international community. After all, "We the Peoples of the United Nations" are determined to "reaffirm faith in fundamental human rights, in the dignity and worth of the human person, in the equal rights of men and women and of nations large and small."[9] An international law of disaster relief and risk reduction is a just manifestation of our commitments to value simultaneously human rights and the sovereign rights of "nations large and small" to help us navigate this windy, rainy, and likely shaky no-analog future.

[9] Charter of the United Nations, concluded June 26, 1945, entered into force October 24, 1945, 1 U.N.T.S. XVI.

APPENDIX A

Guidelines for the Domestic Facilitation and Regulation of International Disaster Relief and Initial Recovery Assistance (IDRL Guidelines)

INTRODUCTION

1. *Purpose and Scope*

1. These Guidelines are non-binding. While it is hoped that States will make use of them to strengthen their laws, policies and/or procedures related to international disaster response, as appropriate, the Guidelines do not have a direct effect on any existing rights or obligations under domestic law.
2. They draw from many existing international instruments, including United Nations General Assembly Resolutions 46/182 of 1991 and 57/150 of 2002, the Measures to Expedite International Relief of 1977 and the Hyogo Framework for Action of 2005.
3. Their purpose is to contribute to national legal preparedness by providing guidance to States interested in improving their domestic legal, policy and institutional frameworks concerning international disaster relief and initial recovery assistance. While affirming the principal role of domestic authorities and actors, they recommend minimum legal facilities to be provided to assisting States and to assisting humanitarian organizations that are willing and able to comply with minimum standards of coordination, quality and accountability. It is hoped that the use of these Guidelines will enhance the quality and efficiency of international disaster relief and initial recovery assistance in order to better serve disaster-affected communities.
4. These Guidelines are not intended to apply to situations of armed conflict or disasters that occur during armed conflicts, or to imply changes in any rules governing relief in those contexts. They are also not intended to recommend any changes to, or affect the meaning or

implementation of, any existing international law or agreements, including but not limited to:

a. International humanitarian, human rights and refugee law;
b. The legal personality and status of States, inter-governmental organizations, the
c. the International Federation of Red Cross and Red Crescent Societies and the International Committee of the Red Cross; International law related to privileges and immunities;
d. The Statutes and regulations of the International Red Cross and Red Crescent Movement and existing legal arrangements between the individual components of the Movement and States; and
e. Existing agreements between States or between States and assisting actors

2. *Definitions*

For the purposes of these Guidelines,

1. "Disaster" means a serious disruption of the functioning of society, which poses a significant, widespread threat to human life, health, property or the environment, whether arising from accident, nature or human activity, whether developing suddenly or as the result of long-term processes, but excluding armed conflict.
2. "Disaster relief" means goods and services provided to meet the immediate needs of disaster-affected communities.
3. "Initial recovery assistance" means goods and services intended to restore or improve the pre-disaster living conditions of disaster-affected communities, including initiatives to increase resilience and reduce risk, provided for an initial period of time, as determined by the affected State, after the immediate needs of disaster-affected communities have been met.
4. "Goods" means the supplies intended to be provided to disaster-affected communities for their relief or initial recovery.
5. "Services" means activities (such as rescue and medical care) undertaken by disaster relief and initial recovery personnel to assist disaster-affected communities.
6. "Equipment" means physical items, other than goods, that are necessary for disaster relief or initial recovery assistance, such as vehicles and radios.
7. "Personnel" means the staff and volunteers providing disaster relief or initial recovery assistance.
8. "Affected State" means the State upon whose territory persons or property are affected by a disaster.

9. "Assisting State" means a State providing disaster relief or initial recovery assistance, whether through civil or military components.
10. "Originating State" means the State from which disaster relief and initial recovery personnel, goods and equipment begin travel to the affected State.
11. "Transit State" means the State through whose territorial jurisdiction disaster relief or initial recovery assistance has received permission to pass on its way to or from the affected State in connection with disaster relief or initial recovery assistance.
12. "Assisting humanitarian organization" means a foreign, regional, intergovernmental or international non-profit entity whose mandate and activities are primarily focused on humanitarian relief, recovery or development.
13. "Eligible assisting humanitarian organization" means an assisting humanitarian organization determined to be eligible to receive legal facilities pursuant to Part V by the originating, transit or affected State, as applicable.
14. "Assisting actor" means any assisting humanitarian organization, assisting State, foreign individual, foreign private company providing charitable relief or other foreign entity responding to a disaster on the territory of the affected State or sending in-kind or cash donations.

PART I: CORE RESPONSIBILITIES

3. *Responsibilities of Affected States*

1. Affected States have the primary responsibility to ensure disaster risk reduction, relief and recovery assistance in their territory. National Red Cross and Red Crescent Societies, as auxiliaries to the public authorities in the humanitarian field, and domestic civil society actors play a key supporting role at the domestic level.
2. If an affected State determines that a disaster situation exceeds national coping capacities, it should seek international and/or regional assistance to address the needs of affected persons.
3. Affected States have the sovereign right to coordinate, regulate and monitor, disaster relief and recovery assistance provided by assisting actors on their territory, consistent with international law.

4. *Responsibilities of Assisting Actors*

1. Assisting actors and their personnel should abide by the laws of the affected State and applicable international law, coordinate with

domestic authorities, and respect the human dignity of disaster-affected persons at all times.

2. Assisting actors should ensure that their disaster relief and initial recovery assistance is provided in accordance with the principles of humanity, neutrality and impartiality, and in particular:
   a. Aid priorities are calculated on the basis of need alone;
   b. Provided without any adverse distinction (such as in regards to nationality, race, ethnicity, religious beliefs, class, gender, disability, age and political opinions) to disaster-affected persons;
   c. Provided without seeking to further a particular political or religious standpoint, intervene in the internal affairs of the affected State, or obtain commercial gain from charitable assistance;
   d. Not used as a means to gather sensitive information of a political, economic or military nature that is irrelevant to disaster relief or initial recovery assistance.
3. To the greatest extent practicable, their disaster relief and initial recovery assistance should also be:
   a. Responsive to the special needs, if any, of women and particularly vulnerable groups, which may include children, displaced persons, the elderly, persons with disabilities, and persons living with HIV and other debilitating illnesses;
   b. Adequate for the needs of affected persons and consistent with any applicable international standards of quality;
   c. Coordinated with other relevant domestic and assisting actors;
   d. Provided and conducted in a manner that is sensitive to cultural, social and religious customs and traditions;
   e. Carried out with adequate involvement of affected persons, including women, youth and the elderly, in their design, implementation, monitoring and evaluation;
   f. Provided by competent and adequately trained personnel;
   g. Commensurate with their organizational capacities;
   h. Building upon and conducted in a manner that strengthens local disaster risk reduction, relief and recovery capacities and reduces future vulnerabilities to disasters;
   i. Carried out so as to minimize negative impacts on the local community, economy, job markets, development objectives and the environment; and
   j. Provided in a transparent manner, sharing appropriate information on activities and funding.

5. *Additional Responsibilities of All States*

1. States providing funding to other assisting actors should encourage them to act in a manner consistent with the provisions of paragraph 4.
2. All States should actively encourage members of the public interested in contributing to international disaster relief or initial recovery to make financial donations where possible or otherwise donate only those types of relief goods expressly requested by the affected State.

6. *Responsibilities Concerning Diversion and the Intended Use of Resources*

1. States and assisting humanitarian organizations should cooperate t prevent unlawful diversion, misappropriation, or fraud concerning disaster relief or initial recovery goods, equipment or resources and initiate proceedings as appropriate.
2. Affected States should use funds and relief goods donated to them, and which they have accepted in relation to a disaster, in a manner consistent with the expressed intent with which they were given.

PART II: EARLY WARNING AND PREPAREDNESS

7. *Early Warning*

1. In order to minimize transboundary impacts and maximize the effectiveness of any international assistance that might be required, all States should have procedures in place to facilitate the expeditious sharing of information about disasters, including emerging hazards that are likely to cause disasters, with other States and assisting humanitarian organizations as appropriate, including the United Nations' Emergency Relief Coordinator.

8. *Legal, Policy and Institutional Frameworks*

1. As an essential element of a larger disaster risk reduction programme, States should adopt comprehensive legal, policy, and institutional frameworks and planning for disaster prevention, mitigation, preparedness, relief and recovery which take full account of the auxiliary role of their National Red Cross or Red Crescent Society, are inclusive of domestic civil society, and empower communities to enhance their own safety and resilience. States, with the support, as appropriate, of relevant regional and international organizations, should devote adequate resources to ensure the effectiveness of these frameworks.

2. These frameworks should also adequately address the initiation, facilitation, transit and regulation of international disaster relief and initial recovery assistance consistent with these Guidelines. They should allow for effective coordination of international disaster relief and initial recovery assistance, taking into account the role of the United Nations Emergency Relief Coordinator as central focal point with States and assisting humanitarian organizations concerning United Nations emergency relief operations. They should also clearly designate domestic governmental entities with responsibility and authority in these areas. Consideration should be given to establishing a national focal point to liaise between international and government actors at all levels.
3. Where necessary and appropriate, national governments should encourage other domestic actors with authority over areas of law or policy pertinent to international disaster relief or initial recovery assistance, such as provincial or local governments and private regulatory bodies, to take the necessary steps at their level to implement the Guidelines.

9. *Regional and International Support for Domestic Capacity*

1. With a view to increasing resilience and reducing the need for international disaster relief and initial recovery assistance, the international community, including donors, regional and other relevant actors, should support developing States, domestic civil society actors and National Red Cross and Red Crescent Societies to build their capacities to prevent, mitigate, prepare for and respond to disasters domestically.
2. The international community should also support developing States to build the capacity to adequately implement legal, policy and institutional frameworks to facilitate international relief and initial recovery assistance. This support should be provided to States in a coordinated manner among the relevant actors.

PART III: INITIATION AND TERMINATION OF INTERNATIONAL DISASTER RELIEF AND INITIAL RECOVERY ASSISTANCE

10. *Initiation*

1. Disaster relief or initial recovery assistance should be initiated only with the consent of the affected State and in principle, on the basis of an appeal. The affected State should decide in a timely manner whether or not to request disaster relief or initial recovery assistance

and communicate its decision promptly. In order to make this decision, the affected State should promptly assess needs. Consideration should be given to undertaking joint needs assessments with the United Nations and other assisting humanitarian organizations.

2. Requests and offers for assistance should be as specific as possible as to the types and amounts of goods as well as the services and expertise available or required, respectively. Affected States may also wish to indicate particular types of goods and services likely to be offered that are not needed.
3. Affected States should make available to assisting actors adequate information about domestic laws and regulations of particular relevance to the entry and operation of disaster relief or initial recovery assistance.

11. *Initiation of Military Relief*

1. Military assets should be deployed for disaster relief or initial recovery assistance only at the request or with the express consent of the affected State, after having considered comparable civilian alternatives. Prior to any such deployment, terms and conditions (including such issues as the duration of deployment, whether they must be unarmed or may be armed the use of their national uniforms, and mechanisms for cooperation with civilian actors) are to be agreed by the affected and assisting States.

12. *Termination*

1. When an affected State or an assisting actor wishes to terminate disaster relief or initial recovery assistance, it should provide appropriate notification. Upon such notification, the affected State and the assisting actor should consult with each other, bearing in mind the impact of such termination on disaster-affected communities.

PART IV: ELIGIBILITY FOR LEGAL FACILITIES

13. *Facilities for Assisting States*

1. It is recommended that transit and affected States grant, at a minimum, the legal facilities described in Part V to assisting States with respect to their disaster relief or initial recovery assistance.

14. *Facilities for Assisting Humanitarian Organizations*

1. Subject to existing international law, it is the prerogative of originating, transit and affected States to determine which assisting humanitarian organizations will be eligible to receive the legal facilities described in Part V with respect to their disaster relief or initial recovery assistance.
2. It is recommended that States establish criteria for assisting humanitarian organizations seeking eligibility for legal facilities. These criteria should include a showing by the organization of its willingness and capacity to act in accordance with the responsibilities described in paragraph 4 of these Guidelines.
3. Any additional requirements imposed on assisting humanitarian organizations should not unduly burden the provision of appropriate disaster relief and initial recovery assistance.
4. Determination of eligibility by the State granting the facilities should be possible in advance of a disaster, or as soon as possible after its onset. Applicable procedures and mechanisms should be as simple and expeditious as possible. They should be clearly described and information about them should be made freely available. They might include the use of a national roster, bilateral agreements or reliance upon international or regional systems of accreditation, if available.
5. Retention of the legal facilities in Part V should be made dependent on ongoing compliance with the provisions of subsection 2 of this paragraph. However, entitlement to legal facilities should not be changed arbitrarily, retroactively or without notice appropriate to the circumstances.

15. *Facilities for Other Assisting Actors*

1. Affected States may also wish to extend, upon request, some of the legal facilities in Part V to assisting actors other than those covered by paragraphs 13 and 14, such as private companies providing charitable relief, provided this does not negatively affect operations of assisting humanitarian organizations or assisting States. Any actor receiving such facilities should be required to abide, at a minimum, by the same conditions described in paragraph 4.

PART V: LEGAL FACILITIES FOR ENTRY AND OPERATIONS

It is recommended that States provide the legal facilities described in paragraphs 16–24 to assisting States and eligible assisting humanitarian organizations. It is understood that the granting of these facilities will be subject to the interests of national security, public order, public and environmental health, and public morals of the concerned affected, originating and transit States. Measures to protect such interests should be tailored to the exigencies of the specific disaster and consistent with the humanitarian imperative of addressing the needs of affected communities.

Where specific facilities recommended here are within the competence of authorities other than the national government, the national government should, where possible and appropriate, encourage those authorities to provide the relevant facilities to assisting States and eligible assisting humanitarian organizations.

16. *Personnel*

1. With regard to disaster relief and initial recovery personnel of assisting States and eligible assisting humanitarian organizations, affected States should:
   a. Grant visas and any necessary work permits, ideally without cost, renewable within their territory, for the time necessary to carry out disaster relief or initial recovery activities;
   b. In disaster relief operations, waive or significantly expedite the provision of such visas and work permits;
   c. Establish expedited procedures for temporary recognition of professional qualifications of foreign medical personnel, architects, and engineers, drivers licenses and other types of licenses and certificates that are necessary for the performance of disaster relief or initial recovery functions and that have been certified as genuine by the concerned assisting State or eligible assisting humanitarian organization, for the time necessary to carry out disaster relief or initial recovery activities;
   d. Facilitate freedom of access to and freedom of movement in and from the disaster-affected area, bearing in mind the safety of disaster relief and initial recovery personnel.
2. Upon request, originating and transit States should likewise waive or promptly issue, ideally without cost, exit or transit visas, as appropriate, for the disaster relief and initial recovery personnel of eligible assisting humanitarian organizations.
3. Assisting States and eligible assisting humanitarian organizations should consider to what degree disaster relief and initial recovery objectives can be met through hiring local staff.

17. *Goods and Equipment*

1. With regard to disaster relief and initial recovery goods and equipment exported or imported by, or on behalf of, assisting States and eligible

assisting humanitarian organizations, originating, transit and affected States should:

a. Exempt them from all customs duties, taxes, tariffs or govern mental fees;
b. Exempt them from all export, transit, and import restrictions;
c. Simplify and minimize documentation requirements for export, transit and import;
d. Permit re-exportation of any equipment or unused goods which the assisting State or assisting humanitarian organization owns and wishes to retain

2. With regard to disaster relief goods and equipment only, originating, transit and affected States should additionally:
   a. Waive or reduce inspection requirements. Where waiver is not possible, clear relief goods and equipment rapidly and as a matter of priority, through a "preclearance" process where feasible; and
   b. Arrange for inspection and release outside business hours and/or at a place other than a customs office as necessary to minimize delay, in accordance with the safety regulations of the affected State. Assisting States and eligible assisting humanitarian organizations should respect any routes and delivery points prescribed by the affected State.
3. In order to benefit from the facilities above, assisting States and assisting humanitarian organizations should, in accordance with agreed international standards, appropriately pack, classify and mark disaster relief and initial recovery goods and equipment, and include detailed manifests with each shipment. They should additionally inspect all such goods and equipment to ensure their quality, appropriateness for the needs in the affected State, and conformity with the national law of the affected State and international standards.
4. Assisting States and eligible assisting humanitarian organizations should assume responsibility for removing or disposing of any unwanted and unused relief and initial recovery goods, particularly if they may pose a threat to human health or safety, or the environment.

18. *Special Goods and Equipment*

In addition to the facilities described in paragraph 17:

1. Affected States should grant temporary recognition to foreign registration and plates with regard to vehicles imported by assisting States and eligible assisting humanitarian organizations or on their behalf in disaster relief and initial recovery assistance.

2. Affected States should waive or expedite the granting of any applicable licenses and reduce any other barriers to the use, import or export of telecommunications and information technology equipment by assisting States and assisting humanitarian organizations or on their behalf in disaster relief and initial recovery assistance. Without discrimination against or negative impact to domestic relief actors, affected States should also grant (or where, appropriate, encourage other domestic actors to grant) assisting States and eligible assisting humanitarian organizations priority access to bandwidth, frequencies and satellite use for telecommunications and data transfer associated with disaster relief operations.
3. Originating, transit and affected States should reduce legal and administrative barriers to the exportation, transit, importation and re-exportation of medications and medical equipment by assisting States and eligible assisting humanitarian organizations or on their behalf in disaster relief and initial recovery assistance, to the extent consistent with public safety and international law. Assisting States and eligible assisting humanitarian organizations should take all reasonable steps to ensure the quality, appropriateness and safety of any such medications and equipment and in particular:
   a. Any medications they import should be approved for use in the originating and affected State;
   b. Medications they use in their own operations should be:
      (i) transported and maintained in appropriate conditions to ensure their quality and;
      (ii) guarded against misappropriation and abuse.
   c. Any medications they donate for use by others in the affected State should be:
      (i) at least twelve months from their expiration date upon arrival, unless otherwise agreed by receiving authorities;
      (ii) transported and maintained in appropriate conditions to ensure their quality until they reach the affected State; and
      (iii) appropriately labeled in a language understood in the affected State with the International Nonproprietary Name or generic name, batch number, dosage form, strength, name of manufacturer, quantity in the container, storage conditions and expiry date.
4. Originating, transit and affected States should consider whether normal requirements regarding fumigation and prohibitions and restrictions on food imports and exports by assisting States and eligible assisting humanitarian organizations in disaster relief operations can be modified or reduced.

19. *Transport*

1. Originating, transit and affected States should grant, without undue delay, permission for the speedy passage of land, marine and air vehicles operated by an assisting State or eligible assisting humanitarian organization or on its behalf, for the purpose of transporting disaster relief or initial recovery assistance and, ideally, waive applicable fees.
2. In particular, permission should be granted for overflight, landing and departure of aircraft. Such aircraft should also be authorized to operate within the territory of the affected State as required for the delivery of assistance.
3. Any applicable exit, transit and entry visas for the operating personnel of such transport vehicles should be promptly issued.

20. *Temporary Domestic Legal Status*

1. Affected States should grant relevant entities of assisting States and eligible assisting humanitarian organizations, upon entry or as soon as possible thereafter, at least a temporary authorization to legally operate on their territory so as to enjoy the rights, inter alia, to open bank accounts, enter into contracts and leases, acquire and dispose of property and instigate legal proceedings, for the purpose of providing disaster relief and initial recovery assistance.
2. Assisting States and eligible assisting humanitarian organizations should also be granted the right to freely bring the necessary funds and currencies in or out of the country through legal means and to obtain legal exchange rates in connection with their disaster relief or initial recovery assistance.
3. Affected States should allow assisting States and eligible assisting humanitarian organizations to legally hire and terminate the contracts of local personnel

21. *Taxation*

1. Affected States should provide exemptions to assisting States and eligible assisting humanitarian organizations from value-added and other taxes or duties directly associated with disaster relief and initial recovery assistance.

22. *Security*

1. Affected States should take appropriate measures to address the safety and security of disaster relief and initial recovery personnel of assisting States and eligible assisting humanitarian organizations and of the

premises, facilities, means of transport, equipment and goods used in connection with their disaster relief or initial recovery assistance. Assisting States and assisting humanitarian organizations should also take appropriate steps in their own planning and operations to mitigate security risks.

23. *Extended Hours*

1. Affected States should endeavor to ensure, when necessary, that State-operated offices and services essential to the timely delivery of international disaster relief function outside of normal business hours.

24. *Costs*

1. The costs of providing international disaster relief or initial recovery assistance pursuant to these Guidelines should normally be borne by the assisting State or assisting humanitarian organization. However, assisting States may agree in advance with the affected State for the reimbursement of certain costs and fees, or for the temporary loan of equipment.
2. Affected States should consider, when it is in their power and to the extent possible under the circumstances, providing certain services at reduced or no cost to assisting States and eligible assisting humanitarian organizations, which may include:
   a. In-country transport, including by national airlines;
   b. Use of buildings and land for office and warehouse space; and
   c. Use of cargo handling equipment and logistic support.

APPENDIX B

UN Draft Articles on Protection of Persons in the Event of Disasters

ARTICLE 1 SCOPE

The present draft articles apply to the protection of persons in the event of disasters.

ARTICLE 2 PURPOSE

The purpose of the present draft articles is to facilitate an adequate and effective response to disasters that meets the essential needs of the persons concerned, with full respect for their rights.

ARTICLE 3 DEFINITION OF DISASTER

"Disaster" means a calamitous event or series of events resulting in widespread loss of life, great human suffering and distress, or large-scale material or environmental damage, thereby seriously disrupting the functioning of society.

ARTICLE 4 RELATIONSHIP WITH INTERNATIONAL HUMANITARIAN LAW

The present draft articles do not apply to situations to which the rules of international humanitarian law are applicable.

ARTICLE 5 DUTY TO COOPERATE

In accordance with the present draft articles, States shall, as appropriate, cooperate among themselves, and with the United Nations and other competent intergovernmental organizations, the International Federation of the Red Cross and Red Crescent Societies and the International Committee of the Red Cross, and with relevant non-governmental organizations.

ARTICLE 5 BIS FORMS OF COOPERATION

For the purposes of the present draft articles, cooperation includes humanitarian assistance, coordination of international relief actions and communications, and making available relief personnel, relief equipment and supplies, and scientific, medical and technical resources

ARTICLE 6 HUMANITARIAN PRINCIPLES IN DISASTER RESPONSE

Response to disasters shall take place in accordance with the principles of humanity, neutrality and impartiality, and on the basis of non-discrimination, while taking into account the needs of the particularly vulnerable.

ARTICLE 7 HUMAN DIGNITY

In responding to disasters, States, competent intergovernmental organizations and relevant non-governmental organizations shall respect and protect the inherent dignity of the human person.

ARTICLE 8 HUMAN RIGHTS

Persons affected by disasters are entitled to respect for their human rights.

ARTICLE 9 ROLE OF THE AFFECTED STATE

1. The affected State, by virtue of its sovereignty, has the duty to ensure the protection of persons and provision of disaster relief and assistance on its territory.
2. The affected State has the primary role in the direction, control, coordination and supervision of such relief and assistance.

ARTICLE 10 DUTY OF THE AFFECTED STATE TO SEEK ASSISTANCE

To the extent that a disaster exceeds its national response capacity, the affected State has the duty to seek assistance from among other States, the United Nations, other competent intergovernmental organizations and relevant non-governmental organizations, as appropriate.

ARTICLE 11 CONSENT OF THE AFFECTED STATE TO EXTERNAL ASSISTANCE

1. The provision of external assistance requires the consent of the affected State.
2. Consent to external assistance shall not be withheld arbitrarily.
3. When an offer of assistance is extended in accordance with the present draft articles, the affected State shall, whenever possible, make its decision regarding the offer known.

ARTICLE 12 OFFERS OF ASSISTANCE

In responding to disasters, States, the United Nations, and other competent intergovernmental organizations have the right to offer assistance to the affected State. Relevant non-governmental organizations may also offer assistance to the affected State.

ARTICLE 13 CONDITIONS ON THE PROVISION OF EXTERNAL ASSISTANCE

The affected State may place conditions on the provision of external assistance. Such conditions shall be in accordance with the present draft articles, applicable rules of international law, and the national law of the affected State. Conditions shall take into account the identified needs of the persons affected by disasters and the quality of the assistance. When formulating conditions, the affected State shall indicate the scope and type of assistance sought.

ARTICLE 14 FACILITATION OF EXTERNAL ASSISTANCE

1. The affected State shall take the necessary measures, within its national law, to facilitate the prompt and effective provision of external assistance regarding, in particular:
   (a) civilian and military relief personnel, in fields such as privileges and immunities, visa and entry requirements, work permits, and freedom of movement; and
   (b) goods and equipment, in fields such as customs requirements and tariffs, taxation, transport, and disposal thereof.
2. The affected State shall ensure that its relevant legislation and regulations are readily accessible, to facilitate compliance with national law.

ARTICLE 15 TERMINATION OF EXTERNAL ASSISTANCE

The affected State and the assisting State, and as appropriate other assisting actors, shall consult with respect to the termination of external assistance and the modalities of termination. The affected State, the assisting State, or other assisting actors wishing to terminate shall provide appropriate notification

Index